Handbuch der Zoologie
Handbook of Zoology
Band/Volume VIII Mammalia

Karl F. Koopman
Chiroptera: Systematics
Teilband/Part 60

Handbuch der Zoologie

Eine Naturgeschichte der Stämme des Tierreiches

Handbook of Zoology

A Natural History of the Phyla of the Animal Kingdom

Gegründet von / Founded by Willy Kükenthal
Fortgeführt von / Continued by M. Beier, M. Fischer, J.-G. Helmcke,
D. Starck, H. Wermuth

Band/Volume VIII Mammalia Teilband/Part 60

Herausgeber/Editors   J. Niethammer, H. Schliemann, D. Starck
Schriftleiter/Managing Editor   H. Wermuth

Walter de Gruyter · Berlin · New York 1994

Karl F. Koopman

Chiroptera: Systematics

Walter de Gruyter · Berlin · New York 1994

Autor/Authors
Dr. Karl F. Koopman
American Museum of Natural History
Central Park West at 79th Street
New York, New York 10024-5192
U.S.A

Herausgeber/Editors

Professor
Dr. Jochen Niethammer
Zoologisches Institut der
Universität Bonn
Poppelsdorfer Schloß
D-53115 Bonn
Federal Republic of Germany

Professor
Dr. Harald Schliemann
Zoologisches Institut und
Zoologisches Museum
Martin-Luther-King-Platz 3
D-20146 Hamburg
Federal Republic of Germany

Schriftleiter/Managing Editor
Dr. Heinz Wermuth
Falkenweg 1
D-71691 Freiberg
Federal Republic of Germany

Professor
Dr. med. Dr, phil. h. c.
Dietrich Starck
Balduinstraße 88
D-60599 Frankfurt
Federal Republic of Germany

Verlag/Publisher

Walter de Gruyter & Co.
Genthiner Straße 13
D-10785 Berlin
Federal Republic of Germany
Tel. (0 30) 2 60 05-0
Telefax (0 30) 2 60 05-2 51

Walter de Gruyter, Inc.
200 Saw Mill River Road
Hawthorne, N.Y. 10532
U.S.A.
Tel. (9 14) 747-01 10
Telefax (9 14) 747-13 26

Das Buch enthält 192 Abbildungen.
With 192 illustrations.

♾ Printed on acid-free paper which falls within the guidelines of the ANSI to ensure permanence and durability.

Die Deutsche Bibliothek – CIP-Einheitsaufnahme

Handbuch der Zoologie – eine Naturgeschichte der Stämme des
Tierreiches / gegr. von Willy Kükenthal. Fortgef. von M. Beier
... – Berlin ; New York : de Gruyter.
  Teilw. mit Parallelt.: Handbook of zoology
NE: Kükenthal, Willy [Begr.]; Beier, Max [Hrsg.]; PT

Bd. 8. Mammalia / Hrsg. J. Niethammer ...
  Teilbd. 60. Chiroptera : systematics / Karl F. Koopman. – 1994
  ISBN 3-11-014081-0
NE: Niethammer, Jochen [Hrsg.]; Koopman, Karl F.

Library of Congress Cataloging-in-Publication Data

Handbuch der Zoologie
  Parallel title: Handbook of zoology.
  Vol. 8, part 56 has: fortgeführt von M. Beier,
... – [et al.].
  Issued in parts.
  German on English.
  Includes bibliographies and indexes.
  Contents: v. 1. Ed. Protozoa. Porifera. Coelenterata. Mesozoa – v. 3. Bd. 1. Hälfte. Tardigrada. Pentastomida. Myzostomida. Arthropoda: Allgemeines. Crustacea. 2. Hälfte. Chelicerata, Pantopoda, Onychophora, Vermes Oligomera. (2 v.) – [etc.] – v. 8. Bd. Mammalia.

  1. Zoology. I. Kükenthal, W. G. (Willy Georg), 1861–1922. II. Krumbach, Thilo, b. 1874. III. Handbook of zoology.
  QL45.H2   591   23-1436
  ISBN 3110140810

Copyright © 1994 by Walter de Gruyter & Co., D-10785 Berlin.
All rights reserved, including those of translation into foreign languages. No part of this book may be reproduced in any form – by photoprint, microfilm or any other means nor transmitted nor translated into a machine language without written permission from the publisher.
Typesetting and printing: Tutte Druckerei GmbH, Salzweg-Passau – Binding: Lüderitz & Bauer, Berlin – Printed in Germany

Table of contents

Preface	VII
Charactarization	1
Appearance	1
Skin	1
Skull	3
Dentition	5
Post-cranial skeleton	6
Muscular System	7
Brain and Spinal Cord	7
Sense Organs	7
Digestive System	8
Circulatory System	8
Urinogenital System	8
Ecology	9
Distribution	12
Phylogeny	14
Systematics	17
Suborder Megachiroptera	18
Family Pteropodidae	18
Subfamily Pteropodinae	18
Suborder Macroglossinae	37
Suborder Microchiroptera	40
Suborder Yinochiroptera	40
Superfamily Emballonuroidea	40
Family Rhinopomatidae	40
Family Craseonycteridae	41
Superfamily Rhinolophoidea	48
Family Nycteridae	49
Family Megadermatidae	50
Family Rhinolophidae	52
Subfamily Rhinolophinae	52
Subfamily Hipposiderinae	60
Infraorder Yangochiroptera	69
Superfamily Noctilionoidea	69
Family Noctilionidae	69
Family Mormoopidae	70
Family Phyllostomidae	72
Subfamily Phyllostominae	72
Subfamily Lonchophyllinae	77
Subfamily Brachyphyllinae	78
Subfamily Phyllonycterinae	79
Subfamily Glossophaginae	80
Subfamily Carolliinae	83
Subfamily Stenodermatinae	85
Subfamily Desmodontinae	94
Superfamily Vespertilionoidea	94
Family Natalidae	95
Family Furipteridae	96
Family Thyropteridae	96
Family Myzopodidae	96
Family Vespertilionidae	97
Subfamily Kerivoulinae	97
Subfamily Vespertilioninae	100
Subfamily Murininae	131
Subfamily Miniopterinae	133
Subfamily Tomopeatinae	135
Family Mystacinidae	135
Family Molossidae	135
Skull photographs	147
Table 1. Characters of Families	189
Table 2. Dental formulae of genera of bats	190
Table 3. Forearm lengths for the genera of bats	190
Literature	192

Preface

The following work should not be regarded as definitive. Rather, it is a progress report on the author's views on systematics and distribution of the order Chiroptera and its taxonomic subdivisions down to species, with subspecies indicated in most cases. Bat systematics is an active field with new taxa being described and old ones synonymized at a surprisingly high rate. Any bat systematist therefore must be resigned to seeing his work modified within a few years if not sooner. This is also not a strictly original work in the sense that the author can personally vouch for all characters and taxonomic arrangements. To have done this would have required at least 10 years and might not have been completed in the author's lifetime. Instead, it has largely been based on the literature (including the author's own work), supplemented by checking specimens (mostly in the more difficult genera) for diagnostic characters. This has mainly been done on specimens at the American Museum of Natural History but to some extent at the Field Museum of Natural History, access to whose collections I am indebted to Drs. Robert Timm and Bruce Patterson and Mr. Robert Izor. However, the most important outside museum has been the British Museum (Natural History). Here, Mr. John Hill has been of tremendous assistance, both in making the collection available and in sharing with me his vast knowledge concerning bat characteristics and relationships. Without him, this would have been a much less adequate work. The Literature section is not intended to be exhaustive. Rather, it consists of those books and papers I found most useful in compiling this work. Many of these contain extensive bibliographies to further important literature. The manuscript was essentially completed late in 1988 and few changes have been made since. It therefore does not incorporate systematic changes made during the subsequent five years.

Order Chiroptera BLUMENBACH 1779 (Bats)

Characterization:
Very small to medium-sized mammals with the pectoral limb modified as a wing for flight (the only truely flying mammals). Digits 2–5 greatly elongated and supporting a fold of skin which connects the fingers with one another and with the body and hind limbs.

Appearance:
Basically that of a flying animal (never flightless) but great diversity in size, the smallest with a head and body length of 3 cm and a weight of 1,7 g, the largest with a head and body length of 43 cm and a weight of 1200 g. Wing proportions vary greatly from short and broad to long and narrow.

Skin:
The most distinctive feature of the integumentary system is the possession of a series of wing membranes which enable the animal to fly (Fig. 1). These consist of a double thickness of skin and are composed of four main parts, the propatagium, chiropatagium or dactylopatagium, the plagiopatagium, and uropatagium. The propatagium lies in front of the forelimb and runs from the neck to the base of the thumb. The dactylopatagium brevis is from the thumb to the second finger. The dactylopatagium minus from the second to the third finger; dactylopatagium medius from the third to the fourth finger; dactylopatagium major from the fourth to the fifth finger. The plagiopatagium runs from the fifth finger to the side of the body (or in a few cases to the mid-dorsal line) and the hind limb. The uropatagium runs between the two hind limbs and usually also involves at least part of the tail when it is present.

Hair normally covers the head and body but not the wing membranes except occasionally the dorsal side of the uropatagium (some Stenodermatinae, *Lasiurus*, Murininae) and proximal portions of the ventral side of the plagiopatagium (*Lasiurus*). There may be a fringe of hair along the edge of the uropatagium (some species of *Myotis* and *Kerivoula*). In a few cases (*Rhinopoma*, some *Taphozous*), hair may be scanty or absent from the rump region. In *Cheiromeles*, there is very little body hair at all. Hair texture is usually fairly long, soft, and lax, but with many exceptions. The hair may be shaggy (*Pteralopex*, some *Rousettus*) or wooly (some *Rhinolophus*, some *Hipposideros*, *Chrotopterus*, *Kerivoula*). The hair may be short and velvety (most *Molossidae*) or short but stiff (*Mystacina*). Body hair color may vary from black to white through various shades of gray, yellow, red, or brown. A white or pale

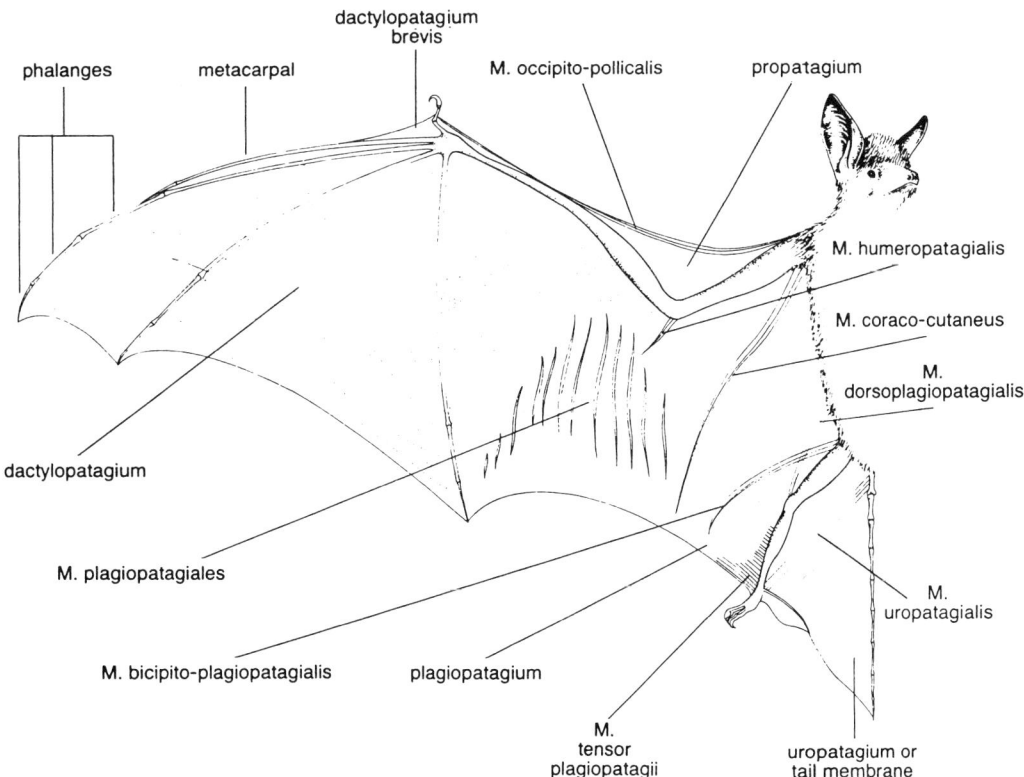

Fig. 1. The wing membranes of bats. Reproduced with the permission of the British Museum (Natural History).

mid dorsal line may be present (*Noctilio*, some Stenodermatinae) or paired white dorsal lines (some *Saccopteryx*). In *Nyctimene*, there is a mid-dorsal black stripe. There may be a sharp distinction between a black or gray dorsum and a white ventrum (some *Saccolaimus*, some *Tonatia*, some *Pipistrellus*). Occasionally there may be white lines on the ventrum next to the attachment of the plagiopatagium.

Specialized skin glands may be found on several parts of the body. These include the forehead (*Hipposideros*), chest (some Phyllostominae and Molossidae, perhaps *Taphozous*), shoulders (many Pteropodidae), propatagium (several genera of neotropical Emballonurinae), and scrotum (*Noctilio*). All these glands are reduced or absent in females. The mammary glands are pectoral (except in *Furipteridae*) and provided with a single pair of teats. In some bats (Rhinopomatidae, Craseonycteridae, Megadermatidae, Rhinolophidae), there are a second pair of teat-like structures in the inguinal region. These are not connected to the mammary glands but are used as holdfasts by the young while the mother is in flight.

Many bats are remarkable for various facial modifications (Figs. 2–6). Most common is some sort of noseleaf (developed several times indepen-

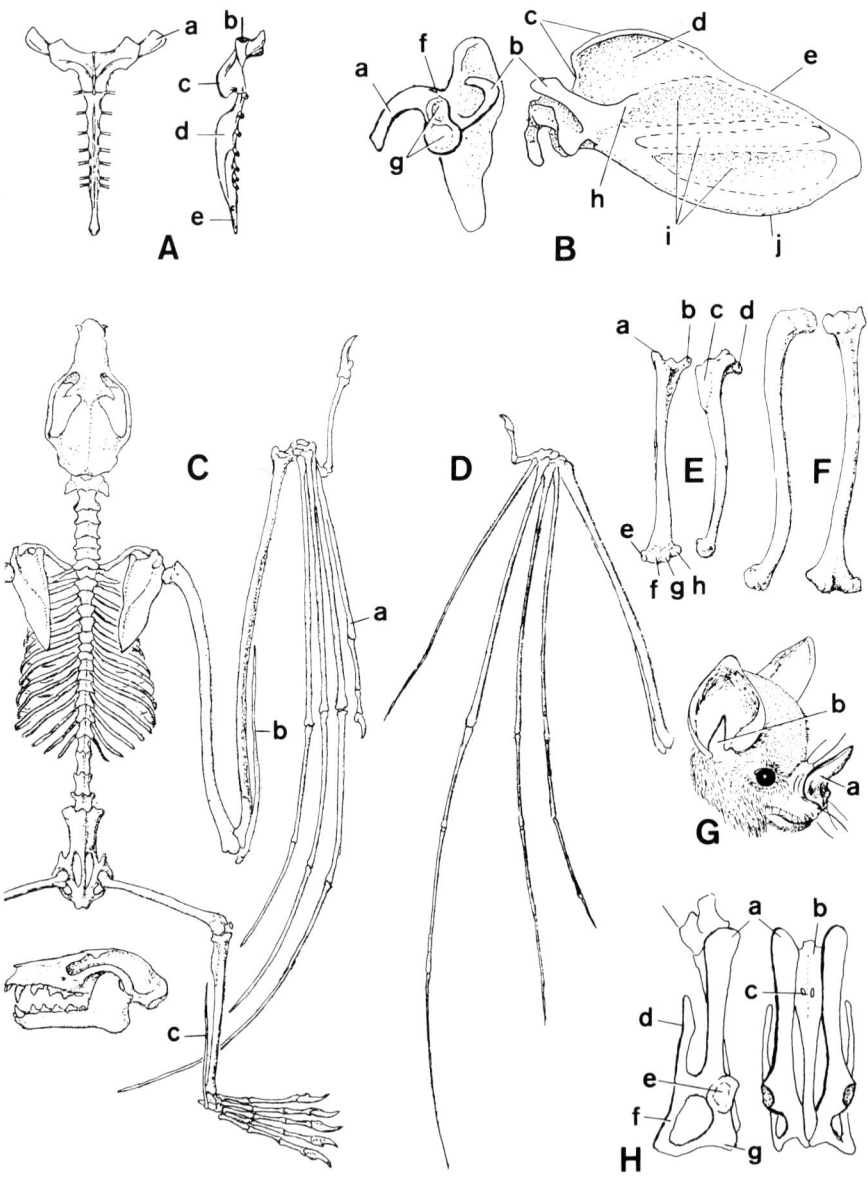

Fig. 2. Features of bats. A, Sternum; B, Scapula of *Chiroderma* showing double articulation with humerus (g); C, Skeleton of *Pteropus*; D, Forearm and manus of *Artibeus*; E, Humerus of *Artibeus* showing trochiter (a) and trochin (b); F, Humerus of *Pteropus* (note relatively feeble development trochiter); G, Head of phyllostomid showing noseleaf (a) and tragus (b); H, Pelvis and sacrum of *Chiroderma*.

Fig. 3. Heads of *Pteropus* (A), *Rousettus* (B), and *Nyctimene* (C). Reproduced with the permission of the British Museum (Natural History).

dently), which is found well developed in Megadermatidae and most Phyllostomidae (where it forms a simple dorsally directed projection from the nasal region) and the Rhinolophidae (where it is much more complex, with several distinct parts). While in *Rhinolophus*, a function related to echolocation has been proposed for the basal portion of the noseleaf (MÖHRES 1953), the explanation is inapplicable to the noseleaves of most bats which have these structures and their function is largely hypothetical at present. Besides well developed noseleaves, a variety of folds, ridges, tubercles, and plates on the muzzle, lips, or chin may be found in various members of the Pteropodidae (particulary *Hypsignathus*), Rhinopomatidae, Craseonycteridae, Nycteridae, Noctilionidae, Mormoopidae (especially *Mormoops*), Phyllostomidae (e.g. *Trachops*), Vespertilionidae (best developed in the Nyctophilini), and Molossidae. In many bats, the ear pinnae have been variously modified. In all but the Pteropodidae and Rhinolophidae, a distinct tragus is developed, which may be very large and occasionally complex (e.g. Megadermatidae). The ear pinnae themselves may be reduced in height (*Lasiurus*), but more often are enlarged as in the Nycteridae, Megadermatidae, some Rhinolophidae, some Phyllostomidae, Myzopodidae, some Vespertilionidae, and some Molossidae. In a number of instances, the ears are joined together; this may be accomplished either by direct fusion of the inner margins of the pinnae (e.g. Megadermatidae) or by a separate band of skin running from the posterior surface of one pinna to that of the other (e.g., various Vespertilionidae).

Skull:

As is characteristic of flying vertebrates (cf. birds) most of the bones are quite thin and, with a few exceptions, are fused together in the adult. In the Megachiroptera, the premaxillaries may be either free, sutured, or fused, but in the Microchiroptera, the premaxillaries are either free and moveable (Infraorder Yinochiroptera) or fused (Infraorder Yangochiroptera). The tympanic bone is usually free and only loosely attached to the remainder of the skull, but in a few (e.g. *Saccolaimus*) it is firmly sutured to the basisphenoid. The periotic is sutured to the rest of the skull in Megachiroptera and in Emballonuridae, but is only loosely attached in other Microchiroptera. This is probably related to echolocation since a loose attach-

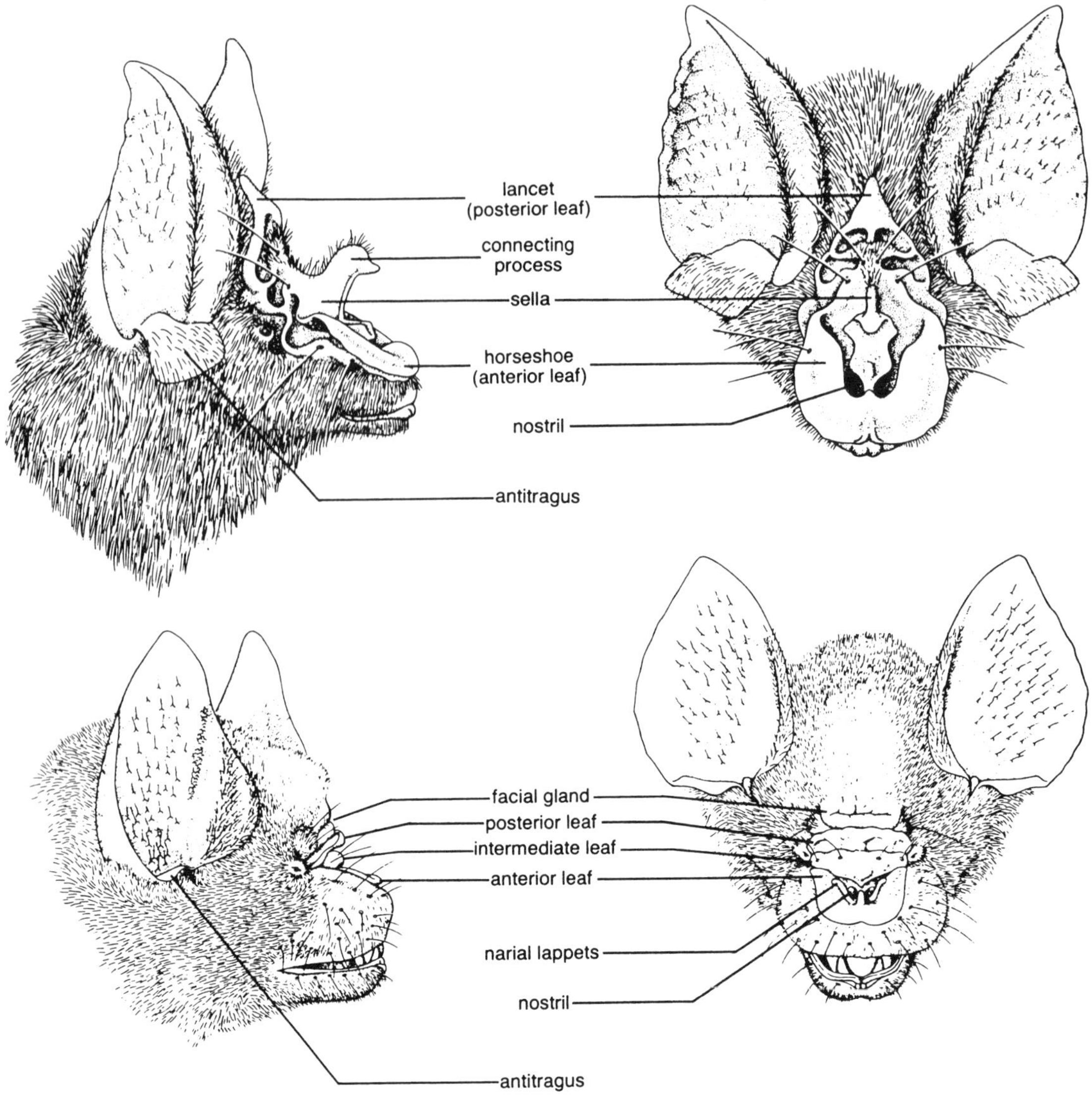

Fig. 4. Heads of *Rhinolophus* (top) and *Hipposideros* (bottom). Reproduced with the permission of the British Museum (Natural History).

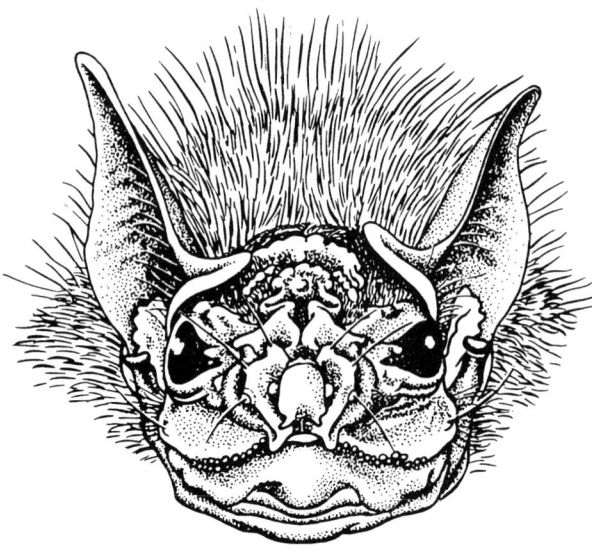

Fig. 5. Face of *Centurio*. Reproduced with the permission of the British Museum (Natural History).

ment prevents the outgoing signal from the larynx from interfering with the incoming echo in the cochlea.

The general proportions of the skull are chiefly related to the nature of the food. Species in which the food is soft (such as moths, overripe fruit, or nectar) tend to have relatively long rostra and weak jaws, whereas those that feed on hard items (such as beetles or tough fruit) tend to have shorter rostra and more powerful jaws. The extremes of this are found in the Phyllostomidae where in *Centurio* the skull is almost as wide as long whereas in *Choeronycteris* (*Musonycteris*) it is four times as long as wide. These dietary differences will also be reflected in the height of the braincase, width of the postorbital constriction, strength of the zygomatic arches (and width across them), and particulary in the development of sagittal and lambdoidal crests (FREEMAN 1981). In some bats

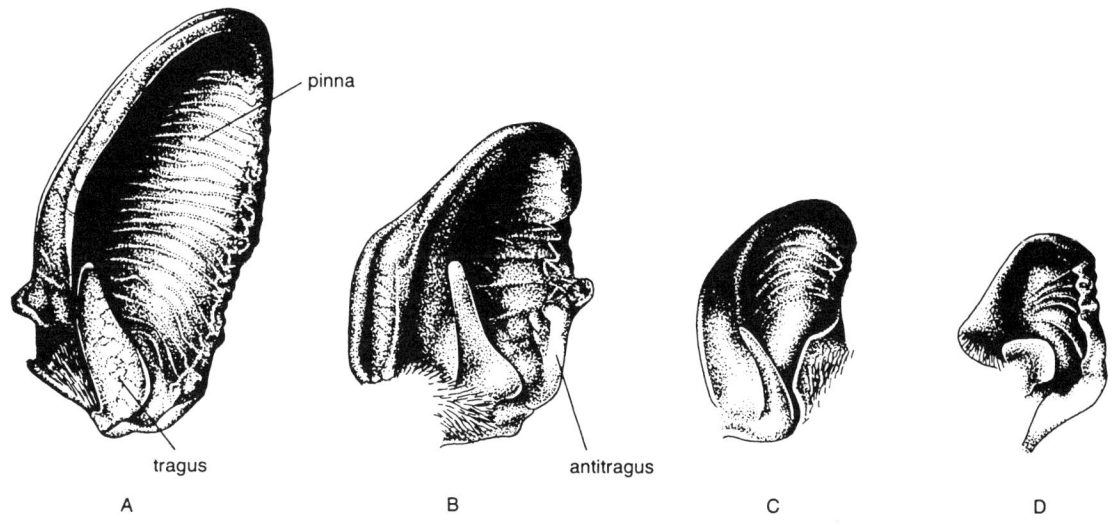

Fig. 6. Ears of *Plecotus* (A), *Barbastella* (B), *Myotis* (C), and *Nyctalus* (D). Reproduced with the permission of the British Museum (Natural History).

which roost in or must pass through narrow crevices, the entire skull is very much flattened, as seen in some species of Vespertilionidae and Molossidae (e.g. *Tylonycteris*, some *Mormopterus*). Another type of skull modification involves modifications of the plane of the palate in relation to the plane of the basicranium. This may involve either an upward rotation of the rostrum (most extreme in *Mormoops*) or, more often, a downward rotation. With long skulled species (*Pteropus*, some *Rousettus*) this gives a sharply angled ventral surface. With a very short skull (e.g. *Centurio*), the result is a very hominoid appearance.

One type of modification of the premaxillary bone has already been mentioned, namely whether or not it is fused to the maxillary. Another type of modification is seen in the reduction or loss of either the nasal or palatal branches. In Nycteridae and Rhinolophidae, the nasal branches have been lost but the palatal branches are retained. In the Emballonuroidea and most of the Vespertilionoidea, the palatal branches have been greatly reduced but the nasal branches retained. In the Megadermatidae, the palatal branches have been lost and the nasal branches greatly reduced. However, at least in *Cardioderma* and *Lavia*, the latter can be seen as slender threadlike structures in skulls which have been cleaned with great care. As mentioned above, the zygomatic arches are well-developed in bats with a powerful bite, but much more delicate in species feeding on soft material and are incomplete in some nectar and soft fruit feeders. Post-orbital processes occur in Pteropodidae and most Emballonuridae. In the former, they are strong and may even (some *Pteropus*) join the zygomatic arches to form a complete bar. In Emballonuridae, they are delicate and may be reduced (some *Emballonura*). In Nycteridae and Megadermatidae, a frontal shield is developed which may be extended (in *Lavia* and *Cardioderma*) into distinct processes behind and in front of the orbit. Another modification involves inflation of the nasal region which is particulary evident in the Rhinolophidae, many Emballonuridae and occasionally elsewhere.

Dentition (Figs. 20–192):
The primitive dental formula for Chiroptera appears to be i 2/3, c 1/1, p 3/3, m 3/3 x 2 = 38. This involves a loss of an upper incisor and at least one upper and lower premolar from the primitive Eutherian mammal dental formula. Originally there were no special peculiarities of the incisors, canines, or premolars, except for partial molarization of the last upper premolars. The molars are derived over the primitive Eutherian pattern in showing the true dilambdodont condition. In the upper molars this involves a relatively medial position of the paracone and metacone with three labial cusps (stylocone, mesostyle, and metastyle). All five of these are connected by a W-shaped ectoloph. Correlated with the more medial position of the metaconid, the lower molars are modified by a lingual shift of the hypoconulid to a position just posterior to the entoconid. This molar pattern has been retained relatively unchanged in all insectivorous, piscivorous, and carnivorous bats, but has been modified, often quite extensively in frugivorous, nectarivorous, and sanguinivorous species.

In the insectivorous-piscivorous-carnivorous bats (most Microchiroptera), there is always at least one pair of upper and lower incisors, except in the Megadermatidae with their greatly reduced premaxillaries. Some Molossidae, however, have reduced the lower incisors to a single vestigial pair. The upper and lower canines are always present and rarely much modified. The premolars are the teeth most likely to be reduced in number when the jaws are shortened. However in the bats of these three animalivorous groups, the number of premolars is never less than 1/2, the semimolarized last upper premolar in particular always being present. In a few cases, on the molars, the W-shaped ectoloph has been modified (most extreme in *Harpiocephalus*). Particularly in species where the jaws have been shortened, the last upper molar is often simplified and reduced in size.

In frugivorous-nectarivorous-sanguinivorous bats (including all Megachiroptera and most Phyllostomidae), tooth modification is often much greater. There are never more than two pairs of lower incisors and these may be further reduced, culminating in the loss of lower incisors in some of the more highly derived Glossophaginae where it facilitates movement of the tongue in taking nectar. Lower incisors are also lost in the highly derived short-faced Nyctimeninae. The upper incisors may also be reduced to a single pair. The upper and lower canines are always present and rarely much modified. The number of premolars is frequently reduced but never to less than 2/2 except in the Desmodontinae, where it is 1/2. There is a tendency in both Phyllostomidae and Pteropodidae for the last premolars to be molarized. Within the Phyllostomidae, there is great diversity in the cusp pattern of the molars. Some (e.g. primitive Glossophaginae) still show much of the primitive pattern including the W-shaped ectoloph. More derived Phyllostomidae, however, show increasing simplification of the original pattern combined with extra cusps and ridges which obscure the original pattern. In extreme cases (*Phyllonycteris*, some Stenodermatinae), it would probably be impossible to homologize cusps if less derived relatives were not known. All Megachiroptera have cusp patterns which are impossible to homologize with any confidence, especially since occlusion of upper with lower molar cusps is the best guide to determination of cusp homologies and in Megachiroptera there is no contact between upper and lower molars. The upper molars of most Megachiroptera are also peculiar since they tend to be quite narrow and look more like lower molars of Microchiroptera than like upper molars. By far the most highly modified dentition in the Phyllostomidae, if not in the entire Chiroptera, is found in the Desmodontinae. Here all the teeth, with the exception of the lower incisors, are either converted into cutting blades, reduced, or lost altogether. This reduction in dental number reaches an extreme in *Desmodus*, where the dental formula is only i 1/2, c 1/1, p 1/2, m 1/1 x 2 = 20. Deciduous teeth in all bats are highly modified hooklike structures whose chief function is to hang on the mother's fur.

Post-cranial skeleton (Fig. 2):

In general, the post-cranial skeleton reflects, directly or indirectly, adaptation for flight. While most of the shaft of the ulna is reduced and often fused with the radius, the radius remains strong and is greatly elongate. Except for the first digit, the metacarpals are all greatly elongated as are many of the phalanges. There is a tendency, however, for loss of ossification of some of the phalanges, particularly on the second digit, which in more derived families loses its independence from the third digit. Primitively in bats, the proximal end of the humerus formed a ball and socket joint with the scapula. The head was the most proximal part and the tuberculum majus (trochiter) and tuberculum minus (trochin) did not project beyond it. This condition is present in Megachiroptera and in some of the more primitive families of the Microchiroptera (Rhinopomatidae, Emballonuridae, Mormoopidae). In more derived families of Microchiroptera, the trochiter becomes enlarged and extends beyond the head of the humerus. In most cases, it makes contact with the scapula when the humerus is raised at end end of the upstroke (recovery stroke) and the beginning of the downstroke (power stroke). This converts the shoulder joint into a hinge joint which can only move through a restricted plane. It is only within this plane that it can move the wing efficiently without relying on stabilizing muscle action. Thus, in the more derived Microchiroptera, flexibility is sacrificed for economy of motion. This presumably makes possible either more rapid or more precisely controlled flight. The scapula is unusually broad and rectangular reflecting the complex muscle attachments involved in raising the wing.

In the hind limb, the femur is directed outward, thus making it more effective for support of the plagiopatagium and uropatagium, but also producing a rather reptilian stance to the hind limbs in walking. The fibula is usually greatly reduced in diameter (threadlike) except where the hindlimb is commonly used in walking (Desmodontinae, *Mystacina*, Molossidae). In many bats there is a cartlaginous, or occasionally bony

(*Noctilio*) spur, the calcar, running from the calcaneum back along the edge of the uropatagium for a variable distance. The hind claws of bats tend to be considerably recurved, since they are usually used for hanging while the bat is in the roost.

The thoracic skeleton of Chiroptera tends to be rather short with a broad rib cage. The individual ribs are considerably broadened and there may be fusions between vertebrae, in the Rhinolophidae involving the first rib and presternum as well. In view of the great diversity of tail length in bats, the number of caudal vertebrae varies greatly from none to an apparent maximum of 18. It might be added that throughout the skeleton, unnecessary bone has been eliminated with the result that a maximum optimization of strength with lightness has been achieved.

Muscular System:
Aside from the muscles involved in flight, these are quite similar to those of other mammals. However, a number of muscles in the shoulder region are highly developed for lowering the wing (power stroke) and raising the wing (recovery stroke). Those involved in the power stroke are the pectoralis, clavodeltoideus, serratus anterior and the subscapularis. Those involved in the recovery stroke include the acromiotrapezius, spinatus, acromiodeltoideus, and spinodeltoideus. Besides these, there are numerous muscles that act to move various parts of the wing on one another. These include the coracocutaneus, humeropatagialis, and the plagiopatagii. Also unique to bats are the occipitopolicaris (which runs from the skull to the thumb) and the depressor ossis styliformis (from the calcar to the ankle). Slow flying bats, with relatively short wings, usually have much of the wing muscle mass in the wing itself, whereas fast flying bats, with relatively long wings, have most of the muscle mass basal to the wing with insertions represented by long tendons.

Brain and Spinal Cord:
Brain proportions vary considerably among bats according to whether the dominant sense is sight or smell (favoring the enlargement of the forebrain) or hearing (favoring the enlargment of the midbrain). In Megachiroptera none of which practice sophisticated echolocation, but locate food by sight and smell, the forebrain (including the neopallium) is much enlarged whereas the midbrain is relatively small. To a lesser degree the same is true of the Phyllostomidae, particularly the fruit and nectar feeders and the vampires. A very different situation is seen among the remaining Microchiroptera, all of which practice sophisticated echolocation and usually use it for catching insect prey. Here the forebrain is relatively small whereas the midbrain is hypertrophied.

The spinal cord is greatly shortened in bats, in its most extreme form (*Artibeus*) extending no further than the ninth thoracic vertebra. This reflects a general shortening of the trunk and the small size of the hind limbs in relation to the wings. An interesting difference (as far as is known) between Megachiroptera and Microchiroptera is in the arrangement of white and gray matter in the spinal cord. In most mammals (including Megachiroptera), there is a large bundle of white matter dorsal to the gray matter (the dorsal funiculus). In Microchiroptera, this funiculus is reduced to a much smaller bundle largely buried in the gray matter on the dorsal side. This affects the way in which the dorsal root must connect with it. The functional significance of this modification is unclear.

Sense Organs:
Functional eyes are present in all bats though they vary considerably in size. They are largest in Megachiroptera (where they are always important in orientation). However, they also tend to be large in Megadermatidae and Phyllostomidae, where they supplement echolocation. In many Microchiroptera, however, they are quite small and play only a minor role in orientation and food-getting.

The ears of bats tend to be far more remarkable, though in the Megachiroptera, the middle and inner ear are essentially similar to those of most other mammals. This is related to the absence of echolocation in almost all Megachiroptera, the only known exception being in *R.* (*Rousettus*) (and perhaps *Eonycteris*), where it has clearly been independently evolved. In Microchiroptera, however, with their dependence on echolocation, there are several important modifications. While this has been worked out in only a few cases, both the stapedius muscle in the middle ear and the cochlea in the inner ear are involved. The stapedius muscle pulls the stapes away from the foramen ovale and this enables the bat to control hearing acuity in relation to sound emission by the larynx. Thus the powerful emission does not confuse the relatively weak returning signal. As mentioned above (in the account of the skull), this is also facilitated by the usually loose attachment of the periotic bone to the remainder of the skull. The cochlea in Microchiroptera tends to be enlarged, the extreme being in the Rhinolophidae, where there can be as many as 3–4 complete turns. The importance of this, of course, is that it allows pitch discrimination over a wide range of frequencies, thus permitting more sophisticated echolocation. In order to allow space for the large cochleae and also to lighten the weight of the large dense periotic bones, there is a tendency toward reduction in the amount of bone around the semicircular canals and cochlea, leaving only a thin semitransparent layer covering the membranous labyrinth.

BHATNAGAR (1980) has discussed the condition of the vomeronasal organ (of Jacobson) in a diversity of bats. While most members of the family Phyllostomidae have the organ well developed (evidently the primitive condition), most other bats show varying degrees of reduction, often total absence.

Digestive System:
The chief modifications are to be seen in the stomach. In insectivorous bats, the stomach is little modified, and in carnivorous and piscivorous bats, the only real change is a tendency to increase the size of the pyloric portion. In frugivores (and to a lesser extent nectarivores), on the other hand, the cardiac portion is enlarged, often to a marked degree, being usually more or less tubular. The greatest modification is to be found among Desmodontinae. Here the cardiac portion of the stomach is very large and sac-like or else more or less tubular, in either case very extensible in order to accomodate the very large blood meal which is characteristically taken by these bats at one feeding. The length of the intestine, as is usual in mammals, is relatively short in insectivores, but relatively long in frugivores.

Circulatory System:
The heart in bats is unusually large with a very high heart rate (up to 1000 beats per minute when in flight) with a large amount of blood pumped per beat. These rates tend to be higher in Microchiroptera than in Megachiroptera and, of course, much higher in flight than at rest (or especially in torpor). Hearts tend to be more elongated in bats than in other mammals, and in Microchiroptera, with a relatively shortened thorax, the heart tends to be rather transversely placed.

The peripheral circulation of bats agrees well with the usual mammalian pattern, except for the larger coronary and pulmonary vessels. These characters, together with the larger heart and rate of blood flow, reflect, of course, the much greater expenditure of energy involved in flight as compared with any other method of locomotion. A special problem is faced by bats in connection with the enormous amount of bare skin exposed on the wing membranes, which makes them very effective in dissipating excess body heat in flight under warm conditions, but can be a severe disadvantage while roosting at low temperatures. Bats, however, have a series of special shunts and valves which enable them to shut off blood to whole segments of the wing, thus reducing heat loss in those regions.

Urinogenital system:
There is nothing very distinctive about the gross morphology of the excretory part of this system. Some species of bats (e.g. *Myotis vivesi*) which feed on marine fish excrete a very concentrated urine. Desmodontinae face a special problem since they take in a large amount of blood at a single feeding. This creates a weight problem in flight and is resolved by rapid absorption and excretion of a large part of the water from the ingested blood. Later, when the concentrated blood is digested, water conservation becomes a problem and a very concentrated urine is excreted. Thus the same kidney must be able to produce either a very dilute or a very concentrated urine as circumstances require.

The reproductive portion of the system shows much more diversity. In the male, the testes frequently descend into a scrotum, at least during the breeding time but not always. The penis shows a great diversity of form from very small to enormous and there may be various peculiar accessory structures of unknown function. The baculum also shows great differences, particularly in size, varying from absent to the size of the tibia.

In the female, though there are always two ovaries, only one may be functional. Thus in some species of Rhinolophidae, Mormoopidae, Natalidae, and Vespertilionidae, all ova are produced by the right ovary, whereas in some species of Megadermatidae and Molossidae, only the left ovary is functional. The form of the uterus is quite diverse, varying from bipartite through bicornuate to simplex.

In many bats that live in climates with marked seasonality, either warm vs. cold or wet vs. dry, a variety of devices have been evolved to take advantage of a relatively short season of optimal development of the young. These include delayed ovulation with sperm storage (either in the male or in the female), delayed implantation, or embryonic diapause (at some time after normal implantation).

Though one young at a time (monotocous) is usual in bats, a number of instances of polytocy (more than one young at a birth) are known in Pteropodidae, Rhinolophidae, Phyllostomidae, and Vespertilionidae, particularly the latter, where three, four, and even five (*Lasiurus*) are known. While in most bats, implantation is relatively superficial, in most of the Phyllostomidae as well as in *Thyroptera*, it is interstitial. The primary amniotic cavity is transitory in most bats, the amniotic cavity being formed later by folds. However, in the Pteropodidae and (independently) in the Phyllostomoidea, the primary amniotic cavity persists. The placenta is variable in its development but usually eventually assumes a more or less discoidal form. For analysis of these and other embryological features, see LUCKETT 1980.

Newborn bats are relatively large at birth though, in the case of most Microchiroptera, they are naked with their eyes still closed. In the Megachiroptera and Phyllostomidae, they are particularly large and well-haired with the eyes open. In most Microchiroptera, development is rapid, adult size and flight ability being attained in a few months. That of Megachiroptera proceeds more slowly.

Ecology:
Bats are unique among mammals in having what almost amounts to a dual ecology, the roosting ecology and the feeding ecology, which may be widely separated spacially. These are relatively independent inasmuch as two species may occupy the same roost yet have different feeding habits and visa versa. This adds another dimension to any comparison of ecologies of related species. These two aspects of bat ecology will be taken up in turn.

Since virtually all bats are nocturnal, they need some place to hang up during the day and this must offer at least some shelter from the weather and predators and usually from at least direct sunlight. However, different bats may vary widely in their preference or even tolerance for a roost site. Some bats, particularly in the families Pteropodidae and Emballonuridae, roost in relatively open places such as tree branches, the sides of tree trunks, cliff faces, etc. Caves of various types, are, of course, another common roost site. Hollow trees are used by a number of species and some hang up under leaves. Some species of the subfamily Stenodermatinae are able to modify leaves in various ways to make them more effective shelters. Many cave and some tree-roosting species are able to utilize various man-made structures, usually those that provide a micro-environment similar to that of the natural roost. Some bats have very specialized roosts. These include crevices of various kinds, either in or under rocks, or occasionally between closely adpressed leaves. Such crevice roosts are particularly favored by many Molossidae. *Tylonycteris* (Vespertilionidae) roosts inside of hollow bamboo stems passing in and out through cracks or holes, in part made by insects. Finally, the Thyropteridae roost inside of large rolled-up leaves and must shift their roosts when the leaves open.

Bats are equally diverse in their feeding ecology. Most are insectivorous, but insects (and other small arthropods) may be obtained either out of the air (aerial insectivory) or by gleaning from the ground or vegetation. Some gleaners take insects off the surface of the water and from such habits, fish catching has been evolved. Some species of the larger insectivorous bat groups (Nycteridae, Megadermatidae, Phyllostominae) have become carnivorous, feeding on birds, rodents, and other bats. *Trachops* (Phyllostominae) is a frog and lizard specialist. Some bats, particularly in the families Pteropodidae and Phyllostomidae are frugivores with, among various species, a variety of fruits being eaten, from very hard (or even green) fruit to very soft over-ripe fruit. Some bats in these same two families obtain energy from the taking of nectar. The problem here is that nectar is very poor in protein and therefore must be supplemented with either insects or pollen. Finally, in the Desmodontinae, the diet consists entirely of blood. While the nature of the evolutionary transition is not clear, Desmodontinae seem most closely related to nectar-feeders, some of which also eat soft juicy fruit.

Bats living in temperate regions face a special problem. Since food (generally insects) is largely unavailable in winter, bats must either lower their metabolism by going into hibernation or else migrate to warmer climates, an option which is availaible to them, since, like migratory birds, they can fly.

Hibernation in bats has been extensively studied, but mostly in a relatively small number of species in Europe and North America (chiefly Vespertilionidae). An important consideration is the existence of temperature and humidity conditions which, for any particular species, must enable it to keep its metabolism low (but above freezing) and without excressive desication. The number of hibernacula (usually caves) with these characteristics may well be a limiting factor for some species.

The other strategy, migration, may well be more widespread, but it is known in detail for only a few cases because it depends on extensive banding and recovery programs. European populations of *Nyctalus noctula* are known to make fairly extensive migrations from northern Europe to more southern areas on the continent. Likewise, the eastern populations of *Tadarida brasiliensis mexicana* in the south-central and southwestern United States migrate (at least the females) from maternity colonies south to various parts of Mexico. Males of these populations may remain in Mexico all year round. In *Lasiurus cinereus*, both males and females largely disappear from Canada and the northern United States during the winter, presumably migrating to warmer areas to the south.

Some bats combine hibernation and migration, inasmuch as suitable hibernacula may be in some distance (up to several hundred kilometers) from summer roosting sites (particularly maternity colonies) which may be too warm to serve as hibernacula. This has been best documented in two North American *Myotis* (*lucifugus* and *grisescens*).

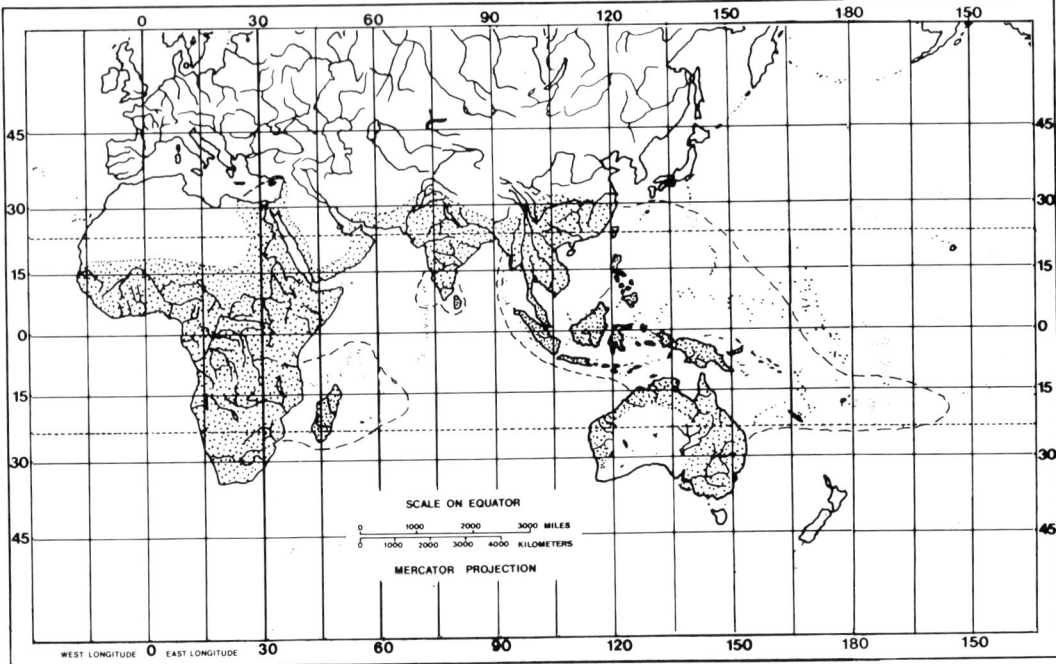

Fig. 7. Distribution of the family Pteropodidae.

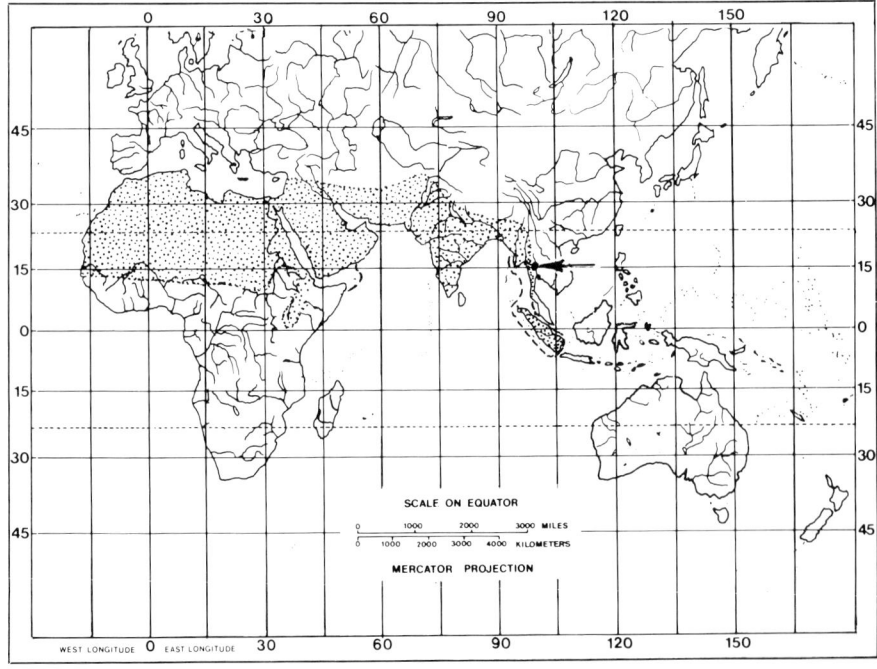

Fig. 8. Distributions of the families Rhinopomatidae (main range) and Craseonycteridae (arrow in Thailand).

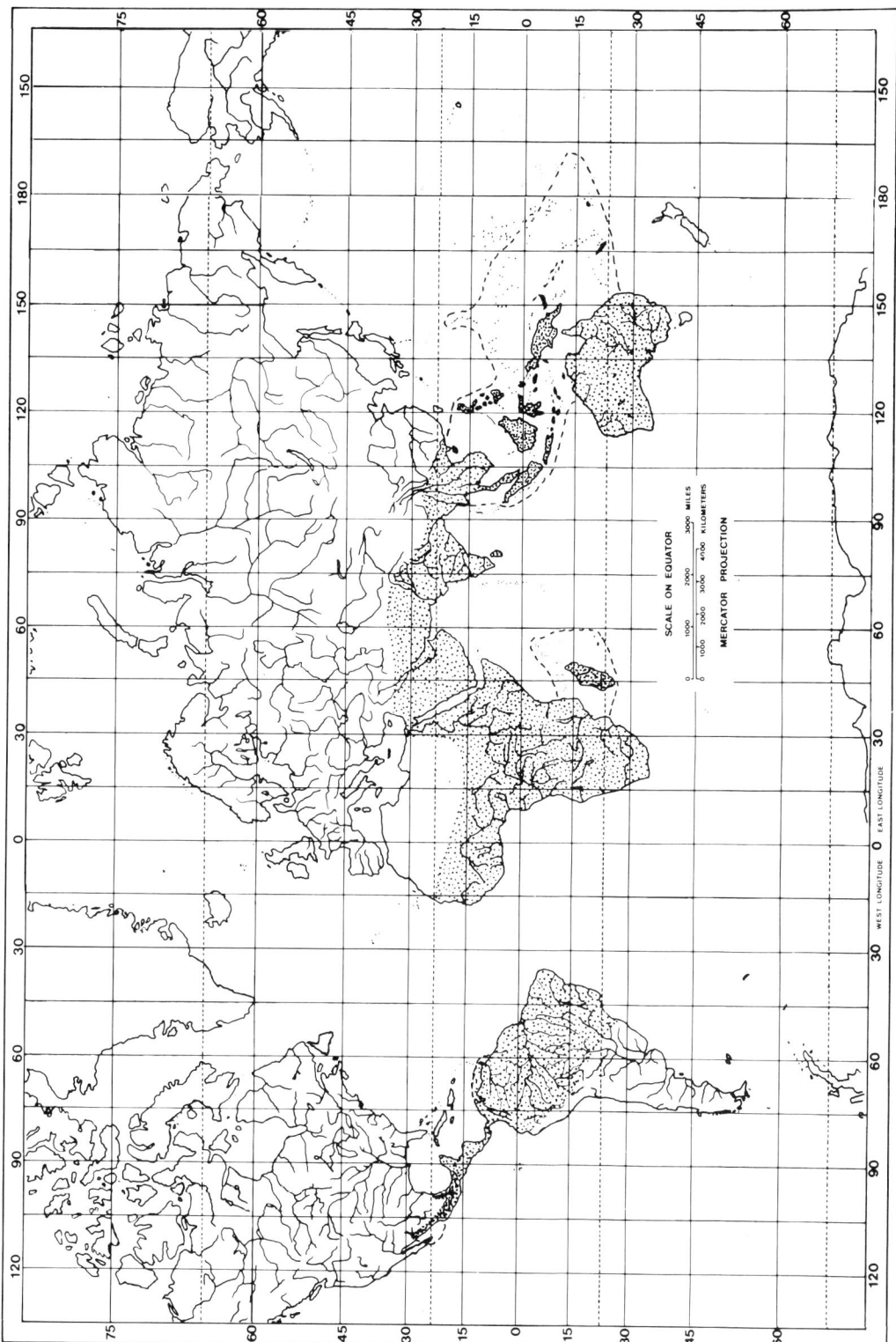

Fig. 9. Distribution of the family Emballonuridae.

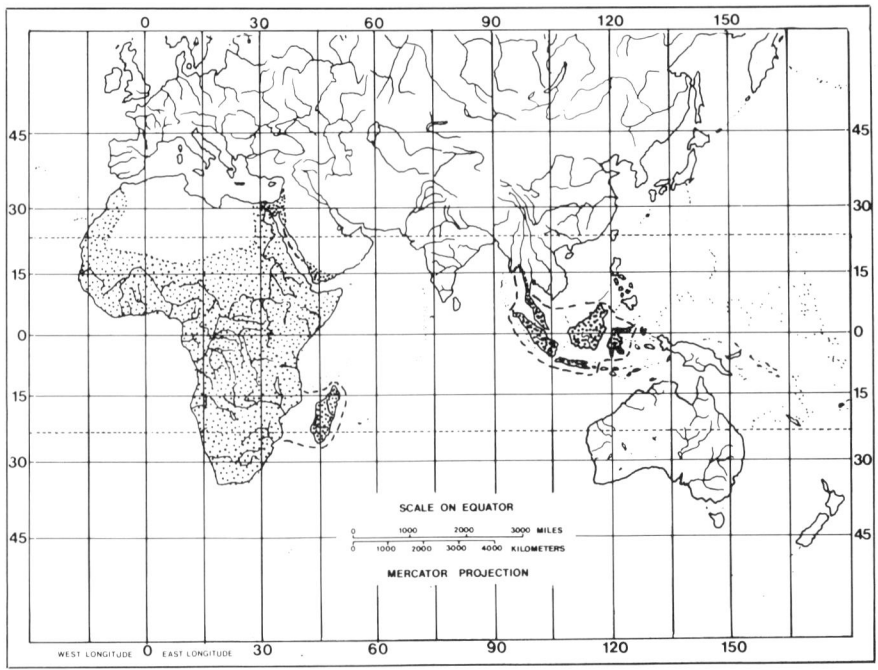

Fig. 10. Distribution of the family Nycteridae.

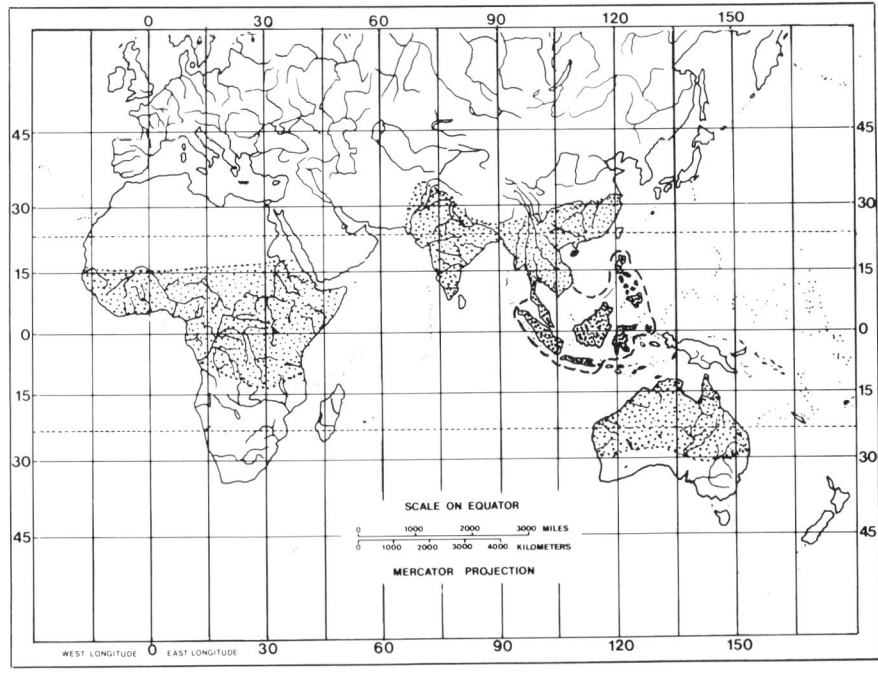

Fig. 11. Distribution of the familiy Megadermatidae.

Distribution:
Except for the primates (including *Homo*), the Chiroptera are the most widespread of mammalian Orders. While a majority of species occur in the wet tropics of the eastern and western hemispheres (Ethiopian, Indo-Malayan, Australian, and Neotropical regions), the bats have colonized dry savanna, and even desert habitats. They extend north to the limit of trees in both Eurasia and North America and south to the southern extremities of Africa, Australia, New Zealand, and South America, thus occurring on all continents except Antarctica. The Chiroptera, as result of their powers of flight, have been able

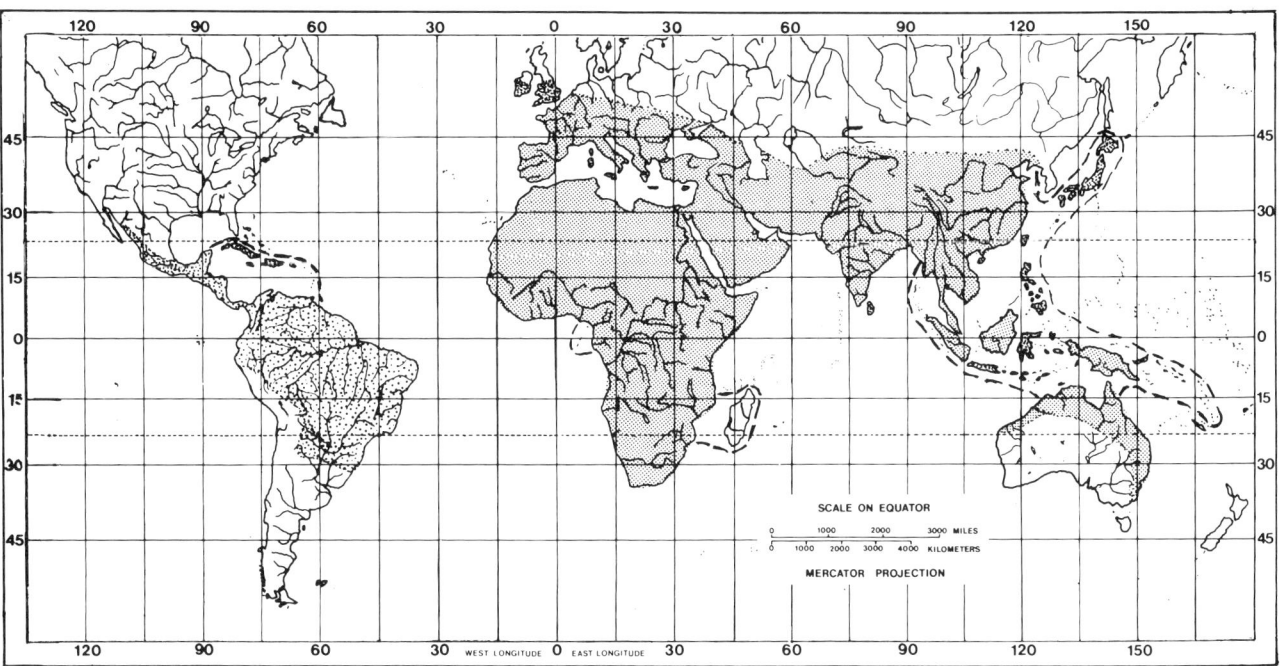

Fig. 12. Distributions of the families Rhinolophidae (Old World) and Noctilionidae (New World).

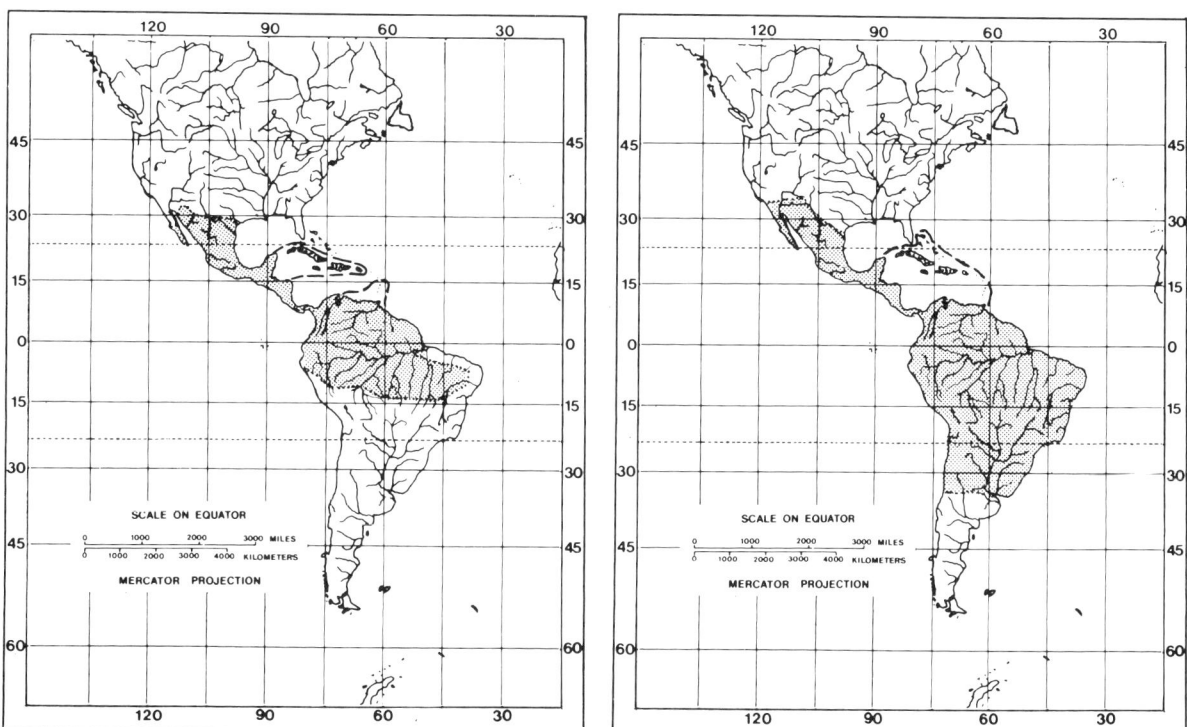

Fig. 13. Distribution of the family Mormoopidae. Fig. 14. Distribution of the family Phyllostomidae.

to colonize many oceanic islands, on some of which they are the only native mammals. These include Bermuda and the Azores in the Atlantic; the Seychelles, Reunion, Mauritius, Rodriguez, and the Maladives in the Indian Ocean. In the North Pacific, these include the Hawaiians, Bonins, Mariannas, and Carolines; in the South Pacific, the Santa Cruz, New Hebrides, New Caledonia, Loyalties, Fijis, Samoa, Cooks, and New Zealand. Various nearer oceanic islands in the

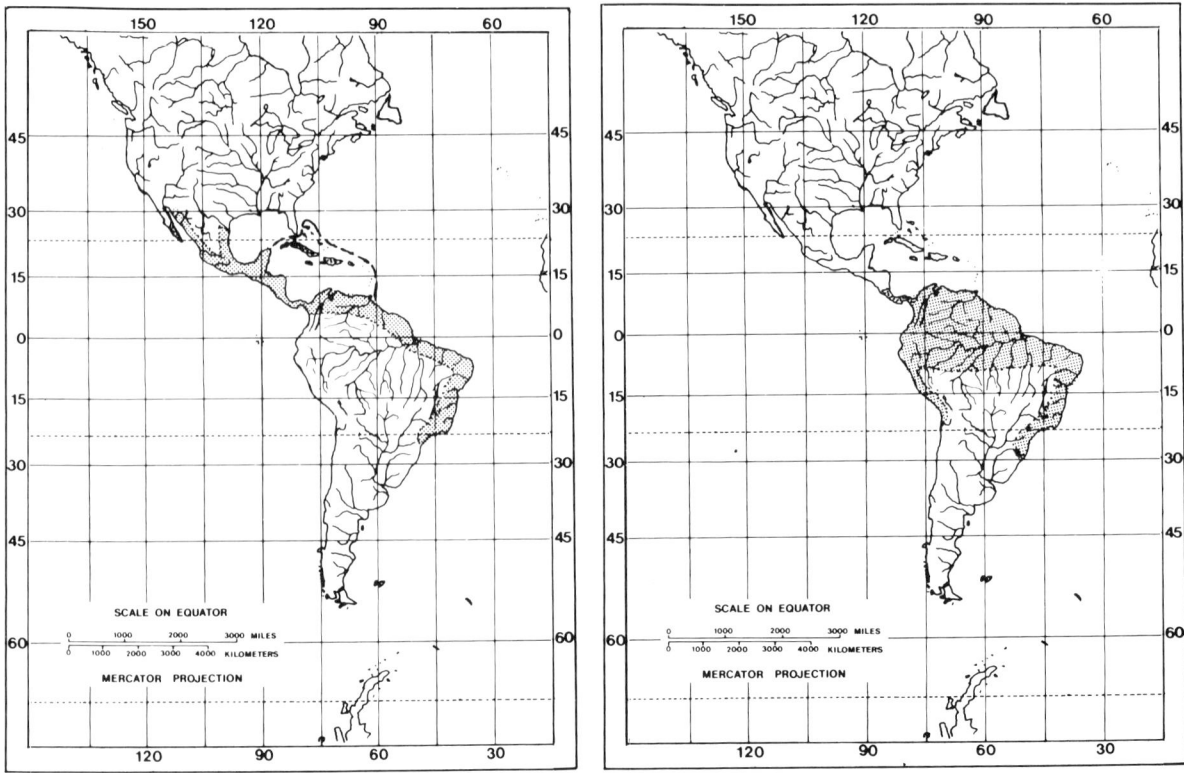

Fig. 15. Distribution of the family Natalidae.   Fig. 16. Distribution of the family Furipteridae.

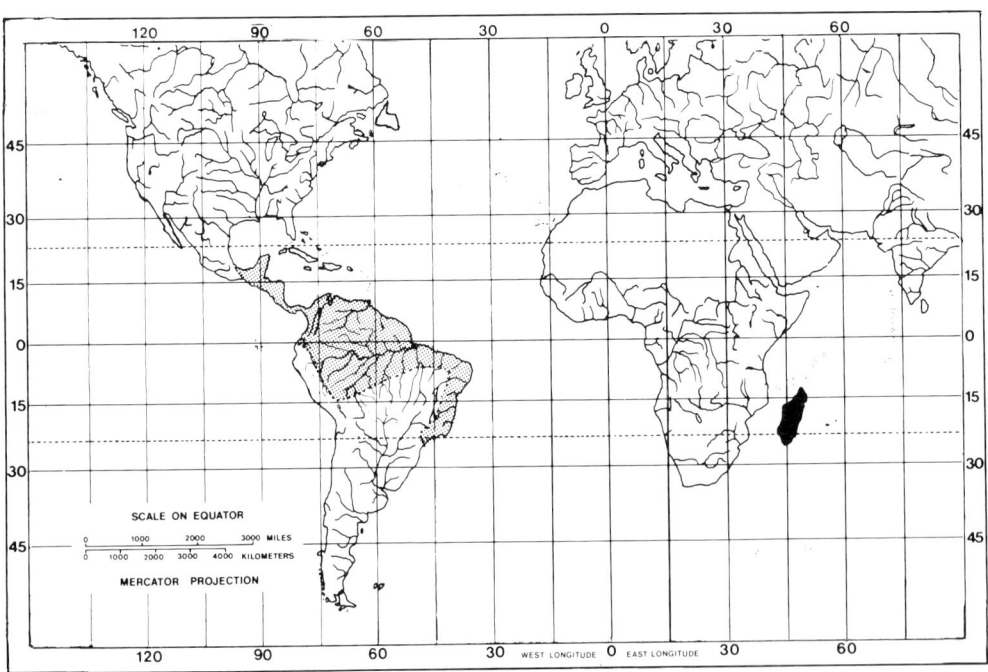

Fig. 17. Distributions of the families Thyropteridae (New World) and Myzopodidae (Madagascar).

tropics support a rich bat fauna. These include Madagascar, the Philippines, Celebes, Lesser Sundas, Moluccas, Bismarcks, Solomons, and West Indies.

Phylogeny:

Several authors (GREGORY 1910, McKENNA 1975, KOOPMAN & MACINTYRE 1980) see the affinities of the Chiroptera within the supraordinal group

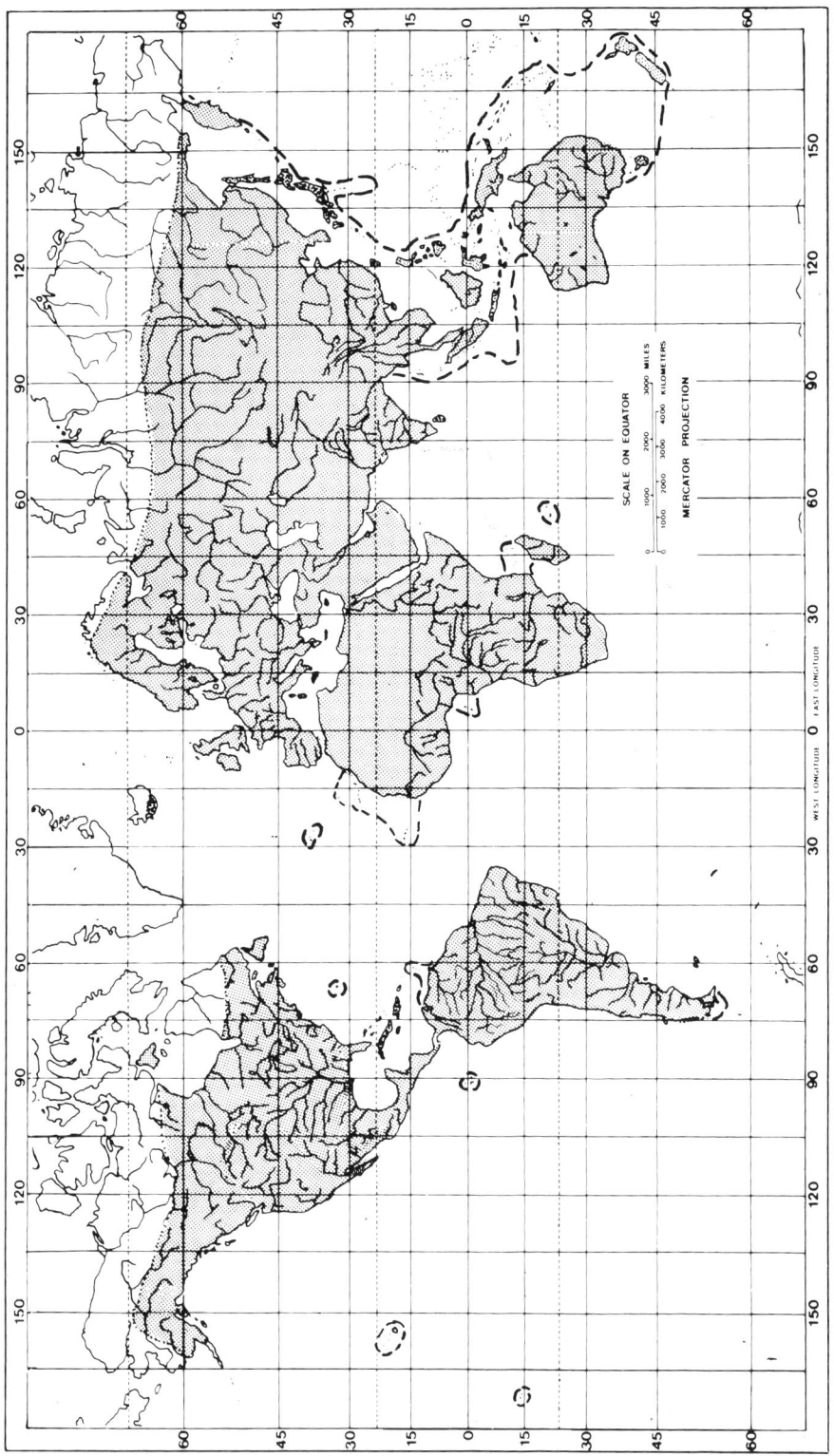

Fig. 18. Distribution of the family Vespertilionidae.

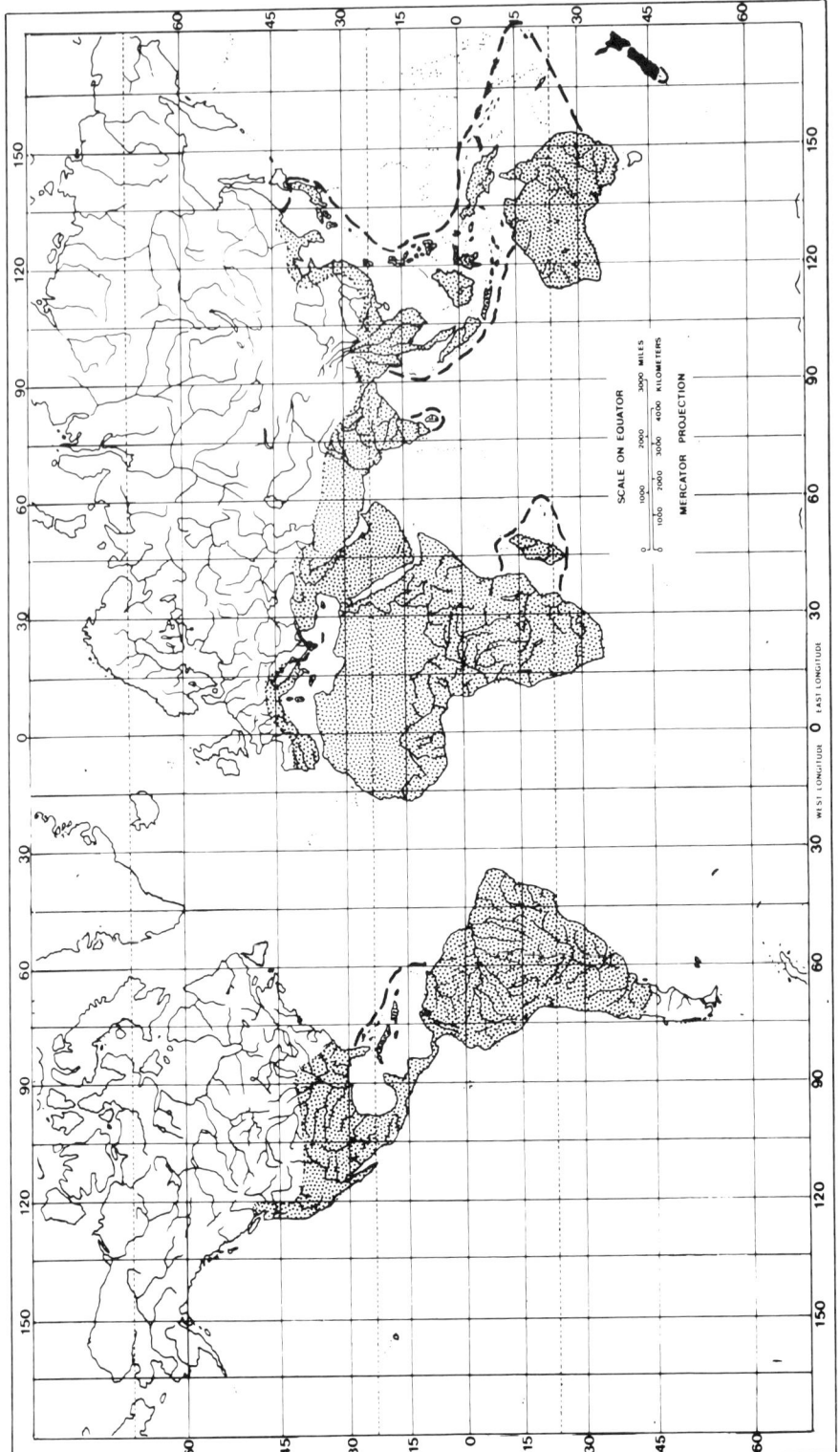

Fig. 19. Distributions of the families Mystacinidae (New Zealand) and Molossidae (main range).

Archonta which comprises Primates, Scandentia, Dermoptera and Chiroptera as well. The taxon Archonta expresses the authors' view that these groups may well have had a common origin; more generally is accepted that these groups originated in an early (perhaps Cretaceous) radiation from tree-shrew-like forms. All these groups posses marked specialisations of their own. This also applies to the Dermoptera for which often a closer relation with the Chiroptera has been proposed. Both orders show extensive structural specialisations of their own; even the patagia of the Chiroptera, especially the chiropatagium, may not be derived from the highly development gliding membranes of the Dermoptera.

In recent years some (e.g. SMITH & MADKOUR 1980) have held that Megachiroptera and Microchiroptera independently evolved flight. While some characters have been cited to show closer relationships between Megachiroptera and Primates than of either with Microchiroptera, the close and unique resemblance in wing structure between the two bat groups make this hypothesis seem unlikely to me.

Systematics:
In the Recent, two suborders, 17 families, 174 genera, 913 species.

Suborder **Megachiroptera** DOBSON 1875
(Old World Fruit and Nectar Bats)

Second finger retaining independence, not clearly associated with third finger, and usually bearing a claw. Humerus with small trochiter and trochin, the former never articulating with the scapula. The external ear is simple, with the margin complete, and without a tragus. The periotic bone small and not compressing the basioccipital. The postorbital processes are well developed. The bony palate is extended behind the last molar, usually narrowing gradually. Incisors never more than 2/2. Arrangement of gray matter in spinal cord essentially as in flightless mammals. Echolocation rarely developed. – One family, 41 genera, 164 species, in the Recent.

Family **Pteropodidae** GRAY 1821
("Flying Foxes")

Structure: Small to large bats (forearm 37–220 mm) with the uropatagium poorly developed and the tail usually short or absent (long only in *Notopteris*). Skeleton with all the basic bat specializations but otherwise generally unmodified. Canines long, and molariform teeth highly specialized for dealing with a fruit or nectar diet, the cusp pattern so modified as not to be clearly homologized with a primitive Eutherian mammal molar. Cardiac portion of stomach elongate. Penis pendent, with a baculum. Except for specializations related to fruit and nectar feeding, these are the most primitive of bats.

Ecology: This familiy occupies all the various fruit and nectar-feeding niches in the tropics and subtropics of the eastern hemisphere. Most species roost in relatively open situations, often in the branches of trees or under leaves. A few species are found in caves, usually near the entrance and except for *Rousettus* (most species of which have a crude echolocation) not in the deeper darker parts of caves. Only rarely are man-made structures used fo roosting.

Distribution (Fig. 7): This family is chiefly tropical and only a few species extend into temperate areas and then not far. In the northern hemisphere it extends north to Turkey in the west and southernmost Japan in the east. Thus, the Pteropodidae occur over most of Africa except for the northwest, across southern Asia and through the East Indies to New Guinea, northern and eastern Australia, but not Tasmania. Bats of this family also occur on most of the islands in the Indian Ocean, in Micronesia east to the eastern Carolines, through Melanesia from the Bismarcks through the Solomons to New Caledonia and Fiji, also Samoa and the Cooks, but only as a rare straggler to New Zealand. The Pteropodidae are, of course, completely absent from the Western Hemisphere.

Systematics: In the Recent, there are two subfamilies, six tribes, 42 genera and 164 species. The basic treatment is that of ANDERSEN (1912).

Subfamily **Pteropodinae** GRAY 1821

Tongue not especially extensible, fixed to floor of mouth by posterior half and without filiform papillae at tip. These are all the Pteropodidae which do not have specialized nectar-feeding tongues (though many are flower feeders). Size ranges from small to largest of all Chiroptera (forearm 56–220 mm). – *Distribution:* Geographical range the same as for the family. – Four tribes, three additional subtribes, 36 genera, 152 species.

Tribe Pteropodini GRAY 1821

Molarform teeth not multicuspidate. Lower canines not proclivous. Facial axis of skull conspicuously deflected against basicranial axis. Never less than two upper molars. Size ranges from medium to largest of all Chiroptera. – *Distribution:* The same as for the family. – Three subtribes, 11 genera, 94 species.

Subtribe Rousettina KOOPMAN & JONES 1970

Two pairs of upper incisors always present. Occipital region of skull not elongate and a short tail present. Two pairs of lower incisors present (and usually two pairs of uppers) and three pairs of upper premolars. Size ranges from medium to large (forearm length, 56–132 mm). – *Distribution:* The geographical range includes virtually all of Africa except the northwest, Madagascar, southern Asia north to southern Turkey and southern China, the East Indies (including the Philippines) east to New Guinea, also the Bismarcks and Solomons. – Four genera, three additional subgenera, 14 species.

Genus *Eidolon* RAFINESQUE 1815 (= *Pterocyon* PETERS 1861) (Fig. 20)

Tympanic bone extended as a short bony auditory meatus. Palate much broader posteriorly than between canines. Length of rostrum much greater than width across lacrimals. Front of orbit vertically above middle or posteroir half of first upper molar. Basicranial axis only moderately deflected in relation to palate. Premaxillaries separated in front. Dental formula i2/2, c1/1, p3/3, m2/3 × 2 = 34. First upper premolar (in cross section) much larger than an upper incisor. First lower molar equal in length to other two lower molars combined. Often roosts from open branches in large colonies, which are frequently migratory. – *Distribution*: The single species is widely distributed in sub-Saharan Africa (including islands in the Gulf of Guinea and off East Africa) as well as southwestern Arabia and Madagascar.

1. *E. helvum* (KERR 1792) (= *stramineus* E. GEOFFROY 1803). – Size relatively large (forearm length, 105–135 mm). – *Distribution:* Same as for genus. – Three currently recognized subspecies:

E. h. sabaeum (southwestern Arabia), *E. h. helvum* (sub-Saharan Africa), *E. h. dupreanum* (Madagascar).

Genus **Rousettus** GRAY 1821 (Fig. 21)

Tympanic bone not extending into auditory meatus. Palate usually clearly broader posteriorly than between canines. Length of rostrum always at least somewhat greater than width across lacrimals. Front of orbit vertically above middle or posterior half of first upper molar. Basicranial deflection variable. Premaxillaries in contact or fused in front. Dental formula i2/2, c1/1, p3/3, m2/3 x 2 = 34. First upper premolar (in cross section) much reduced (equal to or smaller than upper incisor). First lower molar shorter than other two molars combined. This is the most diverse genus of the Rousettina but all species are of more or less medium size (forearm 65–102 mm). – *Distribution:* It ranges through sub-Saharan Africa (except for most of the arid southwest) and Egypt around the eastern end of the Mediterranean as far as Cyprus and southern Turkey. It also occurs on the Comoros and Madagascar and across southern Asia to southern China and Malaya, through the East Indies (including the Philippines) to New Guinea and on the Bismarcks and Solomons. – Three subgenera and nine species.

Subgenus **Rousettus** GRAY 1821

Braincase moderately deflected. Premaxillaries not fused. First lower premolar much larger in bulk than a lower incisor. Width of last upper premolar about one third that of palate between anterior ends of last upper premolars. Edge of plagiopatagium attaches to first toe. Antitragal lobe of ear distinct. – *Distribution*: The same as for the genus. – Seven species are currently recognized, with eight additional subspecies.

1. *R. aegyptiacus* (E. GEOFFROY 1810). – The largest species in the subgenus (forearm length, 82–102 mm; pollex (thumb), 30–38 mm, second phalanx of third digit, 50–62 mm). Molars relatively broad. Median edge of plagiopatagium attached to side of body. – *Distribution:* This is the only species of the subgenus on the African mainland, but it also extends around the eastern end off the Meditteranean to southern Turkey and Cyprus. It also extends across central and southern Arabia and southern Iran to Pakistan, and occurs on the islands of Fernando Poo and São Thomé in the Gulf of Guinea and of Pemba, Zanzibar, and Mafia off the east African coast. – Four subspecies are currently recognized:

R. a. leachi (southern Sudan south to southern Africa), *R. a. unicolor* (= *occidentalis*) (western Africa from Senegal to Angola), *R. a. aegyptiacus* (Egypt north to Cyprus and extreme southern Turkey), *R. a. arabicus* (eastern Ethiopia east to Pakistan).

2. *R. leschenaulti* (DESMAREST 1820). – A medium sized species (forearm length, 75–96 mm; pollex, 23–31 mm; second phalanx of third digit, 41–51 mm). Molars relatively broad. Median edge of plagiopatagium attached to side of body. Uropatagium naked. Body fur relatively short. Last lower molar about twice as long as broad. – *Distribution:* This species ranges from Pakistan across India and the Indo-Chinese region to southeastern China, Vietnam, and peninsular Thailand, also Sri Lanka and, with a gap in Malaya, Sumatra, Simalur, Java and Bali. – Three subspecies may be recognized:

R. l. leschenaulti (mainland range), *R. l. seminudus* (Sri Lanka), *R. l. shortridgei* (Sumatra, Simalur, Java, Bali).

3. *R. amplexicaudatus* (E. GEOFFROY 1810). – A small to medium sized species (forearm length, 66–91 mm; pollex, 24–30 mm; second phalanx of third digit, 39–47 mm). Molars relatively broad. Median edge of plagiopatagium attached to side of body. Uropatagium naked. Body fur relatively short. Last lower molar broader in relation to length than in *leschenaulti* (breadth from five sixths to two thirds of length). – *Distribution:* This species ranges from Thailand through the Malay peninsula and East Indies to the Philippines and the Solomon islands. – While the status of subspecies in this species is confused, four subspecies

are currently recognized (ROOKMAAKER & BERGMANS 1981):

R. a. amplexicaudatus (=*philippinensis*) (Thailand, Malay peninsula, Mentawai islands, Borneo, Philippines, Moluccas, Timor and Sumba in the Lesser Sundas), *R. a. infumatus* (=*minor*) (Sumatra, Java, Lesser Sundas east to Flores), *R. a. stresemanni* (New Guinea), *R. a. brachyotis* (=*hedigeri*) (Bismarcks, Solomons). – Allocation to subspecies of populations from Sulawesi is uncertain.

4. *R. spinalatus* BERGMANS & HILL 1980. – A medium sized species (forearm length, 79–89 mm; pollex, 25–26 mm; second phalanx of third digit, 36–41 mm). Molars relatively broad. Median edge of plagiopatagium attached close to the mid-dorsal line. Uropatagium almost naked. Body fur variable but fairly short. – *Distribution:* The species is known only from Sumatra and Borneo. – No subspecies.

5. *R. celebensis* ANDERSEN 1907. – A medium sized species (forearm length, 67–83 mm; pollex, 28–30 mm; second phalanx of third digit, 40–41 mm). Molars relatively narrow. Median edge of plagiopatagium attached to side of body. Uropatagium well haired. Body fur relatively long. – *Distribution:* This species is known only from Sulawesi and the Sanghir islands. – No subspecies.

6. *R. madagascariensis* GRANDIDIER 1929. – A small species (forearm length, 65–76 mm). Molars relatively narrow. Median edge of plagiopatagium attached to side of body. Uropatagium sparsely haired. Body fur fairly short. – *Distribution:* The species is known only from Madagascar. – No subspecies.

7. *R. obliviosus* KOCK 1978. – A small species (forearm length, 70–75 mm; pollex, 22–25 mm). Molars relatively broad. Median edge of plagiopatagium attached to side of body. Uropatagium relatively naked. Body fur relatively short. Basicranial deflection greater than in other species of *R.* (*Rousettus*), but not as great as in *R.* (*Stenonycteris*). – *Distribution:* This species is confined to the Comoro islands (between eastern Africa and northern Madagascar). – No subspecies.

Subgenus ***Stenonycteris*** ANDERSEN 1912

Braincase strongly deflected. Premaxillaries not fused. First lower premolar much larger in bulk than a lower incisor. Width of last upper premolar about one fifth that of palate between anterior ends of last upper premolars. Molars very narrow. Edge of plagiopatagium attached to second toe. Antitragal lobe of ear obsolete. Several of the characters are approached among some species of *R.* (*Rousettus*) (*celebensis, madagascariensis, obliviosus*), but none show the complete constellation so highly developed. – *Distribution:* A single species confined to eastern Africa.

8. *R. lanosus* THOMAS 1906. – A medium sized species (forearm length, 84–91 mm). – *Distribution:* Confined to mountainous areas of eastern Africa from Tanzania and eastern Zaire to Ethiopia. – Two subspecies:

R. l. lanosus (eastern Zaire, Burundi, Rwanda, Uganda), *R. l. kempi* (Tanzania, Kenya, southeastern Sudan, Ethiopia).

Subgenus ***Lissonycteris*** ANDERSEN 1912

Braincase only slightly deflected. Premaxillaries fused. First lower premolar subequal in bulk to a lower incisor. Molars broad. Edge of plagiopatagium attached to second toe. Antitragal lobe of ear distinct. Orbits larger than in the subgenera *Rousettus* and *Stenonycteris*. – *Distribution:* A single species confined to tropical Africa.

9. *R. angolensis* (BOCAGE 1889). – A medium sized species (forearm length, 66–84 mm). – *Distribution:* ranging from Senegal to Ethiopia and south to Angola and Mozambique. – Three currently recognized subspecies:

R. a. ruwenzorii (Ethiopia to Uganda and Tanzania), *R. a. angolensis* (=*crypticola*) (Cameroon to Zimbabwe, Fernando Poo), *R. a. smithi* (Senegal to Togo).

Genus ***Myonycteris*** MATSCHIE 1899 (Fig. 22)

Tympanic bone not extending into auditory meatus. Palate slightly broader posteriorly than between canines. Length of rostrum only slightly greater than width across lacrimals. Front of orbit vertically above back of last upper premolar or front of first upper molar. Basicranial deflection slight to moderate. Premaxillaries fused. Dental formula i 2/2, c 1/1, p 3/3, m 2/3 x 2 = 34 or i 2/2, c 1/1, p 3/3, m 2/2 x 2 = 32. First upper premolar somewhat larger in cross section than an upper incisor. First lower molar longer than other two molars combined. Size fairly small (forearm 56–75 mm). – *Distribution:* The genus ranges through much of tropical Africa and on at least some of the islands of the Guinea Gulf. – Two subgenera and three species are recognized (BERGMANS 1980).

Subgenus *Myonycteris* MATSCHIE 1899

Molars and premolars relatively weak. Outer and inner ridges of the last lower premolar fused anteriorly. Lower canine at least as high as middle lower premolar. No sagittal crest. Coronoid process of mandible narrow at base. – *Distribution:* The entire distribution of the genus with the exception of Saõ Thomé in the Gulf of Guinea. – Two species are currently recognized.

1. *M. relicta* BERGMANS 1980. – The largest species of the genus (forearm length, 69–75 mm) with a relatively long rostrum and greater cranial deflection than in other species of *Myonycteris*. In these respects, it approaches the genus *Rousettus*, particularly *R.* (*Lissonycteris*). Last lower molar absent. – *Distribution:* A limited coastal forest distribution in southeastern Kenya and northeastern Tanzania. – No subspecies.

2. *M. torquata* (DOBSON 1878). – A relatively small species (forearm length, 54–68 mm) with a relatively short rostrum and relatively little cranial deflection. Last lower molar present. – *Distribution:* An extensive distribution across tropical Africa (but allopatric with *M. relicta*) from Sierra Leone to Angola and Uganda and on Fernando Poo. – Three currently recognized subspecies:

M. t. wroughtoni (western Uganda and at least northeastern Zaire), *M. t. torquata* (Cameroon to Angola and Zambia, Fernando Poo), *M. t. leptodon* (Togo to Sierra Leone).

Subgenus *Phygetis* ANDERSEN 1912

Molars and premolars relatively strong. Outer and inner ridges of last lower premolar well separated. Lower canine not as high as middle lower premolar. Sagittal crest present. Coronoid process of mandible broad at base. – *Distribution:* Known only from São Thomé island in the Gulf of Guinea. – A single known species.

3. *M. brachycephala* (BOCAGE 1889). – A small species (forearm length, 59–64 mm) with a short rostrum and relatively little cranial deflection. Last lower molar present. – *Distribution:* Known only from São Thomé. – No subspecies (FEILER 1984).

Genus *Boneia* JENTINCK 1879 (Fig. 23)

Tympanic bone not extending into auditory meatus. Palate relatively broad anteriorly so that the upper toothrows are almost parallel. Length of rostrum greater than width across lacrimals. Front of orbit vertically above posterior half of first upper molar. Basicranial deflection marked. Premaxillaries separated in front. Dental formula i 1/2, c 1/1, p 3/3, m 2/3 x 2 = 32. First upper molar (in cross section) somewhat larger than the upper incisor. First lower molar shorter than other two molars combined. – *Distribution:* Confined to Celebes. – A single species.

1. *B. bidens* JENTINCK 1879. – Medium sized (forearm length, 92–102 mm). – This species is only known from a small area, the eastern part of the northern peninsula of Celebes. – No subspecies.

Subtribe Pteropodina GRAY 1821

Occipital region of skull elongate, tail absent. Size ranges from medium to the largest of all bats (forearm, 86–220 mm). Two pairs of upper incisors present (and usually two pairs of lowers) and three pairs of upper premolars. – *Distribution:* From islands in the western Indian Ocean east to the central Pacific, north to the Ryukyu and Bonin islands and south to southeastern Australia. – Five genera, 68 species.

Genus *Pteropus* ERXLEBEN 1777 (Fig. 24)

Premaxillaries in simple contact in front. Second lower incisor less than 10 times the bulk of the first lower incisor. Upper canine usually without a secondary cusp. No well-differentiated, anterointernal tubercle on the last upper premolar and the first upper molar. Usually no sharply defined inner basal ledges on the last lower premolar or the first lower molar. Dental formula i 2/2, c 1/1, p 3/3, m 2/3 x 2 = 34, but anterior upper premolar very small und often lost. Claw present on second digit of wing. – *Distribution:* The same as that of the subtribe including Pemba and Mafia islands off the east African coast, the Seychelles, Aldabra, Comoros, Madagascar, all three of the Mascarene islands, the Maldives, the Indian subregion east through southeastern Asia to the Ryukyus, through the entire East Indies to New Guinea and much of Australia; east through islands of the tropical Pacific to the Cooks, also Micronesia from the Bonins to the Palaus and Kusiae (eastern Caroline islands). – Currently, some 57 species and 73 additional subspecies may be recognized, but the status of many is uncertain.

1. *P. hypomelanus* TEMMINCK 1853 [*subniger* group]. – Posteroir basal ledges of larger premolars distinct. Skull and dentition primitive for the genus. Tibia naked dorsally. Uropatagium poorly developed. Ear length moderate. Breadth of last upper premolars about one third that of palate between them. Forearm length, 121–148 mm. Total length of skull, 61–69 mm. – *Distribution:* Ranging (mostly on small islands) from the Maldives *(P. h. maris)*, southeastern Asia (5 subspecies), the West Sumatran islands (*P. h. simalurus, P. h. enganus)*, the Natunas (*P. h. canus, P. h. annectens*), northern Borneo *P. h. tomesi*), Philippines (*P. h. cagayanus*), Celebes (*P. h. macassaricus*), northern Moluccas (*P. h. hypomelanus*), and northern New Guinea to the Bismarcks and Solomons (*P. h. luteus)*. – There are 15 subspecies in all.

2. *P. mearnsi* HOLLISTER 1913 [*subniger* group]. – Smaller than *P. hypomelanus* (forearm length, 123 mm; condylobasal length of skull, 53–54 mm). These contrast with Philippine *P. hypomelanus* (forearm length, 130–148 mm; condylobasal length of skull, 60–62 mm). Probably a synonym of *P. speciosus*. – *Distribution:* Confined to Basilan island and the Zamboanga peninsula of Mindanao in the Philippines. – No subspecies.

3. *P. pumilus* MILLER 1910 (= *balutus* HOLLISTER 1913; *tablasi* TAYLOR 1934) [*subniger* group]. – Smaller than *P. mearnsi* (forearm length, 98–118 mm; condylobasal length of skull, 48–52 mm). – *Distribution:* Confined to the Philippines, but not known from Luzon or the Palawan group. – No subspecies (see KLINGENER & CREIGHTON 1984).

4. *P. speciosus* ANDERSEN 1908 [*subniger* group]. – Smaller than *P. hypomelanus* (forearm length, 118–123 mm; total length of skull, 55–61 mm) and evidently somewhat smaller than *P. mearnsi*, but larger than *P. pumilus*. Ears relatively long (25–26 mm). Distinction from *P. griseus* (the following species) not clear but may be darker in color. – *Distribution:* Known only from the southern Philippines and two islands in the Java Sea, south of Borneo (Solombo Besar, Mata Siri). – No subspecies.

5. *P. griseus* E. GEOFFROY 1810 [*subniger* group]. – Smaller than *P. hypomelanus* (forearm length, 114–128 mm; total length of skull, 56–60 mm). Ears relatively long (23–27 mm). – *Distribution:* Ranges from the southern Moluccas and Timor to Celebes and possibly the Philippines. – Three subspecies:

P. g. mimus (Celebes, ?Luzon), *P. g. griseus* (Timor, small islands south of Celebes), *P. g. pallidus* (Banda islands in the southern Moluccas).

6. *P. faunulus* MILLER 1902 [*subniger* group]. – Size smaller than *P. hypomelanus* (forearm length, 114–118 mm; total length of skull, 54–55 mm). Ears relatively short (21–22 mm). – *Distribution:* Known only from the Nicobars. – No subspecies.

7. *P. admiralitatum* THOMAS 1894 [*subniger* group]. – Size small to medium for its group (forearm length, 104–126 mm; total length of skull, 52–57 mm; maxillary toothrow length, 18–22 mm). Ears relatively short (21–23 mm). Geographically variable in amount of hair on the dorsal side of the tibia and in rostral proportions. – *Distribution:* This species is restricted to the Bismarcks and Solomons. – Four subspecies are currently recognized:

P. a. admiralitatum (Bismarcks), *P. a. colonus* (Choiseul and small nearby islands in the western Solomons), *P. a. solomonis* (several islands in the central and eastern Solomons), *P. a. goweri* (Ndai island on the northern edge of the Solomons). *P. a. solomonis* and *P. a. goweri* have shorter rostra and hairier tibiae than *P. a. admiralitatum* and *P. a. colonus* and as a result have by some been put in a different species or even transferred to the *chrysoproctus* group (FELTEN & KOCK 1972).

8. *P. brunneus* DOBSON 1878 [*subniger* group]. – Medium in size (forearm length, 118 mm; maxillary tooth row length, 24 mm). Tibia hairy dorsally. – *Distribution:* A poorly known form (only one slightly immature specimen collected on the Percy island off the eastern coast of Queensland, Australia and possibly an accidental from somewhere in the Solomon islands) and conspecific with *P. admiralitatum* (KOOPMAN 1984a). – No subspecies.

9. *P. howensis* TROUGHTON 1931 [*subniger* group]. — Size larger than at least the Solomon island subspecies of *P. admiralitatum* (forearm length, 116–122 mm; total length of skull, 53–56 mm; maxillary tooth row length, 19–22 mm). Tibia naked dorsally. Ears relatively short (21–23 mm). May be conspecific with *P. admiralitatum*. – *Distribution:* Apparently confined to Ontong Java (just north of the Solomons). – No subspecies.

10. *P. sanctacrucis* TROUGHTON 1930 [*subniger* group]. – Size medium for the group (forearm length, 112–121 mm; total length of skull, 54 mm; maxillary tooth row length, 20 mm). Tibia naked dorsally. Ears relatively short (20–21 mm). May

be conspecific with *P. admiralitatum*. – *Distribution:* Confined to the Santa Cruz islands. – No subspecies.

11. *P. ornatus* GRAY 1870 [*subniger* group]. – Size large for the group (forearm length, 140–168 mm). Tibia hairy dorsally. – Confined to New Caledonia (*P. o. ornatus*) and the Loyalty islands (*P. o. auratus*). – Two subspecies.

12. *P. dasymallus* TEMMINCK 1825 [*subniger* group]. – Size medium for the group (forearm length, 125–137 mm). Fur unusually thick (almost wooly) for a *Pteropus*, the tibia hairy dorsally. – *Distribution:* Range from Taiwan through the Ryukyus to extreme southern Kyushu and the Daito islands east of the Ryukyus. – Five subspecies are currently recognized:

P. d. formosus (Taiwan), *P. d. dasymallus* and two other subspecies (Ryukyus, extreme southern Kyushu), *P. d. daitoensis* (Daito islands).

13. *P. subniger* (KERR 1792) [*subniger* group]. – Tibia hairy dorsally. Ears very small und hidden in fur. Breadth of last molar less than one fourth that of palate between them, thus cheekteeth reduced in size. Size relatively small (forearm length, 95–113 mm). – *Distribution:* Confined to the Mascarene islands of Mauritius and Reunion, where apparently now extinct. – No subspecies.

14. *P. mariannus* DESMAREST 1822 [*mariannus* group]. – Posterior basal ledges of large premolars distinct. Skull and dentition primitive for the genus. Tibia naked dorsally. Uropatagium poorly developed. Ear length moderate. Differing from members of the *subniger* group in being blackish dorsally and ventrally with a light yellowish mantle. Size tending to be small for the group (forearm length, 106–149 mm; total length of skull, 54–64 mm). – *Distribution:* Occurring in the Ryukyu islands, Marianas, western Carolines, and extreme eastern Carolines, but not central Carolines. – Seven subspecies are currently recognized (KURODA 1938):

P. m. pelewensis (Palaus), *P. m. yapensis* (Yap), *P. m. ulthiensis* (Ulithi), *P. m. ualanus* (Kusiae), *P. m. mariannus* (southern and central Marianas), *P. m. paganensis* (northern Marianas), *P. m. loochoensis* (Ryukyus).

15. *P. tonganus* QUOY & GAIMARD 1830 [*mariannus* group]. – Tending to be larger than *P. mariannus* (forearm length, 133–176 mm; total length of skull, 63–75 mm) with a greater orbital diameter (12.5–13.8 vs. 11.0–11.8 mm). – *Distribution:* Ranging from Karkar island (off the northeast coast of New Guinea) and the eastern Solomons through the Santa Cruz islands, New Hebrides, New Caledonia and the Loyalties, Fijis, Tongas, and Samoa to the Cooks. – There are three currently recognized subspecies:

P. t. basiliscus (Karkar), *P. t. geddiei* (= *heffernani*) (Rennell in the Solomons, Santa Cruz, New Hebrides, Loyalties, New Caledonia), *P. t. tonganus* (Fijis, Tongas, Samoa, Niue, Cooks).

16. *P. caniceps* GRAY 1870 [*argentatus* group]. – Posterior basal ledges of large premolars distinct. Skull and dentition primitive for the genus. Tibia naked dorsally. Uropatagium poorly developed. Ear length relatively long. Dentition relatively heavy. Size similar to *mariannus* group (forearm length, 135–145 mm). – *Distribution:* Ranging from Celebes to the Sanghirs and northern Moluccas. – Two subspecies are currently recognized:

P. c. dobsoni (Celebes), *P. c. caniceps* (Sanghirs, Sulas, Halmahera group).

17. *P. argentatus* GRAY 1844 [*argentatus* group]. – Dentition weaker than in *P. caniceps*. Forearm length would probably be about 136 mm. – *Distribution:* Known only from a single immature specimen supposedly from Amboina in the Moluccas. The Celebes record is erroneous (MUSSER & al. 1982). – No subspecies.

18. *P. rufus* E. GEOFFROY 1803 [*niger* group]. – Posterior basal ledges of large premolars distinct. Skull and dentition primitive for the genus. Uropatagium relatively well developed. Ears rather sharply pointed, relatively long (34–41 mm) and exposed. Tibia naked dorsally. Size relatively large (forearm length, 158–171 mm). – *Distribution:* Confined to Madagascar. – Two subspecies.

19. *P. seychellensis* MILNE EDWARDS 1877 [*niger* group]. – Ears shorter (30–35 mm) and size smaller (forearm length, 130–157 mm) than *P. rufus*. – *Distribution:* Ranging from the Seychelles, Aldabra, and the Comoros to Mafia island (south of Zanzibar). – Three subspecies are recognized (HILL 1971):

P. s. seychellensis (Seychelles), *P. s. aldabrensis* (Aldabra), *P. s. comorensis* Comoros, (Mafia).

20. *P. voeltzkowi* MATSCHIE 1909 [*niger* group]. – Ears shorter (20–30 mm) than in *P. seychellensis*, but size intermediate between it and *P. rufus* (forearm length, 151–161 mm). – *Distribution:* Confined to Pemba island (north of Zanzibar). – No subspecies.

21. *P. niger* (KERR 1792) [*niger* group]. – Ears very short and nearly concealed in fur. Tibia hairy dorsally. Size relatively large (forearm length, 159–171 mm). – *Distribution:* Confined to the Mascarene islands where known only subfossil from Rodriguez, apparently now extinct in Reunion, but still surviving on Mauritius. – No subspecies.

22. *P. melanotus* BLYTH 1863 [*melanotus* group]. – Posterior basal ledges of large premolars distinct. Skull and dentition primitive for the genus. Uropatagium relatively well developed. Ears broadly rounded at tips. Size medium (forearm length, 125–163 mm). – *Distribution:* Six subspecies are currently recognized:

Ranging through a series of small islands from the Andamans *(P. m. satyrus, P. m. tytleri)* through the Nicobars *(P. m. melanotus)*, Nias *(P. m. niadicus)*, and Enggano *(P. m. modiglianii)* to Christmas island *(P. m. natalis)*.

23. *P. melanopogon* PETERS 1867 [*livingstonei* group]. – Posterior basal ledges of large premolars distinct. Skull and dentition primitive for the genus except that the dentition is unusually heavy. Tibia naked dorsally. Size very large (forearm length, 179–204 mm). Ears reasonably typical for the genus. – *Distribution:* Ranging from the Sanghir islands through the central and southern Moluccas and Keis to the Arus (the alleged New Guinea form has been shown to be a synonym of *P. neohibernicus*). – Three subspecies are currently recognized:

P. m. melanopogon (Sanghirs, Buru, Ceram and surrounding islands, Banda, and Tanimbar), *P. m. keyensis* (Keis), *P. m. aruensis* (Arus).

24. *P. livingstonei* GRAY 1866 [*livingstonei* group]. – Ears semicircularly rounded. Smaller than *P. melanopogon* (forearm length, 162–172 mm). – *Distribution:* Confined to the Comoros (northwest of Madagascar). – No subspecies.

25. *P. rayneri* GRAY 1870 [*chrysoproctus* group]. – Posterior basal ledges of large premolars distinct. Rostrum moderately shortened. Dentition heavy. Tibia densely hairy dorsally. First lower incisors not reduced and second not enlarged. Anterior lower premolar not enlarged. Posterior lower molar reduced. Size medium to large (forearm length, 121–180 mm). – *Distribution:* Confined to the Solomons. – Seven currently recognized subspecies:

P. r. grandis (Bougainville, Shortland, Choiseul, Ysabel), *P. r. monoensis* (Mono), *P. r. lavellanus* (Vella, Lavella, Ghizo, Ganongga), *P. r. rubianus* (Simbo, Kolombangara, Rendova), *P. r. rayneri* (Guadalcanal, Malaita), *P. r. cognatus* (San Cristobal, Ugi), *P. r. rennelli* (Rennell).

26. *P. fundatus* FELTEN & KOCK 1972 [*chrysoproctus* group]. Differs from *P. rayneri* in the interorbital breadth being less than the postorbital breadth and in much smaller size (forearm length, 95–103 mm). – *Distribution:* Confined to the Banks islands (northern end of the New Hebrides). – No subspecies.

27. *P. chrysoproctus* TEMMINCK 1837 [*chrysoproctus* group]. – Tibia thinly haired dorsally. Size large (forearm length, 163–177 mm). – *Distribution:* Occurring in the central Moluccas (Ceram, Buru, and surrounding small islands) and Sanghirs. – No subspecies.

28. *P. lombocensis* DOBSON 1878 [*molossinus* group]. – Posterior basal ledges of large premolars distinct. Rostrum greatly shortened. Dentition heavy. First lower incisor reduced, but second not enlarged. Anterior lower premolar not enlarged. Ears exposed. Size medium for group (forearm length, 108–122 mm). – *Distribution:* Confined to the Lesser Sundas. – Two currently recognized subspecies:

P. l. lombocensis (Lombok, Flores), *P. l. solitarius* (Alor near Timor).

29. *P. rodricensis* DOBSON 1878 [*molossinus* group]. – Tibia hairy dorsally. Ears nearly concealed in fur. Size large for group (forearm length, 124–127 mm). – *Distribution:* Confined to the Mascarene islands. Known subfossil from Mauritus, but still living on Rodriguez. – No subspecies.

30. *P. molossinus* TEMMINCK 1853 [*molossinus* group]. Tibia naked dorsally. Ears nearly concealed in fur. Size very small (forearm length, 94–99 mm). – *Distribution:* Known only from Mortlock and Ponape in the eastern Carolines. – No subspecies.

31. *P. samoensis* PEALE 1848 [*samoensis* group]. – Posterior basal ledges of large premolars distinct. Rostrum greatly shortened. Dention heavy. First lower incisor unreduced and second enlarged. Anterior lower premolar enlarged. Last lower molar unreduced. Posterior ledges of upper incisors medium in breadth. No inner basal ledges on lower premolars and molars. Size large for group (forearm length, 124–144 mm). – *Distribution:* Confined to the Fijis and Samoa. – Two subspecies (HILL & BECKON 1978):

P. s. nawaiensis (Fiji), *P. s. samoensis* (Samoa).

32. *P. anetianus* GRAY 1870 [*samoensis* group]. – Broad inner basal ledges on last lower premolar

and anterior two lower molars. Size relatively small (forearm length, 114–136 mm). – *Distribution:* Confined to the New Hebrides. – There are seven currently recognized subspecies (FELTEN & KOCK 1972):

P. a. anetianus (Aneityum, Erromanga), *P. a. bakeri* (Efate and nearby small islands), *P. a. pastoris*, *P. a. eotinus*, and *P. a. aorensis* (northern New Hebrides), *P. a. banksianus* and *P. a. motalavae* (Banks islands).

33. *P. tokudae* TATE 1934 [*pselaphon* group]. – Posterior basal ledges of large premolars distinct. Rostrum greatly shortened. Dentition heavy. First lower incisor unreduced and second enlarged. Last lower molar unreduced. Posterior ledges of upper incisors unusually broad. Tibia naked dorsally. Size very small (forearm length, 90–95 mm). Probably only a subspecies of *P. insularis*. – *Distribution:* Confined to Guam (southern Marianas) where probably now extinct. – No subspecies.

34. *P. insularis* HOMBRON & JACQUINOT 1853 [*pselaphon* group]. – Tibia naked dorsally. Size small (forearm length, 99–109 mm). Hair on back dark brown. – *Distribution:* Confined to Truk (eastern Carolines). – No subspecies.

35. *P. phaeocephalus* THOMAS 1883 [*pselaphon* group]. – Tibia naked dorsally. Size small (forearm length, 105 mm). Hair on back golden cream buff. Almost certainly only a subspecies of *P. insularis*. – *Distribution*: Confined to Mortlock (eastern Carolines). – No subspecies.

36. *P. pselaphon* LAY 1829 [*pselaphon* group]. – Tibia and foot hairy dorsally. Upper molariform teeth not shortened. Upper canine without secondary cusp. Rostrum unusually broad. Large for group (forearm length, 123–141 mm). – *Distribution:* Confined to the Bonin and Volcano islands. – No subspecies.

37. *P. pilosus* ANDERSEN 1908 [*pselaphon* group]. – Tibia hairy but foot naked dorsally. Upper molariform teeth not shortened. Upper canine without secondary cusp. Rostrum not unusually broad. Very large species of that group (forearm length, 150–151 mm). – *Distribution:* Confined to the Palau islands, where probably extinct. – No subspecies.

38. *P. tuberculatus* PETERS 1869 [*pselaphon* group]. – Tibia hairy dorsally. Upper molariform teeth not shortened. Upper canine with small secondary cusp. Size medium within the group (forearm length, 115–125 mm). – *Distribution:*. Apparently confined to Vanikoro island in the Santa Cruz group. – No subspecies.

39. *P. nitendiensis* SANBORN 1930 [*pselaphon* group]. – Tibia hairy dorsally. Upper molariform teeth not shortened. Upper canine without a secondary cusp. Size medium within the group (forearm length, 118–121 mm). – *Distribution:* Apparently confined to Ndeni Island in the Santa Cruz group. – Probably only a subspecies of *P. tuberculatus.*

40. *P. vetulus* JOUAN 1863 (=*macmillani* TATE 1942) [*pselaphon* group]. – Tibia hairy dorsally. Upper molariform teeth not shortened, but instead narrowed, and the anterior upper molar is notched. Upper canine apparently without a secondary cusp. Size rather small (forearm length, 100–114 mm). – *Distribution:* Confined to New Caledonia. – No subspecies.

41. *P. leucopterus* TEMMINCK 1853 [*pselaphon* group]. Tibia hairy dorsally. Upper molariform teeth shortened and subquadrate. Postorbital processes forming a complete postorbital bar. Size large for group (forearm length, 136–143 mm). – *Distribution:* Confined to the Philippines. – No subspecies.

42. *P. temmincki* PETERS 1867 [*personatus* group]. – Posterior basal ledges of large premolars distinct. Rostrum shortened. Dentition weak, but premolars and molars not greatly narrowed. Tibia hairy dorsally. Size relatively large for group (forearm length, 94–116 mm). – *Distribution:* Known from the central Moluccas and the Bismarcks. The Timor record is probably erroneous (GOODWIN 1979). – Three subspecies are currently recognized:

P. t. liops (Buru), *P. t. temmincki* (Amboina, Ceram), *P. t. capistratus* (Bismarcks).

43. *P. personatus* TEMMINCK 1825 [*personatus* group]. – Premolars and molars greatly narrowed. Size very small (forearm length, 86–96 mm). – *Distribution:* Known only from the northern Moluccas (Gilolo group) and northern Celebes. – No subspecies.

44. *P. lylei* ANDERSEN 1908 [*vampyrus* group]. – Posterior basal ledges of premolars practically obliterated. Tibia naked dorsally. Premolars and molars not extremely narrow (breadth of last upper premolar at least one fourth width of palate between them). Last two upper premolars clearly elongate. Inner edges of plagiopatagia well separated. Ears long and sharply pointed. Size small

for group (forearm length, 145–160 mm). – *Distribution:* Known only from Thailand, Cambodia, and Vietnam. – No subspecies.

45. P. giganteus BRUNNICH 1782 [*vampyrus* group]. Ventral hair clearly paler than dorsal hair. Size medium for group (forearm length, 160–177 mm). – *Distribution:* Confined to the Indian subcontinent, Burma, Andaman and Maldive islands. – Three subspecies:

P. g. ariel (Maldives), *P. g. giganteus* and *P. g. leucocephalus* (remainder of range).

46. P. vampyrus LINNAEUS 1758 [*vampyrus* group]. Ventral hair nearly the same color as dorsal hair. Large to very large (forearm length, 179–220 mm). – *Distribution:* Ranges from Indo-China through the Malay peninsula, Sumatra, and Borneo to the Philippines, Java, and the Lesser Sundas. – Seven subspecies are currently recognized:

P. v. intermedius (extreme southern Burma and adjacent Thailand), *P. v. malaccensis* (Indo China, Thailand, Malaya, Sumatra, and adjacent small islands), *P. v. natunae* (Borneo and adjacent small islands), *P. v. lanensis* (Philippines), *P. v. vampyrus* (Java), *P. v. pluton* (Bali, Lombok, Sumbawa), *P. c. edulis* (Savu, Timor).

47 P. alecto TEMMINCK 1837 [*alecto* group]. – Posterior basal ledges of premolars practically obliterated. Tibia naked dorsally. Premolars and molars not extremely narrow. Last two upper premolars clearly elongate. Inner edges of plagiopatagia well separated. Ears moderate and rounded. No mantle of contrasting colored hair present. Size medium to large (forearm length, 141–182 mm). – *Distribution:* Ranging from Bawean and Kangean islands (in Java Sea on Sunda shelf) and Celebes through the Lesser Sundas to tropical Australia and extreme southern New Guinea. – Four subspecies are recognized:

P. a. aterrimus (Bawean, Kangean), *P. a. alecto* (Celebes, Saleyer, Lombok), *P. a. morio* (Sumba, Savu), *P. a. gouldi* (coastal regions of tropical Australia, barely extending across Torres Straits to New Guinea).

48. P. conspicillatus GOULD 1850 [*conspicillatus* group]. Posterior basal ledges of premolars practically obliterated. Tibia naked dorsally. Premolars and molars not extremely narrow. Last two upper premolars clearly elongate. Inner edges of plagiopatagia well separated. Ears moderate and rounded. A sharply defined mantle of yellowish hair present. Size large for group (forearm length, 155–183 mm). – *Distribution:* Ranging form northern Moluccas (Gilolo group) through New Guinea and adjacent small islands to Cape York in northeastern Australia. – Two currently recognized subspecies:

P. c. chrysauchen (Gilolo group, northeastern New Guinea and surrounding small islands), *P. c. conspicillatus* (eastern New Guinea, east Papuan islands, Cape York).

49. P. ocularis PETERS 1867 [*conspicillatus* group]. – Size small for group (forearm length, 131–139 mm). – *Distribution:* Known only from Ceram and Buru. – No subspecies.

50. P. neohibernicus PETERS 1876 (=*papuanus* PETERS & DORIA 1881; *sepikensis* SANBORN 1931) [*neohibernicus* group]. – Posterior basal ledges of premolars practically obliterated. Tibia naked dorsally. Premolars and molars not extremely narrow. Last two upper premolars clearly elongate. Inner edges of plagiopatagia almost meeting, leaving only a narrow spinal line of furred back exposed. Size very large (forearm length, 189–203 mm). – *Distribution:* Occurring in New Guinea (and a few small surrounding islands) and the Bismarcks. – Two currently recognized subspecies:

P. n. neohibernicus (entire range except the Admiralty islands), *P. n. hilli* (Admiralty islands in the Bismarcks).

51. P. macrotis PETERS 1867 [*poliocephalus* group]. – Posterior basal ledges of premolars practically obliterated. Tibia naked dorsally. Premolars and molars not extremely narrow. Last two upper premolars subsquarish. Ears long and pointed (33–36 mm). Size small for group (forearm length, 121–141 mm). – *Distribution:* Known from the Aru islands and New Guinea. – Two currently recognized subspecies:

P. m. macrotis (Aru islands), *P. m. epularius* (New Guinea mainland).

52. P. pohlei STEIN 1933 [*poliocephalus* group]. – Tibia naked dorsally. Ears relatively short (26–28 mm). Size small for group (forearm length, 126–136 m). – *Distribution:* Known only from Yapen island off the northern coast of western New Guinea. – No subspecies.

53. P. poliocephalus TEMMINCK 1825 [*poliocephalus* group]. Tibia hairy dorsally. Ears long and pointed (33 mm). Size large for group (forearm length, 138–164 mm). – *Distribution:* Confined to coastal eastern Australia from southeastern Queensland to islands in Bass Strait (but not Tasmania proper) – No subspecies.

54. P. gailliardi VAN DEUSEN 1969 [*scapulatus* group]. – Posterior basal ledges of premolars practically obliterated. Premolars and molars ex-

tremely narrow (breadth of last upper premolar less than one fourth width of palate between them). Tibia naked dorsally. Ears short (20 mm). Size fairly small for group (forearm length, 114 mm). Pelage distinctly woolly. – *Distribution:* Known from a single somewhat immature specimen collected in the mountains of New Britain (Bismarcks). – No subspecies.

55. *P. woodfordi* THOMAS 1888 (= *austini* LAWRENCE 1945) [*scapulatus* group]. – Ears short (14–20 mm). Size very small (forearm length, 86–102 mm). Pelage not woolly. – *Distribution:* Confined to the central and southeastern Solomons (Fauro to Guadalcanal). – No subspecies.

56. *P. mahaganus* SANBORN 1931 [*scapulatus* group]. – Ears medium in length (23–25 mm). Pelage not woolly. Size large for group (forearm length, 131–144 mm). Last upper and lower molars unusually large for group. – *Distribution:* Known only from Bougainville and Santa Isabel in the Solomons. – No subspecies.

57. *P. scapulatus* PETERS 1862 [*scapulatus* group]. – Ears long and pointed (27–31 mm). Pelage not woolly. Size large for group (forearm length, 120–143 mm). Last upper and lower molars relatively small. – *Distribution:* Ranging over the whole of tropical Australia and much of the southeast (accidental in New Zealand); also barely across Torres Strait to extreme southern New Guinea. – No subspecies.

Genus *Acerodon* JOURDAN 1837 (Fig. 25)

Premaxillaries in simple contact in front. Second lower incisor less than 10 times the bulk of the first lower incisor. Upper canine without a secondary cusp. A well-differentiated antero-internal tubercle on the last upper premolar and the first upper molar. Sharply defined inner basal ledges on the last lower premolar and the first two lower molars. Dental formula i2/2, c1/1, p3/3, m2/3 x 2 = 34. Claw present on second digit of wing. – *Distribution:* The geographical distribution includes the Philippines, Talauts, Celebes, and Lesser Sundas. – Currently six species and six additional subspecies are recognized.

1. *A. celebensis* (PETERS 1867) (= *arquatus* MILLER & HOLLISTER 1921). – No antero-internal cusp on the middle lower premolar. Ears relatively long and pointed (29 mm). Dentition relatively weak. Size relatively small (forearm length, 125–142 mm). – *Distribution:* Confined to Celebes and surrounding small islands. – No subspecies.

2. *A. mackloti* (TEMMINCK 1837). – No antero-internal cusp on the middle lower premolar. Ears relatively long and pointed (32–34 mm). Dentition relatively heavy. Size relatively small (forearm length, 135–156 mm). – *Distribution:* Confined to the Lesser Sundas. – Five subspecies are currently recognized:

A. m. prajae (Lombok), *A. m. floresi* (Sumbawa, Flores), *A. m. alorensis* (Alor), *A. m. gilvus* (Sumba), *A. m. mackloti* (Timor), but there is some doubt concerning the validity of these.

3. *A. humilis* ANDERSEN 1909. – A distinct antero-internal cusp on the middle lower premolar. Ears relatively short and rounded (24 mm). Size relatively small (forearm length, 140 mm). – *Distribution:* Confined to the Talaut islands (between Philippines and Moluccas). – No subspecies.

4. *A. jubatus* (ESCHSCHOLTZ 1831). – A distinct antero-internal cusp on the middle lower premolar. Ears relatively short and rounded (28–40 mm). Size large to very large (forearm length, 182–210 mm). – *Distribution:* Confined to the main islands of the Philippines (Luzon, Visayans, Mindanao). – Two subspecies.

5. *A. lucifer* (ELLIOT 1896). – A distinct antero-internal cusp on the middle lower premolar. Ears relatively short and rounded. Size fairly large (forearm length, 165 mm). – *Distribution:* Known only from Panay island in the Philippines, where probably extinct. – No subspecies.

6. *A. leucotis* (SANBORN 1950). – A distinct antero-internal cusp on the middle lower premolar. Ears relatively short and rounded (23–32 mm). Size relatively small (forearm length, 137–143 mm). Distinction from *A. humilis* not clear, though of course, separated geographically by the larger Philippine species. – *Distribution:* Confined to the Palawan group in the southeastern Philippines. – Two subspecies.

Genus *Pteralopex* THOMAS 1888 (Fig. 26)

Premaxillaries co-ossified in front. Second lower incisor more than 10 times the bulk of the first lower incisor. Upper canine with a large external secondary cusp. Upper molariform teeth with prominent anterior and posterior basal ledges. Dental formula i2/2, c1/1, p3/3, m2/3 x 2 = 34.

Claw present on second digit of wing. – *Distribution:* Solomons and Fijis. – Three species are currently recognized (HILL & BECKON 1978).

1. *P. anceps* ANDERSEN 1909. – Labial margin of last upper premolar and first upper molar each forming a single large cusp. Middle lower premolar with a prominent lingual cusp. Last lower premolar and anterior two lower molars with long lingual ridges. Size relatively large (forearm length, 160–171 mm). – *Distribution:* Known only from Bougainville and Choiseul in the western Solomons. – No subspecies.

2. *P. atrata* THOMAS 1888. – Labial margin of last upper premolar and first upper molar each forming a single large cusp. Middle lower premolar with a prominent lingual cusp. Last lower premolar with a subconical lingual cusp. Lingual elevations of first and second lower molars short, on first forming a subconical antero-internal cusp. Size medium (forearm length, 139–144 mm). – *Distribution:* Known only from Santa Isabel and Guadalcanal in the eastern Solomons. – No subspecies.

3. *P. acrodonta* HILL & BECKON 1978. – Labial margin of last upper premolar and first upper molar each divided into two cusps, the anteriormost very small. Middle lower premolar lacking a lingual cusp. First and second lower molars similar. Size relatively small (forearm length, 116–120 mm). – *Distribution:* Known only from Taveuni island in the Fijis. – No subspecies.

Genus **Styloctenium** MATSCHIE 1899 (Fig. 27)

Dental formula i2/1, c1/1, p3/3, m2/2 x 2 = 30. Claw present on second digit of wing. Molars and premolars considerably shortened and simplified. Tooth rows more or less parallel. – *Distribution:* Confined to Celebes and the nearby Togian islands. – A single species.

1. *S. wallacei* (GRAY 1866). – Size relatively small (forearm length, 90–96 mm). – *Distribution:* Same as genus. – No subspecies.

Genus **Neopteryx** HAYMAN 1945 (Fig. 28)

Dental formula i2/2, c1/1, p2/3, m2/3 x 2 = 32. Claw absent on second digit of wing. Molars and premolars greatly shortened and simplified. Toothrows widely diverging posteriorly. – *Distribution:* Confined to Celebes. – A single species.

1. *N. frosti* HAYMAN 1945. – Size medium (forearm length, 110 mm). – *Distribution:* Same as for genus. – No subspecies.

Subtribe Dobsoniina KOOPMAN & JONES 1970

Facial axis of skull conspicuously deflected against basicranial axis. Dental formula never more than i1/1, c1/1, p2/3, m2/3 x 2 = 28 and may be further reduced. Second digit of wing without a claw. A short tail present. Inner edges of plagiopatagia meeting along the mid-dorsal line. (Most of these characters are only known in *Dobsonia*, since available material of *Aproteles* is defective). – *Distribution:* From the Philippines, Celebes, and the Lesser Sundas to the Solomons, east Papuan islands, and Cape York in northeastern Queensland, Australia. – Two genera and 12 species.

Genus **Aproteles** MENZIES 1977 (Fig. 29)

Rostrum fairly elongate. Lower incisors absent. Molars simplified and almost circular in section. Braincase relatively low. The genus was originally described on fossil material and only a single specimen prepared from a living animal is known. This unfortunately is represented by a skull only, lacking the premaxillaries and upper incisors (if existing). – *Distribution:* Known only from the highlands of central New Guinea. – A single species.

1. *A. bulmerae* MENZIES 1977. – Total length of skull, ca. 68 mm. – *Distribution:* Same as for genus. – No subspecies.

Genus **Dobsonia** PALMER 1898 (Fig. 30)

A single pair of lower incisors present. Molars relatively complex and elongate. High braincase. Rostrum relatively shortened. – *Distribution:* Same as for subtribe. – There are currently recognized 11 species and seven additional subspecies.

1. *D. minor* (DOBSON 1878) [*minor* group]. – Premolars and molars simple with no well-marked antero-internal and posterior basal ledges and no surface ridges. Size very small (forearm length, 74–86 mm). – *Distribution:* Known from western and central New Guinea (including Yapen and Bagabag islands) as well as Celebes. – No subspecies.

2. *D. chapmani* RABOR 1952 [*moluccensis* group]. – A well-marked antero-internal basal ledge or cusp on the last upper and lower premolars, but not on the first lower molar. A posterior basal ledge on the anterior upper, middle and posterior lower premolars. Size medium for group (forearm length, 124–131 mm; total length of skull, 47–56 mm; maxillary tooth row length 20–23 mm). Forehead relatively flat. Originally described as a subspecies of *D. viridis* and distinction from *D. exoleta* not clear (BERGMANS 1978). – *Distribution:* Known only from the central Philippines. – No subspecies.

3. *D. exoleta* ANDERSEN 1909 [*moluccensis* group]. – Size small for group (forearm length, 105–125 mm; total length of skull, 46–55 mm; maxillary tooth row length 19–23 mm). Surface ridge present on first lower molar. Forehead relatively flat. Maxillary dentition relatively heavy. – *Distribution:* Known only from Celebes and the nearby Togian islands. – No subspecies.

4. *D. emersa* BERGMANS & SARBINI 1985 [*moluccensis* group]. – Size small for group (forearm length, 113–114 mm; total length of skull, 47–50 mm; maxillary tooth row length, 20–21 mm). Weak surface ridge present on first lower molar. Forehead distinctly concave. Maxillary dentition composed of small and narrow teeth. – *Distribution:* Confined to Biak and Owii islands (an unnamed close relative on Numfoor), all in Geelvink Bay in northwestern New Guinea. – No subspecies.

5. *D. moluccensis* (QUOY & GAIMARD 1830) [*moluccensis* group]. – Size relatively large (forearm length, 113–160 mm; total length of skull, 50–65 mm; maxillary tooth row length, 19–28 mm). Dentition relatively heavy. – *Distribution:* From the northern and central Moluccas through the west Papuan and Aru islands and the mainland of New Guinea to Cape York (northeastern Queensland) and the Bismarcks. – Three currently recognized subspecies (KOOPMAN 1979):

D. m. moluccensis (Halmahera and Buru through Ceram to the Aru islands), *D. m. magna* (West Papuan islands through New Guinea to Cape York), *D. m. anderseni* (Bismarcks).

6. *D. pannietensis* (DE VIS 1905) [*moluccensis* group]. – Size small for group (forearm length, 97–125 mm; total length of skull, 41–56 mm; maxillary toothrow length, 17–24 mm). Forehead relatively flat. Maxillary dentition relatively weak. – *Distribution:* Confined to the east Papuan islands. – Two currently recognized subspecies (KOOPMAN 1982):

D. p. pannietensis (Louisiade and D'Entrecasteaux archipelagos), *D. p. remota* (Kiriwina and Woodlark islands).

7. *D. peroni* (E. GEOFFROY 1810) [*peroni* group]. – A well-marked antero-internal basal ledge or cusp on the last upper and lower premolars as well as the first lower molar but not on the first upper molar. Longitudinal ridges of molars and premolars without notches. Size relatively small (forearm length, 105–130 mm; total length of skull, 46–54 mm). – *Distribution:* Confined to the Lesser Sundas from Nusa Penida (just east of Bali) to Babar island (between Timor and Timorlaut). – Three subspecies are currently recognized (BERGMANS 1978):

D. p. peroni (Timor, probably also nearby Alor, Wetar, and Babar), *D. p. sumbana* (Sumba), *D. p. grandis* (western Lesser Sundas from Nusa Penida to Komodo and probably Flores).

8. *D. viridis* (HEUDE 1896) [*viridis* group]. – A well-marked antero-internal basal ledge or cusp on the last upper and lower premolars as well as the first upper and lower molars. Notches generally present on longitudinal ridges of molars and premolars. Size medium to large for group (forearm length, 108–129 mm; total length of skull, 45–54 mm). Rostrum relatively lighter (lacrimal breadth, 11.5–12.8 mm). Teeth variable in size (maxillary toothrow length, 18–21 mm). Interorbital width medium (9–10 mm). – *Distribution:* Known from Celebes, the Moluccas, and nearby small islands, including Misol in the west Papuan islands. – Two currently recognized subspecies:

D. v. viridis (Celebes, central Moluccas, Misol); *D. v. crenulata* (Sanghir and Togian islands, near Celebes, to the northern Moluccas, Halmahera group).

9. *D. beauforti* BERGMANS 1975 [*viridis* group]. – Size small for group (forearm length, 99–112 mm; total length of skull, 41–46 mm). Teeth relatively small (maxillary toothrow length, 16–17 mm). Interorbital width relatively great (7.9–8.8 mm). – *Distribution:* Confined to Waigeo in the west Papuan islands and to Biak and Owii islands in Geelvink Bay. – No supspecies.

10. *D. praedatrix* ANDERSEN 1900 [*viridis* group]. – Size medium for group (forearm length, 111–122 mm; total length of skull 49–52 mm). Rostrum relatively heavier (lacrimal breadth, 12.8–13.8 mm). Teeth relatively large (maxillary toothrow length, 18–21 mm). Interorbital width great (10.2–10.5 mm). – *Distribution:* Confined to the Bismarcks. – No subspecies.

11. *D. inermis* ANDERSEN 1909 [*viridis* group]. – Size small for group (forearm length, 98–112 mm; total length of skull 40–48 mm). Rostrum relatively heavier (lacrimal breadth, 11.0–11.7 mm). Teeth medium in size (maxillary toothrow length, 15–19 mm). Interorbital width relatively small (6.7–8.2 mm). – *Distribution:* Confined to the Solomons. – Two currently recognized subspecies:

D. i. inermis (=*nesea*) (entire Solomon chain from Buka to San Cristobal and Rennell, except Choiseul and Santa Ysabel), *D. i. minimus* (Choiseul, Santa Ysabel).

Tribe Harpyionycterini MILLER 1907

Molariform teeth multicuspidate. Lower canines strongly proclivous (crossing upper canines at nearly right angles). Facial axis of skull conspicuously deflected against basicranial axis. Size fairly small. – *Distribution:* Confined to Celebes and the main Philippine islands (excluding the Palawan group). – One genus and species.

Genus *Harpyionycteris* THOMAS 1896 (Fig. 31)

Dental formula i1/1, c1/1, p3/3, m2/3 x 2 = 30. Second digit of wing with a claw. Tail absent. Inner edges of plagiopatagia attached to sides of body. – *Distribution:* Same as for tribe. – One species with three subspecies.

1. *H. whiteheadi* THOMAS 1896. – Forearm length, 82–93 mm. – *Distribution:* Known from Celebes and the Philippine islands of Mindanao, Camiguin, Negros, and Mindoro. – Three subspecies are currently recognized:

H. w. celebensis (Celebes), *H. w. whiteheadi* (Philippines except Negros), *H. w. negrosensis* (Negros).

Tribe Epomophorini GRAY 1866

Molariform teeth not multicuspidate. Lower canines not proclivous. Facial axis of skull usually very little deflected against basicranial axis. Braincase flattened posteriorly. Small whitish hair tufts usually present anterior and posterior to ear pinna. Size small to large. – *Distribution:* Confined to sub-Saharan Africa and a few near-shore islands. – Eight genera, 20 species.

Genus *Plerotes* ANDERSEN 1910 (Fig. 32)

Dental formula variable, but at least four upper cheekteeth. Molariform teeth sublinear with flattened crowns. All palate ridges simple. – *Distribution:* Known only from Angola, Zambia, and southern Zaire. – A single species.

1. *P. anchietai* (SEABRA 1900). – Size small (forearm length, 53 mm). Since no adult males are known, it is uncertain wheather epaulettes are present, but an immature male seems to show the beginnings of them. – *Distribution:* Same as for genus. – No subspecies.

Genus *Hypsignathus* H. ALLEN 1861 (Fig. 33)

Dental formula i2/2, c1/1, p2/3, m1/2 x 2 = 28. Posterior palatal ridges modified. Rostrum long (orbit to tip of nasals much more than lacrimal breadth), broad, and greatly deepened. Post-dental palate flattened posteriorly. Premaxillae ankylosed anteriorly. Lower incisors and canines closing anterior to uppers. Outer ridges of lower molars bilobed or trilobed. Upper lip with cutaneous leaves. Tail absent. No epaulettes. – *Distribution:* Forest zone of tropical Africa from Gambia to southwestern Ethiopia and western Kenya, south to Angola and southern Zaire, Fernando Poo. – A single species.

1. *H. monstrosus* H. ALLEN 1861. – Size very large for group (forearm length, 118–137 mm). – *Distribution:* Same as for genus. – No subspecies.

Genus *Epomops* GRAY 1870 (Fig. 34)

Dental formula usually i2/2, c1/1, p2/3, m1/2 x 2 = 28 (second upper incisor may be lost). Palatal ridges more or less modified. Rostrum long (orbit to tip of nasals much more than lacrimal breadth) and broad, but not greatly deepened. Post-dental palate flattened posteriorly. Premaxillae in simple contact anteriorly. Lower incisors and canines biting against uppers. Outer ridges of lower molars simple. Upper lip without cutaneous leaves. Tail rudimentary. Epaulettes present. – *Distribution:* Widely distributed, chiefly in the forest zone, of tropical Africa from Guinea to southern Sudan and south to northern Botswana. – Three species with one additional subspecies.

1. *E. dobsoni* (BOCAGE 1889). – Three thick and prominent interdental palatal ridges and two pairs of thick post-dental ridges each with two triangular projections. Zygomatic breadth not quite half of total skull length. Size relatively small (forearm length, 81–88 mm). – *Distribution:*

Southern tropical Africa from Angola and northern Botswana to Malawi, western Tanzania, Rwanda and southern Zaire. – No subspecies.

2. E. buettikoferi (MATSCHIE 1899). – Four inter-dental and five to seven post-dental palatal ridges (latter thin, serrate, arcuate). Third inter-dental palatal ridge broadly divided in center. Zygomatic breadth about half of total skull length. Size relatively large (forearm length, 88–103 mm). – Distribution: Confined to northwestern Africa from Guinea and Sierra Leone to Nigeria. – No subspecies.

3. E. franqueti (TOMES 1860). – Four inter-dental and five to seven post-dental palatal ridges (latter thin, serrate, arcuate). Third inter-dental palatal ridge nearly always undivided. Zygomatic breadth about three fifths of total skull length. Size relatively small to fairly large (forearm length, 76–102 mm). – Distribution: The most widely distributed Epomops, from Sierra Leone east to southern Sudan and south to Angola and Zambia. – Two subspecies:

E. f. strepitans (west of the Niger river), E. f. franqueti (east of the Niger river).

Genus **Epomophorus** BENNETT 1836 (Fig. 35)

Dental formula i2/2, c1/1, p2/3, m1/2x2 = 28. Palatal ridges more or less modified. Rostrum long (orbit to tip of nasals much more than lacrimal breadth) and narrow, but not deepened. Post-dental palate deeply depressed posteriorly. Epaulettes present. – Distribution: Over virtually all of sub-Saharan Africa, including Pemba and Zanzibar islands off the east coast. – Eight species with two additional subspecies.

1. E. wahlbergi (SUNDEVALL 1846) [wahlbergi group]. – A single main post-dental palatal ridge. Size relatively large (forearm length, 69–89 mm; total length of skull, 44–56 mm). – Distribution: Widely distributed from Cameroon and Somalia south to Cape province in South Africa, including the islands of Pemba and Zanzibar. – Two poorly defined subspecies:

E. w. haldemani and E. w. wahlbergi.

2. E. pousarguesi TROUESSART 1904 [gambianus group]. – Two main post-dental palatal ridges. Fourth palatal ridge much nearer third than fifth. Sixth palatal ridge posterior to middle of post-dental palate. Size large (forearm length, 91–100 mm; total length of skull 54–64 mm). – Distribution: Known only from the Central African Republic. – No subspecies.

3. E. reii AELLEN 1950 [gambianus group]. – Fourth palatal ridge much nearer third than fifth. Sixth palatal ridge posterior to middle of post-dental palate. Size fairly large (forearm length, 82 mm; total length of skull, 50–51 mm). – Distribution: A poorly known species recorded only in Cameroon. – No subspecies.

4. E. angolensis GRAY 1870 [gambianus group]. Fourth palatal ridge much nearer third than fifth. Sixth palatal ridge anterior to middle of post-dental palate. Size relatively large (forearm length, 82–91 mm; total length of skull, 58–62 mm). – Distribution: Known only from western Angola and northwestern Namibia. – No subspecies.

5. E. gambianus (OGILBY 1835) [gambianus group]. – Fourth palatal ridge midway between third and fifth. Size large (forearm length, 76–93 mm; total length of skull, 51–62 mm). – Distribution: As here defined (not including parvus), this species ranges across northern tropical Africa (for the most part north of the forests) from Senegal to Ethiopia. – No subspecies.

6. E. crypturus PETERS 1852 (= parvus ANSELL 1960) [gambianus group]. – Fourth palatal ridge midway between third and fifth. Size fairly large (forearm length, 75–86 mm; total length of skull, 46–55 mm). – Distribution: Chiefly southeastern tropical Africa from Angola, southern Zaire, and southern Tanzania to the Cape province of South Africa. – No subspecies.

7. E. labiatus HEUGLIN 1864 [gambianus group]. – Fourth palatal ridge midway between third and fifth. Size medium (forearm length, 62–81 mm; total length of skull, 40–49 mm). – Distribution: Chiefly northern tropical Africa from Senegal to Ethiopia and south to southern Congo Republic, northern Zaire, Burundi, and northern Tanzania. – Two subspecies (KOOPMAN 1975):

E. l. labiatus (Ethiopia and northeastern Sudan), E. l. anurus (remaining range).

8. E. minor DOBSON 1880 [gambianus group]. – Fourth palatal ridge midway between third and fifth. Size small (forearm length, 57–67 mm; total length of skull, 35–37 mm). – Distribution: Chiefly an eastern African species ranging from Ethiopia to Malawi and eastern Zambia (including Zanzibar), but extending across southern Zaire to the lower Congo river. – No subspecies.

Genus *Micropteropus* MATSCHIE 1899 (Fig. 36).

Dental formula i2/2, c1/1, p2/3, m1/2x2 = 28. Palatal ridges modified. Rostrum short (orbit to tip of nasals about the same as lacrimal breadth) and broad. Post-zygomatic palate about the same width anteriorly and posteriorly and at least as broad as long. All palatal ridges divided by a median groove. Epaulettes present. – *Distribution:* Over most of tropical Africa from Gambia to Ethiopia and south to northern Angola, northern Zambia, and northern Tanzania. – Three species.

1. *M. grandis* SANBORN 1950. – Six palatal ridges, second to sixth narrowly divided medially, arranged in two parallel lines of separate pairs. Size relatively large (forearm length, 63–66 mm; total length of skull, 35–36 mm). – *Distribution:* Known by a few specimens from southern Congo Republic and northern Angola. – No subspecies.

2. *M. intermedius* HAYMAN 1963. – Five palatal ridges, second to fifth divided medially, converging posteriorly in clearly separate pairs. Size medium (forearm length, 58–64 mm; total length of skull, 32–34 mm). – *Distribution:* Known by a few specimens from southwestern Zaire and northern Angola. – No subspecies.

3. *M. pusillus* (PETERS 1867). – Five palatal ridges, second to fifth widely separated anteriorly by a deep V-shaped groove, not arranged in separate pairs. Size small (forearm length, 46–55 mm; total length of skull, 28–31 mm). – *Distribution:* Same as for genus. – No subspecies.

Genus *Nanonycteris* MATSCHIE 1899 (Fig. 37)

Dental formula i2/2, c1/1, p2/3, m1/2x2 = 28. Palatal ridges modified. Rostrum short (orbit to tip of nasals about the same as lacrimal breadth) but slender. Post zygomatic palate about the same width anteriorly and posteriorly, but almost twice as broad as long. Maxillary tooth row considerably shortened. Only posterior group of palatal ridges divided. Epaulettes present. – *Distribution:* Restricted to western tropical Africa (chiefly forests) from Guinea to western Central African Republic and northern Congo Republic. – A single species.

1. *N. veldkampi* (JENTINCK 1887). – Size small (forearm length, 45–51 mm; total length of skull, 24–28 mm). – *Distribution:* Same as for genus. – No subspecies.

Genus *Scotonycteris* MATSCHIE 1894 (Fig. 38)

Dental formula i2/2, c1/1, p2/3, m1/2x2 = 28. Palatal ridges more or less modified. Cranial rostrum short (orbit to tip of nasals about the same as lacrimal breadth) and broad. Postzygomatic palate long, its lateral margins straight and converging posteriorly. The only genus of epomophorines which may lack whitish tufts at the base of each ear. No epaulettes. – *Distribution:* Restricted to the tropical African forest belt from Liberia to eastern Zaire and southern Congo Republic. – Two species and one additional subspecies.

1. *S. zenkeri* MATSCHIE 1894. – Tufts at bases of ears absent. Size relatively small (forearm length, 45–55 mm; total length of skull, 24–27 mm). – *Distribution:* Same as for genus, including Fernando Poo. – Two poorly marked subspecies.

2. *S. ophiodon* POHLE 1943. – Tufts at bases of ears present. Size relatively large (forearm length, 74–78 mm; total length of skull, 38–40 mm). – *Distribution:* Known only from a few localities in Liberia, Ghana, Cameroon, and Congo Republic. – No currently recognized subspecies.

Genus *Casinycteris* THOMAS 1910 (Fig. 39).

Dental formula, i2/2, c1/1, p2/3, m1/2x2 = 28. Palatal ridges modified. Cranial rostrum very short (orbit to tip of nasals less than lacrimal breadth) and broad. Postzygomatic palate greatly emarginated so as to be virtually absent. No epaulettes. – *Distribution:* Restricted to the forest region of central Africa. – A single species.

1. *C. argynnis* THOMAS 1910. – Size fairly small (forearm length, 50–62 mm; total length of skull, 23–29 mm). – *Distribution:* Known only from Cameroon and Zaire. – No subspecies.

Tribe Cynopterini GRAY 1866

Molariform teeth not multicuspidate. Lower canines not proclivous. Facial axis of skull very little deflected against basicranial axis. Braincase not flattened posteriorly. No whitish tufts at bases of ear pinna. A single upper molar present. Size small to medium. Rostrum much shortened (orbit to tip of nasals equal to or less than lacrimal breadth). – *Distribution:* From Pakistan to the Santa Cruz islands (east of the Solomons) and northeastern Australia. – 16 genera, 37 species.

Subtribe Cynopterina GRAY 1866

One or two pairs of lower incisors. Lower canines separated. Nostrils not tubular. Tongue with three circumvallate papillae. – *Distribution:* From Pakistan to the Philippines, Moluccas, and Lesser Sundas. – 14 genera, 22 species.

Genus *Cynopterus* F. CUVIER 1824 (Fig. 40)

Postorbital foramen (through base of postorbital process) large. Premaxillaries in simple contact anteriorly. Dental formula i2/2, c1/1, p3/3, m1/2 x 2 = 30. Edge of plagiopatagium attached to first toe. Cranial rostrum lower at canine than at middle premolar. Upper canine with secondary cusp at inner edge, but without a groove on the front face. Tail present. Posterior upper incisor not shortened. – *Distribution:* Same as for subtribe, with the exception of the Moluccas. – Five species and 17 additional subspecies.

1. *C. brachyotis* (MÜLLER 1838). – Premolars and molars relatively narrow and oval in outline. Surface cusp on last lower premolar and first lower molar small or absent. Size relatively small (forarm length, 54–70 mm; total length of skull, 13–21 mm). – *Distribution:* From Ceylon, the Andamans, southern Burma, and Vietnam to the Philippines, Celebes, and the Lesser Sundas. – Eight currently recognized subspecies (HILL 1983).

C. b. ceylonensis (Ceylon), *C. b. brachysoma* (Andamans), *C. b. brachyotis* (= *archipelagus*) (southern Burma and Vietnam to Sumatra, Boreo, Celebes, Talauts, and Philippines, except Malayan highlands), *C. b. altitudinis* (Malayan highlands), *C. b. minutus* (Nias island off western Sumatra), *C. b. concolor* (Enggano island off Western Sumatra), *C. b. javanicus* (Java, Bali, and probably Lombok), *C. b. insularum* (Kangean and Mata Siri islands in the Java Sea.

2. *C. minor* REVILLIOD 1911. – Presumably with all the characters of *C. brachyotis* except for slightly smaller size (forearm length, 53 mm; ear length, 13–14 mm). – *Distribution:* Known only by a single immature specimen from Celebes and probably a synonym of *C. brachyotis.* – No subspecies.

3. *C. sphinx* (VAHL 1797). – Molars and premolars relatively narrow and oval in outline. Surface cusp on last lower premolar and first lower molar small or absent. Size medium (forearm length, 64–78 mm; ear length, 16–24 mm). – *Distribution:* Pakistan and Ceylon east to southern China and south to Sumatra and probably Borneo. – Probably six currently recognized subspecies (HILL 1983):

C. s. sphinx (Pakistan and Ceylon to Burma), *C. s. angulatus* (southern China to Sumatra and probably Borneo), *C. s. scherzeri* (Nicobars), *C. s. serasani* (Natunas), *C. s. babi* (Babi island off western Sumatra), *C. s. pagensis* (Mentawi islands off western Sumatra).

4. *C. titthaecheileus* (TEMMINCK 1825). – Molars and premolars relatively narrow and oval in outline. Surface cusp on last lower premolar and fist lower molar small or absent. Size large (forearm length, 70–83 mm; ear length, 16–23 mm). – *Distribution:* Confined to Sumatra, Java, Lesser Sundas, and small nearby islands. – Three currently recognized subspecies (HILL 1983):

C. t. titthaecheileus (Sumatra, Java, Lombok), *C. t. major* (Nias island off western Sumatra), *C. t. terminus* (Timor).

5. *C. horsfieldi* GRAY 1843. – Molars and premolars relatively broad and subrectangular in outline. Surface cusp on last lower premolar and first lower molar always well developed. Size medium to large (forearm length, 64–90 mm; ear length, 17–18 mm). – *Distribution:* From Thailand to Java and Borneo. – Four currently recognized subspecies:

C. h. horsfieldi (Java), *C. h. persimilis* (Borneo), *C. h. harpax* (= *minor, lyoni*) (Sumatra, Malaya, Thailand), *C. h. princeps* (Nias island off western Sumatra).

Genus *Ptenochirus* PETERS 1861 (Fig. 41)

Postorbital foramen large. Premaxillae in simple contact anteriorly. Dental formula, i2/1, c1/1, p3/3, m1/2 x 2 = 28. Edge of plagiopatagium attached to first toe. Cranial rostrum lower at canine than at middle premolar. Upper canine with a secondary cusp at the inner edge and a distinct vertical groove on the antero-medial surface. Tail present. Posterior upper incisor much shortened. – *Distribution:* Confined to the Philippines. – Two species.

1. *P. jagorii* (PETERS 1861). – Size relatively large (forearm length, 76–87 mm; greatest length of skull, 36–39 mm). – *Distribution:* Known only from Luzon, Mindanao, and a number of smaller islands in between (but not from the Palawan group) in the Philippines. – No subspecies.

2. *P. minor* YOSHIYUKI 1979. – Size relatively small (forearm length, 62–77 mm; greatest length of skull, 29–35 mm). – *Distribution:* Known only from Mindanao, Dinagat and Palawan in the Philippines. – No subspecies.

Genus *Megaerops* PETERS 1865 (Fig. 42)

Postorbital foramen large. Premaxillae in simple contact anteriorly. Dental formula i2/1, c1/1, p3/3, m1/2 x 2 = 28. Egde of plagiopatagium attached to first toe. Slope of cranial rostrum variable. Upper canine without a well-developed secondary cusp or antero-medial groove. Tail absent or greatly reduced. – *Distribution:* From northeastern India and Vietnam to Java, Borneo, and the Philippines. – Four species.

1. *M. wetmorei* TAYLOR 1934. – A short tail usually present. Cranial rostrum lower at canine than at middle premolar, relatively short and slender. Size small (forearm length, 46–53 mm). Vomer extending posteriorly beyond palate. – *Distribution:* Known only from Borneo and the island of Mindanao in the Philippines. – No subspecies.

2. *M. niphanae* YENBUTRA & FELTEN 1983. – Tail absent. Cranial rostrum lower at canine than at middle premolar, and short. Size relatively large (forearm length, 52–60 mm). Vomer extending posteriorly beyond palate. – *Distribution:* Known from northeastern India, Thailand and Vietnam. – No subspecies.

3. *M. ecaudatus* (TEMMINCK 1837). – Tail absent. Cranial rostrum as deep at canine as at middle premolar, relatively long and broad. Size relatively large (forearm length, 51–56 mm). Vomer extending posteriorly beyond palate. – *Distribution:* Ranging from Thailand and Vietnam through Malaya to Sumatra and Borneo. The specimen from northeastern India is referable to *niphanae*. – No subspecies.

4. *M. kusnotoi* HILL & BOEADI 1979. – Tail absent. Cranial rostrum as deep at canine as at middle premolar, short and very broad. Size fairly large (forearm length, 49–54 mm). Vomer does not extend posteriorly beyond palate. – *Distribution:* Confined to Java. – No subspecies.

Genus *Dyacopterus* ANDERSEN 1912 (Fig. 43)

Postorbital foramen vestigial. Premaxillae ankylosed anteriorly. Dental formula i2/2, c1/1, p2/3, m1/2 x 2 = 28. Edge of plagiopatagium attached to second toe. Tail absent. – *Distribution:* Confined to Malaya, Sumatra, Borneo, and the Philippines. – A single species.

1. *D. spadiceus* (THOMAS 1890). – Size relatively large (forearm length, 77–93 mm). – *Distribution:* Same as for genus. – Two subspecies:

D. s. spadiceus (Malaya, Borneo), *D. s. brooksi* (Sumatra).

Genus *Balionycteris* MATSCHIE 1899 (Fig. 44)

Postorbital foramen absent. Upper and lower incisors simple subvertical pegs. Molars and premolars not narrow, but middle upper premolar with a large antero-external basal lobe. Uropatagium not greatly reduced, calcar present. Wings spotted. Premaxillaries in simple contact anteriorly. Dental formula i2/1, c1/1, p3/3, m2/2 x 2 = 30. – *Distribution:* Confined to the Malay peninsula and Borneo. – A single species and one additional subspecies.

1. *B. maculata* (THOMAS 1893). – Size small (forearm length, 39–42 mm). – *Distribution:* Same as for genus. – Two subspecies:

B. m. seimundi (Malaya, including extreme southern Thailand and the Rhio archipelago), *B. m. maculatus* (Borneo).

Genus *Chironax* ANDERSEN 1912 (Fig. 45)

Postorbital foramen absent. Upper und lower incisors simple subvertical pegs. Molars and premolars not narrow, but middle upper premolar with a well-defined antero-external basal cusp. Uropatagium not greatly reduced, calcar present. Premaxillaries solidly united anteriorly. Dental formula i2/2, c1/1, p3/3, m1/2 x 2 = 30. – *Distribution:* Ranges form the Malay peninsula through Sumatra, Java, and Borneo to Celebes. – A single species.

1. *C. melanocephalus* (TEMMINCK 1825). – Size small (forearm length, 41–50 mm). – *Distribution:* Same as for genus. – No subspecies.

Genus *Thoopterus* MATSCHIE 1899 (Fig. 46)

Postorbital foramen absent. Upper and lower incisors simple subvertical pegs, lateral upper incisor not shortened. Molars and premolars not narrow and middle upper premolar without a clear antero-external lobe or cusp. Uropatagium not greatly reduced, calcar present. Dental formula i2/2, c1/1, p3/3, m1/2 x 2 = 30. Last lower premolar and anterior upper molar with surface cusps. Tail vestigial. Posterior edge of plagiopatagium attached to second toe. – *Distribu-*

tion: Known only from Celebes and Morotai island (northern Moluccas) with a doubtful record from Luzon (Philippines). – A single species.

1. *T. nigrescens* (GRAY 1870). – Size relatively large (forearm length, 73–79 mm). – *Distribution:* Same as for genus. – No subspecies.

Genus ***Sphaerias*** MILLER 1906 (Fig. 47)

Postorbital foramen absent. Upper and lower incisors proclivous, crowns triangularly pointed. Molars and premolars unusually narrow. Uropatagium greatly reduced, calcar absent. Tail absent. Dental formula i 2/2, c 1/1, p 3/3, m 1/2 x 2 = 30. – *Distribution:* Confined to northern India, northern Burma, northern Thailand, and southwestern China. – A single species.

1. *S. blanfordi* (THOMAS 1891). – Size fairly small (forearm length, 50–52 mm). – *Distribution:* Same as for genus. – No subspecies.

Genus ***Aethalops*** THOMAS 1923 (Fig. 48)

Postorbital foramen absent. Inner upper incisors usually smaller than outer ones. Uropatagium and calcar greatly reduced. Dental formula i 2/1, c 1/1, p 3/3, m 1/2 x 2 = 28. Molars and premolars not narrow, but middle lower premolar with or without a clear antero-external basal cusp. Premaxillaries in simple contact anteriorly. – *Distribution:* Confined to Malaya, Sumatra, Java, and Borneo. – A single species with two additional subspecies.

1. *A. alecto* (THOMAS 1923). – Size small (forearm length, 43–51 mm). – Distribution same as for genus. – Three subspecies.

A. a. alecto (Malaya, Sumatra), *A. a. ocypete* (Java), *A. a. aequalis* (Borneo).

Genus ***Penthetor*** ANDERSEN 1912 (Fig. 49)

Postorbital foramen absent. Upper and lower incisors simple subvertical pegs, lateral upper incisors shortened. Molars and premolars not narrow and middle upper premolar without a clear antero-external lobe or cusp. Uropatagium not greatly reduced, calcar present. Dental formula i 2/1, c 1/1, p 3/3, m 1/2 x 2 = 28. No surface cusps on molars or premolars. Tail half as long as foot. Posterior edge of plagiopatagium attached to first toe. – *Distribution:* Confined to Malaya and Borneo. – A single species.

1. *P. lucasi* (DOBSON 1880). – Size medium (forearm length, 58–63 mm). – *Distribution:* Same as for genus. – No subspecies.

Genus ***Latidens*** THONGLONGYA 1972 (Fig. 50)

Postorbital foramen absent. Upper and lower incisors simple subvertical pegs. Molars and premolars not narrow. A small calcar is present, but no tail. Dental formula i 1/1, c 1/1, p 3/3, m 1/2 x 2 = 26. Posterior edge of plagiopatagium attached to first toe. – *Distribution:* Confined to southern India. – A single species.

1. *L. salimalii* THONGLONGYA 1972. – Size fairly large (forearm length, 67–68 mm). – *Distribution:* Known only by the type specimen from the Madura district. – No subspecies.

Genus ***Alionycteris*** KOCK 1969 (Fig. 51)

Postorbital foramen absent. Upper incisors somewhat caniniform, lowers simple pegs, both subvertical. Molars and premolars not narrow, but middle upper premolar with an antero-external basal ledge. Uropatagium, calcar, and tail absent. Dental formula i 1/1, c 1/1, p 3/3, m 1/2 x 2 = 26. Premaxillaries in simple contact. Posterior edge of plagiopatagium attached to first toe. – *Distribution:* Confined to the Philippines. – A single species.

1. *A. paucidentata* KOCK 1969. – Size small (forearm length, 43–46 mm). – *Distribution:* Known only from Mindanao. – No subspecies.

Genus ***Otopteropus*** KOCK 1969 (Fig. 52)

Postorbital foramen absent. Upper and lower incisors simple subvertical pegs. Molar and premolars not narrow. Uropatagium and calcar somewhat reduced, tail absent. Dental formula i 1/1, c 1/1, p 3/3, m 1/1 x 2 = 24. Premaxillaries in simple contact. Posterior edge of plagiopatagium attached to second toe. Orbit unusually large. A peculiar extra lobe on the hind border of the ear pinna. – *Distribution:* Confined to the Philippines. – A single species.

1. *O. cartilagonodus* KOCK 1969. – Size small (forearm length, 46–48 mm). – *Distribution:* Known only from Luzon. – No subspecies.

Genus ***Haplonycteris*** LAWRENCE 1939 (Fig. 53)

Postorbital foramen absent. Upper and lower incisors fairly trenchant. Molars and premolars not narrow. Uropatagium fairly well developed, but calcar and tail absent. Dental formula i 1/1, c 1/1, p 3/3, m 1/1 x 2 = 24. Premaxillaries in simple contact. – *Distribution:* Confined to the Philippines, where it is widely distributed. – A single species.

1. *H. fischeri* LAWRENCE 1939. – Size small (forearm length, 49 mm). – *Distribution:* Same as for genus. – No subspecies.

Subtribe Nyctimenina MILLER 1907

Lower incisors absent with canines close together. Nostrils elongated into cylindrical tubes. Tongue with four circumvallate papillae. Dental formula i 1/0, c 1/1, p 3/3, m 1/2 x 2 = 24. – *Distribution:* Ranging from the Philippines, Celebes, and Timor to the Santa Cruz islands (east of the Solomons) and northeastern Australia. – Two genera, 15 species.

Genus *Paranyctimene* TATE 1942 (Fig. 54)

The middle and last upper and lower premolars (as well as the canines) unusually tall (upper and lower middle premolars each at least twice as tall as long). Post-dental palate elongate. No dorsal stripe. – *Distribution:* Confined to New Guinea. – A single species.

1. *P. raptor* TATE 1942. – Size small (forearm length, 47–56 mm). – *Distribution:* Same as for genus. – No subspecies.

Genus *Nyctimene* BORKHAUSEN 1797 (Fig. 55)

Middle and last upper and lower premolars not unusually tall (upper and lower middle premolars each less than twice as tall as long). Post-dental palate not elongate. Dorsal stripe present. – *Distribution:* Same as that of subtribe. – 14 species and 7 additional subspecies.

1. *N. minutus* ANDERSEN 1910 [*albiventer* group]. – Dorsal stripe narrow (much less than one third breadth of furred area of back). Premaxillae not proclivous. Size relatively small (forearm length, 51–55 mm; maxillary tooth row length, 8.8–10.0 mm). Upper molar subequal in size to last upper premolar. Ears not unusually broad and somewhat pointed. Inner cusp of middle upper premolar completely fused with outer. – *Distribution:* Known only from Celebes and Buru (Moluccas). – Two subspecies:

N. m. minutus (Celebes), *N. m. varius* (Buru).

2. *N. draconilla* THOMAS 1922 [*albiventer* group]. – Size very small (forearm length, 47–53 mm; maxillary tooth row length, 7.8–8.6 mm). Inner cusp of middle upper premolar not completely fused with outer. Upper canine at least twice middle upper premolar in height. – *Distribution:* Confined to southern New Guinea (the northern New Guinea record is in error, see HILL 1983). – No subspecies.

3. *N. albiventer* (GRAY 1863) [*albiventer* group]. – Size fairly small (forearm length, 50–59 mm; maxillary tooth row length, 8.7–10.3 mm). Inner cusp of middle upper premolar not completely fused with outer. – *Distribution:* Ranging from the Moluccas through New Guinea to the Bismarcks (the Australian record is in error). – Two subspecies.

N. a. albiventer (northern Moluccas), *N. a. papuanus* (Kei islands, New Guinea, Bismarcks).

4. *N. cyclotis* ANDERSEN 1910 [*cyclotis* group]. – Dorsal stripe narrow (much less than one third breadth of furred area of back). Premaxillae not proclivous. Size relatively small (forearm length, 53–58 mm; maxillary tooth row length, 8.5–9.8 mm). Upper molar clearly smaller than last upper premolar. Ears unusually broad and rounded. – *Distribution:* Confined to the mountains of New Guinea and New Britain. – Two subspecies:

N. c. cyclotis (extreme northwestern New Guinea), *N. c. certans* (remainder of New Guinea and, presumeably, New Britain).

5. *N. vizcaccia* THOMAS 1914 [*cephalotes* group]. – Dorsal stripe narrow (much less than one third breadth of furred area of back). Premaxillae not proclivous. Size small to medium (forearm length, 54–60 mm; maxillary tooth row length, 8.7–10.1 mm). Rostrum relatively long. Inner cusp of middle upper and lower premolars completely fused with outer. – *Distribution:* Ranges from Umboi island (off the northeast coast of New Guinea), New Britain, and New Ireland through the Solomons as far as Guadalcanal. – Two subspecies are currently recognized (SMITH & HOOD 1983).

N. v. vizcaccia (Umboi, Bismarcks), *N. v. bougainville* (= *minor*) (Solomons).

6. *N. cephalotes* (PALLAS 1767) [*cephalotes* group]. – Size medium (forearm length, 60–70 mm; maxillary tooth row length, 9.3–11.2 mm). Inner cusp of middle upper and lower premolars distinct. – *Distribution:* Ranging from Celebes and Timor

through the central and southern Moluccas to Numfoor island (off northwestern New Guinea); also recorded from extreme southern New Guinea. – No subspecies.

7. <u>N. rabori</u> HEANEY & PETERSON 1984 [*cephalotes* group]. – Size fairly large (forearm length, 73–81 mm; maxillary tooth row length, 11.2/12.1 mm). Sagittal crest and coronoid process unusually well developed. – *Distribution:* Known only from Negros island in the central Philippines. – No subspecies.

8. <u>N. masalai</u> SMITH & HOOD 1983 [*cephalotes* group]. – Size medium (forearm length, 63–68 mm; maxillary tooth row length, 10.4–10.9 mm). Rostrum unusually long. Braincase unusually broad. – *Distribution:* Known only from New Ireland in the Bismarcks. – No subspecies.

9. <u>N. malaitensis</u> PHILLIPS 1968 [*cephalotes* group]. – Size medium (forearm length, 65 mm; maxillary tooth row length, 10.5 mm). Rostrum relatively short and broad. – *Distribution:* Known only from Malaita island in the eastern Solomons. – No subspecies.

10. <u>N. major</u> (DOBSON 1877) [*major* group]. – Dorsal stripe narrow (much less than one third breadth of furred area of back). Premaxillae not proclivous. Size medium to large (forearm length, 67–86 mm; maxillary tooth row length, 10.2–14.2 mm). Ear somewhat rounded and clearly shorter than hind foot (with claws). Cranium relatively flat. – *Distribution:* Some small islands off the northern and eastern sides of New Guinea, also the Bismarcks and Solomons. – Four subspecies.

<small>*N. m. lullulae* (Schouten islands, Karkar, and Bagabag, all off the northern coast of New Guinea, also Woodlark and possibly Kiriwina in the Trobriands), *N. m. geminus* (D'Entrecasteaux and Louisiade archipelagos), *N. m. major* (Bismarcks), *N. m. scitulus* (Solomons east to Guadalcanal).</small>

11. <u>N. robinsoni</u> THOMAS 1904 [*major* group]. – Size medium (forearm length, 60–70 mm; maxillary tooth row length, 10.0–12.0 mm). Ears definitely pointed and equal in length to hind foot (with claws). – *Distribution:* Restricted to the Pacific coast of Queensland (Australia). – No subspecies.

12. <u>N. sanctacrucis</u> TROUGHTON 1931 [*major* group]. – Size fairly large (forearm length, 75 mm; maxillary tooth row length, 12.9 mm). Ears unusually small. Rostrum relatively long. – *Distribution:* Known only from the Santa Cruz islands (east of the Solomons). – No subspecies.

13. <u>N. aello</u> (THOMAS 1900) [*aello* group]. – Dorsal stripe broad (about one third breadth of furred area of back). Premaxillae relatively proclivous. Size large (forearm length, 77–85 mm; maxillary tooth row length, 12.2–14.2 mm). Frontal region of skull relatively depressed. Rostrum not greatly shortened. – *Distribution:* Confined to New Guinea (where apparently widely distributed), including Misol (off its western end). – No subspecies.

14. <u>N. celaeno</u> THOMAS 1922 [*aello* group]. – Size relatively large (forearm length, 83 mm; maxillary tooth row length, 13.2 mm). Frontal region of skull relatively elevated. Rostrum greatly shortened. – *Distribution:* Known only from the region of Geelvink Bay in northwestern New Guinea. (HILL 1983), for status. – No subspecies.

Subfamiliy **Macroglossinae** GRAY 1866

Tongue definitely extensible, fixed to floor of mouth only by its posterior third, its terminal fourth or fifth covered above with unfringed filiform papillae. These are the Pteropodidae with specialized nectar-feeding tongues. Size ranges from small to medium (forearm length, 37–80 mm) – *Distribution:* Ranging in the forested regions of tropical Africa and from northern India east to the Fiji islands, northern and eastern Australia. – Two tribes, six genera, 12 species.

Tribe <u>Macroglossini</u> GRAY 1866

Premaxillary about as broad in upper as in lower half. Infraorbital canal relatively short. Terminal phalanx of third digit definitely shorter than third metacarpal. Anterior upper and lower premolars always functional teeth. Size ranges from small to medium (forearm length, 37–80 mm). – *Distribution:* Ranging in the forested regions of tropical Africa and from northern India east to the Solomons, northern and eastern Australia. – Four genera, eight species.

Genus ***Eonycteris*** DOBSON 1873 (Fig. 56)

Premaxillaries separate from one another anteriorly. Molars and premolars relatively unreduced. No claw on second digit of wing. Tail relatively well-developed (as long as hind foot). Facial axis not strongly deflected on braincase. Dental formula normally i2/2, c1/1, p3/3, m2/3 x 2 = 34 (last lower molar occasionally lost) – *Distribution:* Ranging from India and the Andaman islands to the Philippines, Celebes, and Timor. – Two species are here recognized with three additional subspecies.

1. *E. major* ANDERSEN 1910. – Size relatively large (forearm length, 75–82 mm). Anal glands absent or poorly developed. – *Distribution:* Known only from Borneo and the Philippines. – Two subspecies (which may be specifically distinct):

E. m. major (Borneo), *E. m. robusta* (Philippines).

2. *E. spelaea* DOBSON 1871. – Size relatively small (forearm length, 61–77 mm). Anal glands well-developed. – *Distribution:* Range same as for genus (including Mentawai islands). – Three poorly defined subspecies, one of which, *E. s. rosenbergi* (northern Celebes), is generally treated as a separate species.

Genus *Megaloglossus* PAGENSTECHER 1885 (Fig. 57)

Premaxillaries solidly fused together anteriorly. Molars and premolars reduced in size. Claw present on second digit of wing. Tail vestigial. Upper incisors minute, lower incisors subequal in size and bilobate. Uropatagium and calcar relatively unmodified. Dental formula i 2/2, c 1/1, p 3/3, m 2/3 x 2 = 34. Facial axis not strongly deflected on braincase. Premaxillaries not greatly proclivous. Middle upper premolar much higher than anterior upper premolar. Fifth metacarpal much shorter than third. Differentiated neck-ruff in males. – *Distribution:* Restricted to forested regions of tropical Africa from Liberia to Uganda and south to northern Angola and southern Zaire including Fernando Poo. – A single species with an additional subspecies.

1. *M. woermanni* PAGENSTECHER 1885. – Size relatively small (forearm length, 37–50 mm). – *Distribution:* Same as for genus. – Two poorly marked subspecies.

Genus *Macroglossus* F. CUVIER 1824 (Fig. 58)

Premaxillaries solidly fused together anteriorly. Molars and premolars reduced in size. Claw present on second digit of wing. Tail vestigial or absent. Upper incisors minute, lower incisors subequal in size and simple widely spaced pegs. Uropatagium and calcar relatively unmodified. Dental formula normally i 2/2, c 1/1, p 3/3, m 2/3 x 2 = 34. Facial axis strongly deflected on braincase. Premaxillaries greatly proclivous. Middle upper premolar reduced, little higher than anterior upper premolar. Fifth and third metacarpals subequal. No differentiated neck-ruff in males. – *Distribution:* From Burma to the Solomon islands and northern Australia. – Two species with three additional subspecies.

1. *M. sobrinus* ANDERSEN 1911. – Size relatively large (forearm length, 42–51 mm; condylobasal length of skull, 25–29 mm) with a relatively long rostrum (length from orbit to nares, 9.5–11.7 mm). Mandible projecting forward to form a definite chin. – *Distribution:* From Burma and Vietnam to Java, including the Mentawai islands. – Two well-marked subspecies:

E. s. fraternus (Mentawais) *E. s. sobrinus* (remaining range).

2. *M. minimus* (GEOFFROY 1810). – Size relatively small (forearm length, 36–45 mm; condylobasal length of skull, 22–26 mm) with a relatively short rostrum (length from orbit to nares, 7.5–9.2 mm). Mandible slopes forward without forming a definite chin. – *Distribution:* From Thailand and Vietnam south to Java and extending east to the Solomons islands and northern Australia, including the Philippines. – Three subspecies are here recognized:

M. m. minimus (Java and surrounding islands), *M. m. lagochilus* (= *fructivorus*) (Thailand and Vietnam to Nias (off the west coast of Sumatra), Borneo, Philippines, Celebes, central Moluccas, and the Lesser Sundas), *M. m. nanus* (= *pygmaeus, microtus*) (Northern and eastern Moluccas, New Guinea and surrounding islands, northern Australia, Bismarcks, and Solomons).

Genus *Syconycteris* MATSCHIE 1899 (Fig. 59)

Premaxillaries solidly fused together anteriorly. Molars and premolars reduced in size. Claw present on second digit of wing. Tail vestigial or absent. Upper incisors large and narrowly chisel-shaped, lateral incisors much larger than inner and with triangular crowns. Uropatagium unusually narrow or absent, calcar vestigial or absent. Dental formula normally i 2/2, c 1/1, p 3/3, m 2/3 x 2 = 34, but a molar above and/or below may be lost. – *Distribution:* Moluccas, New Guinea, Bismarcks, and the eastern coast of Australia as far south as northeastern New South Wales. – Three species and six additional subspecies are here recognized.

1. *S. australis* (PETERS 1867). – Size relatively small (forearm length, 38–49 mm). Uropatagium

present. Terminal phalanx of third digit of wing less than 34 mm. Ear pinna relatively long and pointed. Metatarsus and digits of foot relatively naked. – *Distribution:* Same as for genus except for northern Moluccas. – Seven subspecies recognized here:

S. a. australis (eastern Australia), *S. a. papuana* (New Guinea and some nearby islands; including the Louisiades), *S. a. finschi* (Bismarcks), *S. a. crassa* (Fergusson and Kiriwina islands in the East Papuan group), *S. a. naias* (Woodlark island in the East Papuan group), *S. a. keyensis* (Keis), *S. a. major* (Ceram and nearby islands).

2. *S. hobbit* ZIEGLER 1982. – Size medium (forearm length, 45–50 mm). Uropatagium absent. Terminal phalanx of third digit of wing usually more than 34 mm. Ear pinna short and rounded. Metatarsus and digits of foot relatively hairy. – *Distribution:* Known only from a small area above 2000 meters on Mt. Kaindi in eastern New Guinea. – No subspecies.

3. *S. carolinae* ROZENDAAL 1984. – Size relatively large (forearm length, 60 mm). Uropatagium present. – *Distribution:* Known only from Halmahera in the northern Moluccas. – No subspecies.

Tribe Notopterini KOOPMAN & JONES 1970

Premaxillary two or three times as broad in upper as in lower half. Infraorbital canal relatively long. Terminal phalanx of third digit subequal to or longer than third metacarpal. Anterior upper and lower premolars vestigial. Size medium (forearm length, 52–69 mm). – *Distribution:* Bismarcks (a New Guinea record is almost certainly erroneous), Solomons, New Hebrides, Fijis, New Caledonia, and a record from the Carolines. – Two genera, four species.

Genus *Melonycteris* DOBSON 1877 (Fig. 60)

Premaxillaries separate or in simple contact anteriorly. Angular process of dentary relatively large. Anterior upper and lower premolars present, middle lower premolar smaller than anterior or middle lower molar. Inner edges of plagiopatagia attached to sides of body. Tail absent. Tibia much less than half length of forearm. – *Distribution:* Bismarcks and Solomons. – Two subgenera and three species.

Subgenus *Melonycteris* DOBSON 1877

Dental formula normally i 2/2, c 1/1, p 3/3, m 2/3 x 2 = 34. Claw on second digit of wing. – *Distribution:* Bismarcks (a New Guinea record almost certainly erroneous). – A single species.

1. *M. melanops* DOBSON 1877. – Size relatively large (forearm length, 57–63 mm). Ventral fur nearly black, strongly contrasting with dorsum. – *Distribution:* As for subgenus. – No subspecies.

Subgenus *Nesonycteris* THOMAS 1877

Dental formula normally i 2/1, c 1/1, p 3/3, m 2/3 x 2 = 32. No claw on second digit of wing. – *Distribution:* Confined to the Solomons. – Two species.

2. *M. aurantius* PHILLIPS 1966. – Postorbital region of skull, relatively broad (8.0–8.9 mm). Size relatively small (forearm length, 42–54 mm). Pelage bright reddish-brown, little contrast between dorsum and ventrum. – *Distribution:* Known only from Choiseul and Florida islands in the Solomons. – No subspecies.

3. *M. woodfordi* (THOMAS 1887). – Postorbital region of skull relatively narrow (7.2–7.5 mm). Size medium (forearm length, 52–58 mm). Pelage dark brown, little contrast between dorsum and ventrum. – *Distribution:* Known from a number of islands from Bougainville to Guadalcanal in the Solomons. – No subspecies.

Genus *Notopteris* GRAY 1859 (Fig. 61)

Premaxillaries co-ossified anteriorly. Angular process of dentary relatively small. Anterior upper and lower premolars absent, middle lower premolar larger than any other lower premolar or molar. Inner edges of plagiopatagia meeting in mid-dorsal line. Tail very long, subequal to forearm. No claw on second digit of wing. Tibia half length of forearm. Dental formula i 2/1, c 1/1, p 2/2, m 2/3 x 2 = 28, but minute anterior upper incisor lost in adults. – *Distribution:* New Hebrides, Fijis, and New Caledonia, also recorded from Ponape island in the Carolines. – A single species, one additional subspecies.

1. *N. macdonaldi* GRAY 1859. – Size relatively large (forearm length, 60–72 mm). – *Distribution:* The same as for genus. – Two subspecies:

N. m. macdonaldi (New Hebrides, Fijis, possibly Ponape), *N. m. neocaledonica* (New Caledonia).

Suborder **Microchiroptera** Dobson 1875
(Typical Bats)

Second finger closely associated with third finger and lacking a claw in all recent genera. Humerus with a large trochiter and trochin, the former often articulating with the scapula. The external ear is often complicated with the margin not forming a complete ring, tragus usually present. The cochlea of the ear is large and usually compresses the basioccipital. The postorbital processes are usually absent or rudimentary. The bony palate is usually not extended behind the last molar, narrowing abruptly. Incisors often 2/3. Gray matter in spinal cord greatly extended dorsally. Echolocation universal. – *Distribution:* Virtually coextensive with that of order. – Two infraorders, four superfamilies, 16 families, 133 genera, 749 species.

Infraorder **Yinochiroptera** Koopman 1985
(Bats with movable premaxillae)

Premaxillaries almost never fused with maxillaries. – *Distribution:* Predominately an Old World group but also occurring in tropical South and Middle America. – Two superfamilies, six families, 29 genera, 198 species.

Superfamily **Emballonuroidea** Gervais 1855
(Mouse-tailed, Sheath-tailed, and Bumblebee Bats)

Last cervical vertebra not fused with first thoracic. Reduction of palatal, but not nasal, branch of the premaxillary. Usually no special modifications of the muzzle or rhinarium. – *Distribution:* From Africa and Madagascar through southern Asia and the east Indies to Australia and many western Pacific islands. Also the whole of mainland tropical America. – Three families, one additional subfamily, 14 genera, 51 species.

Family **Rhinopomatidae** Bonaparte 1838
(Mouse-tailed bats)

Structure: Small to medium-sized bats (forearm length, 45–75 mm) with the uropatagium poorly developed, no calcar, but the tail long. Second digit of wing retaining two bony phalanges. Muzzle with a thickened narial pad surmounted by a ridge-like dermal outgrowth. Skull lacking postorbital processes and lacrimal region swollen. Premaxillaries fused neither to each other nor to other parts of the skull, not meeting dorsal to the nasal aperture. Humerus with trochiter well-developed but not as large as trochin and not articulating with scapula. Post-cranial skeleton in general primitive for Microchiroptera. Inguinal false nipples present.

Ecology: These are insect-eating bats, chiefly of arid and semi-arid regions. They roost in caves, rock crevices and similar man-made structures and may remain torpid at times when insects are scarce. During these times they utilize fat stored in the rump region.

Distribution (Fig. 8): The family ranges across northern Africa, both north and south of the Sahara (and in Kenya almost to the Equator), through southern Asia as far east as Thailand and south to Sumatra.

Systematics: A single genus and three species are currently recognized.

Genus **Rhinopoma** E. Geoffroy 1818 (Fig. 62)

Ears joined by their inner margins, tragi fairly small. Dental formula i2/2, c1/1, p1/2, m3/3 x 2 = 28. – *Distribution:* Same as for family. – Three species and eight additional subspecies are recognized (Hill 1977a).

1. *R. microphyllum* (Brunnich 1782). – Size relatively large (forearm length, 57–75 mm). Tail usually shorter than forearm. Sagittal crest prominent. Supraorbital ridges high and enclosing a prominent recess. Rostrum with narial wellings not pronounced. – *Distribution:* From Morocco and Senegal to the eastern end of the Mediterranean and central Sudan, also western Saudi Arabia and from Iran at least to India and possibly Thailand, also Sumatra. – Six subspecies are currently recognized:

R. m. tropicalis (Nigeria and south-central Sudan), *R. m. microphyllum* (= *cordofanicum*) (remainder of African range, eastern end of the Mediterranean south to northwestern Saudi Arabia, southern Iran, except southwestern coastal portion, east to western India), *R. m. asirensis* (southwestern Saudi Arabia), *R. m. harrisoni* (southwestern Iran), *R. m. kinneari* (central India), *R. m. sumatrae* (Sumatra).

2. *R. hardwickei* Gray 1831. – Size small to medium (forearm length, 45–74 mm). Tail usually longer than forearm. Sagittal crest low. Supraorbital ridges low and not enclosing a prominent

recess. Rostrum with pronounced globose narial swellings, which do not project laterally much beyond anterior ends of nasals. Muzzle with a well-developed transverse dermal ridge. – *Distribution:* From Morocco and Mauretania across northern Africa to Egypt and Ethiopia (south to Kenya), widely distributed on the Arabian peninsula and east to western Iran; also Afghanistan and Pakistan east to India and possibly Thailand. – Four poorly defined subspecies are currently recognized:

R. h. cystops, R. h. arabium (= *sennaariense*), *R. h. macinnesi, R. h. hardwickei.*

3. *R. muscatellum* THOMAS 1903. – Size small (forearm length, 45–55 mm). Tail usually longer than forearm. Sagittal crest low. Supraorbital ridges low and not enclosing a prominent recess. Rostrum with pronounced and rather angular swellings, which project laterally considerably beyond anterior ends of nasals. Muzzle with low dermal ridge. – *Distribution:* Restricted to Oman, southern and eastern Iran, and southern Afghanistan. – Two currently recognized subspecies:

R. m. muscatellum (= *pusillum*) (Oman and southern Iran),
R. m. seianum (confined to a small area of extreme eastern Iran and southern Afghanistan).

Family **Craseonycteridae** HILL 1974 (Bumblebee bats)

Structure: The smallest of all bats (forearm length, 22–26 mm) with the uropatagium well-developed, but no calcar or tail. Second digit of wing retaining a single small bony phalanx. No ridge-like dermal outgrowth above the thickened narial pad. Skull lacking postorbital processes and lacrimal region only slightly inflated. Premaxillaries fused to each other both ventral and dorsal to the nasal aperture. Humerus with trochiter larger than trochin and probably articulating with scapula. Post-cranial skeleton primitive except for vertebral fusion in posterior thoracic and anterior lumbar regions. Inguinal false nipples present.

Ecology: The single species is known to be insectivorous and cave-dwelling and appears to be both an aerial insectivore and foliage gleaner.

Distribution (Fig. 8): Known only from a small area in western Thailand.

Systematics: A single genus and species are known.

Genus *Craseonycteris* HILL 1974 (Fig. 63)

Ears large but not joined together, tragi very large and distinctive. Dental formula i 1/2, c 1/1, p 1/2, m 3/3 x 2 = 28. – *Distribution:* Same as for family. – A single species.

1. *C. thonglongyai* HILL 1974. – Characters and *Distribution:* Same as for genus. – No subspecies.

Family **Emballonuridae** GERVAIS 1855 (Sheath-tailed bats)

Structure: Small to fairly large bats (forearm length, 35–95 mm) with the uropatagium and calcar well-developed. The tail is present, but does not extend the full width of the uropatagium, its end protruding a short distance above its dorsal surface. Second digit of wing without phalanges. Muzzle without any special modifications. Skull usually with well-developed postorbital processes. Lacrimal region not swollen. Premaxillaries not fused with one another and usually not to other parts of the skull, not meeting dorsal to the nasal aperture. Humerus with trochiter well-developed but not as large as trochin and not articulating with scapula. Postcranial skeleton in general primitive for Microchiroptera. Wings unusually long, the proximal phalanx of the third digit of the wing flexed back on the metacarpal when not in flight.

Ecology: As far as is known, always insectivorous. While confined to tropical and warm temperate regions, they may be found in a variety of macrohabitats from rain forests to deserts. They usually hang or prop themselves up on vertical surfaces, often in open situations, but these may be on rocks or trees, in caves or cave-like structures or even under vegetation.

Distribution (Fig. 9): Most of Africa (except for the Sahara and the northwest), Madagascar, Mascarenes, Seychelles, across southern Asia to southern China, through the entire Malay archipelago to Micronesia, the central Pacific and Australia. Also the whole of mainland tropical America.

Systematics: In the Recent, there are, 12 genera, 47 species.

Genus *Taphozous* E. GEOFFROY 1818 (Fig. 72)

Dental formula i 1/2, c 1/1, p 2/2, m 3/3 x 2 = 30. Postorbital processes long, curved, and slender, not obscured by supraorbital ridges. No wing sacs. Dorsal surface of rostrum short and flat, narrowed anteriorly. Clavicle not expanded. Tibia without a longitudinal groove. Ventral side of dentary usually concave anteriorly. Tympanic bulla incomplete medially and separated from basioccipital. A radio-metacarpal pouch present and at least fairly well developed in the wing. – *Distribution:* An extensive Old World distribution including most of Africa (except the Sahara and northwest). Madagascar and the Mascarenes, across southern Asia to southern China, through the Malay archipelago to the Philippines and Australia, but only very marginal in New Guinea. – Two subgenera, 13 species.

Subgenus *Taphozous* E. GEOFFROY 1818

Frontal region of skull strongly concave. Rump and pygal areas well-haired. No occipital „helmet" on skull.

Distribution: Essentially same as that of the genus, but in Arabia and Iran confined to the southern portions. – Eleven species, 14 additional subspecies (KOOPMAN 1984a for discussion of Australian species).

1. *T. hildegardeae* THOMAS 1909. – Gular sac absent. A black beard-like throat patch in males. Skull relatively broad. Fur relatively pale in color. Size medium (forearm length, 63–70 mm). – *Distribution:* Confined to a small area of southeastern Kenya and northeastern Tanzania including Zanzibar island. – No subspecies.

2. *T. melanopogon* TEMMINCK 1841. – Gular sac absent. A black beard-like throat patch variably developed. Braincase relatively slender. Fur relatively dark in color. Size medium (forearm length, 60–69 mm). – *Distribution:* Widely distributed from India and Ceylon east through southeastern Asia and the Malay archipelago to the Philippines, Celebes, the Kei islands (eastern Moluccas) and Timor. – Six subspecies may be recognized:

T. m. bicolor (India and Ceylon east to southern China and Vietnam, including the Andaman islands), *T. m. fretensis* (Malay peninsula and Borneo), *T. m. cavaticus* (Sumatra), *T. m. melanopogon* (Java, Bali), *T. m. achates* (Lombok, Sumbawa, Savu, and Timor in the Lesser Sundas), *T. m. philippinensis* (= *solifer*) (main Philippine islands). Celebes and Kei populations have not been allocated to subspecies.

3. *T. theobaldi* DOBSON 1872. – Gular sac absent. A blackish or reddish beard-like throat patch variably developed. Size relatively large (forearm length, 69–73 mm). – *Distribution:* Confined to India, Burma, Thailand, Vietnam, and Java (a Malay record is evidently erroneous). – Two subspecies are currently recognized:

T. t. secatus (central India), *T. t. theobaldi* (remainder of distribution).

4. *T. perforatus* E. GEOFFROY 1818. – Gular sac absent or poorly developed. No differentiated throat patch present. Ears relatively short. Size medium (forearm length, 57–67 mm). – *Distribution:* Widely distributed in Africa from Mauretania and Egypt south to Zimbabwe, also southern Asia from southwestern Arabia to northwestern India. – Six poorly defined subspecies are here recognized:

T. p. senegalensis (Mauretania, Senegal), *T. p. swirae* (Mali and Ghana to Central African Republic), *T. p. sudani* (central and southern Sudan, eastern Zaire), *T. p. rhodesiae* (Botswana and Zimbabwe), *T. p. perforatus* (Egypt and northern Sudan), *T. p. haedinus* (Tanzania north to Ethiopia and east across southern Asia to India).

5. *T. longimanus* HARDWICKE 1825. – Gular sac present and throat area virtually naked. Size relatively small (forearm length, 55–62 mm). Basisphenoid pits broad. Anterior ventral mandibular emargination weak. – *Distribution:* From India and Ceylon to Cambodia and Malaya, also Sumatra, Borneo, Java and east to Flores in the Lesser Sundas. – Four currently recognized subspecies:

T. l. longimanus (India and Ceylon to Cambodia), *T. l. albipinnis* (Malay peninsula, Sumatra, Borneo), *T. l. kampenii* (Java, Bali), *T. l. leucopleura* (Flores).

6. *T. kapalgensis* MCKEAN & FRIEND 1979. – Gular sac present and throat area virtually naked. Size medium (forearm length, 58–63 mm). Basisphenoid pits medium in width. Anterior ventral mandibular emargination weak. – *Distribution:* Confined to a small area on the coast of the Northern Territory of Australia. – No subspecies.

7. *T. hilli* KITCHENER 1980. – Gular sac present and throat area virtually naked. Size medium (forearm length, 63–72 mm), though the skull is unusually small. Basisphenoid pits narrow. Anterior ventral mandibular emargination strong. – *Distribution:* Known only from arid areas in Western Australia and the Northern Territory. – No subspecies.

8. *T. australis* GOULD 1854. – Gular sac present and throat area virtually naked. Size medium (forearm length, 63–67 mm). Basisphenoid pits broad. Anterior ventral mandibular emargination

strong. – *Distribution:* Confined to Cape York, with a single record from southeastern New Guinea (Port Moresby). – No subspecies.

9. *T. georgianus* THOMAS 1915. – Gular sac absent, but no differentiated throat patch. Size medium (forearm length, 64–71 mm). Basisphenoid pits medium in width. Anterior ventral mandibular emargination strong. – *Distribution:* Tropical and sub-tropical Australia (except for most of Cape York). – No subspecies.

10. *T. troughtoni* TATE 1952. – Gular sac absent, but no differentiated throat patch. Size relatively large (forearm length 70–75 mm). Basisphenoid pits medium in width. Anterior ventral mandibular emargination strong. – *Distribution:* Definitely known only from central Queensland, but may extend into the Northern Territory. – No subspecies.

11. *T. mauritianus* E. GEOFFROY 1818. – Gular sac present and throat area virtually naked. Size medium (forearm length, 57–65 mm). A distinctive color pattern, grizzled gray dorsally and white ventrally. – *Distribution:* Over most of sub-Saharan Africa from at least Sierra Leone on the west to Ethiopia in the east and south to Cape Province, South Africa. Also Madagascar, Aldabra, Reunion, and Mauritius, all in the western Indian Ocean. – No subspecies.

Subgenus ***Liponycteris*** THOMAS 1922

Frontal region of skull only weakly concave. Rump and pygal areas at least partly naked. Occipital "helmet" on skull more or less developed. – *Distribution:* Over sub-Saharan Africa from Senegal and Somalia south to Tanzania and north to Egypt and Israel. Also east across Arabia, Iran, Afghanistan, Pakistan, and India, to Burma. Recently recorded from the Cape Verde islands. – Two species, four additional subspecies.

12. *T. hamiltoni* THOMAS 1920. – Size relatively small (forearm length, 61–69 mm). A slight but definite concavity in the frontal region. Hairless rump and pygal areas relatively small. Occipital "helmet" of skull poorly developed. – *Distribution:* Known only from southern Chad, southern Sudan, and northern Kenya. – No subspecies. (KOOPMAN 1975, for placement and distribution of this species).

13. *T. nudiventris* CRETZSCHMAR 1830. – Size relatively large (forearm length, 66–79 mm). Frontal concavity virtually absent. Hairless rump and pygal areas relatively large. Occipital "helmet" well developed. – *Distribution:* Same as for subgenus. – Five subspecies are currently recognized:

T. n. nudiventris (African, southwestern Arabian, and Israeli range), *T. n. zayidi* (Oman), *T. n. magnus* (Iraq, Iran), *T. n. kachensis* (Afghanistan, Pakistan, India), *T. n. nudaster* (Burma).

Genus ***Saccolaimus*** TEMMINCK 1841 (Fig. 73)

Dental formula i 1/2, c 1/1, p 2/2, m 3/3 x 2 = 30. Postorbital processes long, curved, and slender, not obscured by supraorbital ridges. No wing sacs. Dorsal surface of rostrum short and flat, narrowed anteriorly. Clavicle not expanded. Tibia without a longitudinal groove. Ventral side of dentary usually convex anteriorly. Tympanic bulla complete medially and sutured to basioccipital. Radio-metacarpal pouch absent or poorly developed. – *Distribution:* Through the forested regions of tropical Africa, also India and Ceylon through southeastern Asia and the Malay archipelago to the Philippines, Solomons and Australia. – Five species, three additional subspecies.

1. *S. mixtus* TROUGHTON 1925. – Posterior floor of mesopterygoid fossa deeply grooved. Basisphenoid pits separated by a high septum. Size relatively small (forearm length, 61–68 mm). Radio-metacarpal pouch small but distinct. Sagittal crest relatively low and not forming an occipital "helmet". – *Distribution:* Confined to southern and eastern New Guinea and the northern end of Cape York peninsula (Australia). – No subspecies.

2. *S. flaviventris* PETERS 1867. – Posterior floor of mesopterygoid fossa deeply grooved. Basisphenoid pits separated by a high septum. Size relatively large (forearm length, 70–80 mm). Radio-metacarpal pouch absent. Sagittal crest relatively low with occipital "helmet" poorly developed. – *Distribution:* Widely distributed in northern and eastern Australia (except Tasmania). – No subspecies.

3. *S. saccolaimus* (TEMMINCK 1841). – Posterior floor of mesopterygoid fossa not deeply grooved. Basisphenoid pits separated by a relatively low septum. Size medium (forearm length, 67–76 mm). Radio-metacarpal pouch virtually absent. Sagittal crest relatively high with occipital "helmet" usually well developed. – *Distribution:* From India and Ceylon through southeastern Asia and the Malay archipelago (but not the Philippines) to Guadalcanal in the Solomons and to

northeastern Queensland and northern Northern Territory in Australia. – Four currently recognized subspecies:

S. s. crassus (Asian mainland and Sumatran range), S. s. saccolaimus (Java), S. s. affinis (= flavomaculatus) (Borneo), S. s. nudicluniatus (New Guinea, Guadalcanal, northeastern Queensland). The Celebes, Timor, and Northern Territory populations have not been allocated to subspecies.

4. S. pluto (MILLER 1910) (= capito HOLLISTER 1913). – Apparently similar to S. saccolaimus in all respects except for slightly smaller size (forearm length, 69–73 mm). – Distribution: Confined to the Philippines. – No subspecies. Probably only a subspecies of S. saccolaimus.

5. S. peli (TEMMINCK 1853). – Posterior floor of mesopterygoid fossa deeply grooved. Basisphenoid pits separated by a relatively low septum. Size very large (forearm length, 84–95 mm). Radio-metacarpal pouch absent. Sagittal crest relatively high with occipital "helmet" well developed. – Distribution: Forested regions of tropical Africa from Liberia to western Kenya and south to Angola. – No subspecies.

Genus *Emballonura* TEMMINCK 1838 (Fig. 64)

Dental formula i2/3, c1/1, p2/2, m3/3x2 = 34. Postorbital processes usually long, curved, and slender, not obscured by supraorbital ridges. No wing sacs. Clavicle not expanded. Tibia without a longitudinal groove. Anterior lower border of orbit so expanded that toothrows are not visible from above. – Distribution: Occurs in Madagascar and from the Malay peninsula through the Malay archipelago and beyond to the Philippines, Caroline and Mariana islands, and Samoa. – Two subgenera, nine species.

Subgenus *Mosia* GRAY 1844

Rostrum relatively short. Size small (forearm length, 32–37 mm). Nostrils widely separated, opening by elliptical apertures placed obliquely. Tragus relatively long and narrow. Upper lip not projecting beyond lower. Lower incisors filling up space between canines. – Distribution: From Celebes through the Moluccas, New Guinea, and the Bismarcks to the Solomons. – A single species, two additional subspecies.

1. E. nigrescens (GRAY 1843). – Postorbital crests confluent with sagittal crest. Basisphenoid pits deep but not extending into alisphenoids, median septum absent. – Distribution: Same as subgenus. – Three currently recognized subspecies:

E. n. nigrescens (known only from central Moluccas), E. n. papuana (Celebes, northern Moluccas, Keis, New Guinea), E. n. solomonis (Bismarcks, Solomons, perhaps Woodlard island in the Trobriands).

Subgenus *Emballonura* TEMMINCK 1838

Rostrum relatively long. Size medium (forearm length, 37–53 mm). Nostrils usually close together, opening by circular apertures directed forwards. Upper lip projecting more or less beyond the lower. Lower incisors separated from canines. Upper molar tooth rows parallel. – Distribution: Coextensive with that of genus. – Eight species and nine additional subspecies.

2. E. atrata PETERS 1874 [atrata group]. – Inner margin of tragus convex. Calcar longer than tibia. Size relatively small (forearm length, 37–39 mm). Postorbital crests not confluent with sagittal crest. Basisphenoid pits of medium depth and separated by a median septum, but not extending into alisphenoids. – Distribution: Confined to Madagascar. – No subspecies.

3. E. monticola TEMMINCK 1838 [alecto group]. – Inner margin of tragus convex. Calcar shorter than tibia. Size medium (forearm length, 42–46 mm). Postorbital crests not confluent with sagittal cest. Basisphenoid pits relatively deep and separated by a median septum, extending, to some extent, into alisphenoids, but not recessed into basioccipital. – Distribution: From southern Burma through the Malay peninsula, Sumatra, Java, and Borneo to southern Celebes. – No subspecies.

4. E. alecto (EYDOUX & GERVAIS 1836) [alecto group]. – Size relatively large (forearm length, 43–49 mm). Postorbital crests more or less confluent with sagittal crest. Basisphenoid pits separated by a median septum. – Distribution: Occuring in Borneo, Philippines, Celebes and Moluccas. – Three subspecies are currently recognized:

E. a. rivalis (Borneo), E. a. palawanensis (Palawan in the southwestern Philippines), E. a. alecto (main Philippine islands, Celebes and Moluccas).

5. E. beccarii PETERS & DORIA 1880. [alecto group]. – Inner margin of calcar straight. Size relatively small (forearm length, 37–45 mm). Postorbital and sagittal crests virtually absent. Basisphenoid pits relatively deep and extending to

some extent, into alisphenoids, but median septum absent or poorly developed. Rostrum greatly inflated and postorbital processes poorly developed. – *Distribution:* Known from New Guinea, the Kei islands, the Trobriands (east of New Guinea) and New Ireland in the Bismarcks. – Three subspecies are currently recognized:

E. b. clavium (Keis), *E. b. beccarii* (New Guinea), *E. b. meeki* (Trobriands).

6. *E. raffrayana* DOBSON 1879 [*raffrayana* group]. – Tragus considerably shortened with inner margin straight. Calcar shorter than tibia. Size medium (forearm length, 38–47 mm). Postorbital crests confluent with sagittal crest. Basisphenoid pits deep and extending far into alisphenoids, recessed into basioccipital, and median septum usually well developed. – *Distribution:* Ranging from the Moluccas, New Guinea, and the Bismarcks to the Solomons. – Three subspecies are currently recognized:

E. r. stresemanni (Ceram in the central Moluccas), *E. r. raffrayana* (New Guinea), *E. r. cor* (Tabar islands in the Bismarcks, Solomons). A New Ireland record has not been allocated subspecifically.

7. *E. dianae* HILL 1956 [*raffrayana* group]. – Size relatively large (forearm length, 41–48 mm). Postorbital crests not confluent with sagittal crest. Basisphenoid pits very deep and extending well into alisphenoids, with the median septum moderately developed. Ears unusually broad. Rostrum greatly inflated and postorbital processes poorly developed. – *Distribution:* Known from New Guinea, New Ireland in the Bismarcks and from Malaita and Rennell in the Solomons. – No subspecies.

8. *E. furax* THOMAS 1911 [*raffrayana* group]. – Size large (forearm length, 45–50 mm). Postorbital crests virtually absent. Basisphenoid pits deep and extending well into alisphenoids, but median septum poorly developed. – *Distribution:* A poorly known species which occurs on New Guinea and on New Ireland in the Bismarcks. – No subspecies.

9. *E. semicaudata* (PEALE 1848) [*semicaudata* group]. – Inner margin of tragus concave. Calcar shorter than tibia. Size medium to large (forearm length, 40–53 mm). Development of supraorbital and sagittal crests variable, confluent if well-developed. Basisphenoid pits relatively shallow more or less separated by a median septum, but not extending into alisphenoids. – *Distribution:* Known from the Palaus, Marianas, eastern Carolines, and from the New Hebrides east to Samoa. – Four subspecies are here recognized of which *sulcata* has almost always been considered a separate species (but with *rotensis* intermediate):

E. s. semicaudata (Samoa, Tongas, Fijis, New Hebrides), *E. s. palauensis* (Palaus), *E. s. rotensis* (Marianas), *E. s. sulcata* (Truk and Ponape in the Carolines).

Genus ***Coleura*** PETERS 1867 (Fig. 65)

Dental formula i 1/3, c 1/1, p 2/2, m 3/3 x 2 = 32. Postorbital processes long, curved, and slender, not obscured by supraorbital ridges. No wing sacs. Clavicle not expanded. Tibia without a longitudinal groove. Anterior upper premolar styliform. Premaxillary bent strongly inward dorsally. Basisphenoid pits very deep and coalesced. Muzzle not extending anterior to incisors. – *Distribution:* Confined to tropical Africa from Guinea (Bissau) to Somalia and south to Angola and northern Mozambique, also South Yemen and the Seychelles. – Two species and one additional subspecies.

1. *C. afra* (PETERS 1852) (= *gallarum* THOMAS 1915: = *kummeri* MONARD 1939). – Size relatively small (forearm length, 43–53 mm). Forearm length ca. three times total skull length. – *Distribution:* Same as that of genus excluding the Seychelles. – No subspecies.

2. *C. seychellensis* PETERS 1869. – Size relatively large (forearm length, 52–57 mm). Forearm length more than three and a half times total skull length. – *Distribution:* Confined to the Seychelles, including the Amirante islands. – Two subspecies are currently recognized:

C. s. seychellensis (Amirante islands, Mahe and Praslin in the main Seychelle group), *C. s. silhouettae* (Silhouette and La Digne in the main Seychelle group).

Genus ***Rhynchonycteris*** PETERS 1867 (Fig. 66)

Dental formula i 1/3, c 1/1, p 2/2, m 3/3 x 2 = 32. Postorbital processes long, curved, and slender, not obscured by supraorbital ridges. No wing sacs. Clavicle not expanded. Tibia without a longitudinal groove. Frontal region of skull not conspicuously concave. Anterior upper premolar flat and triangular. Basisphenoid pits deep and coalesced. Muzzle extending anterior to incisors. Free edge of plagiopatagium attached to tibia. – *Distribution:* Ranging from tropical Mexico through Central America and South America (West of the Andes not south of Colombia) to Trinidad and eastern Brazil. – A single species.

1. *R. naso* (WIED-NEUWIED 1820). – Size relatively small (forearm length, 35–41 mm). – *Distribution:* Same as for genus. – No subspecies.

Genus **Centronycteris** GRAY 1838 (Fig. 68)

Dental formula i 1/3, c 1/1, p 2/2, m 3/3 x 2 = 32. Postorbital processes long, curved, and slender, not obscured by supraorbital ridges. No wing sacs. Clavicle not expanded. Tibia without a longitudinal groove. Premaxillaries well developed dorsally. Anterior root of zygoma so little expanded that upper molars and premolars are visible in dorsal view. Basisphenoid pits deep and separated by a median septum. Anterior upper premolar round with anterior and posterior cusps. Rostrum relatively narrow with no angle between it and the forehead. Free edge of plagiopatagium attached to metatarsus. – *Distribution:* Ranging from tropical Mexico through Central America to South America (west of the Andes south to Ecuador) east to eastern Brazil. – A single species with one additional subspecies.

1. *C. maximiliani* (FISCHER 1829). – Size relatively large (forearm length, 43–48 mm). – *Distribution:* Same as for genus. – Two currently recognized subspecies:

C. m. centralis (Middle America and northwestern South America. *C. m. maximiliani* (eastern South America).

Genus **Saccopteryx** ILLIGER 1811 (Fig. 67)

Dental formula i 1/3, c 1/1, p 2/2, m 3/3 x 2 = 32. Postorbital processes long, curved, and slender, not obscured by supraorbital ridges. Wing sac (particularly evident in males) close to forearm near elbow. Clavicle not expanded. Tibia without a longitudinal groove. Anterior root of zygoma so expanded that upper molars and premolars are hidden in dorsal view. Anterior upper premolar a structureless spicule. Interpterygoid fossa strongly narrowed anteriorly, the palate without a median projection. Upper surface of rostrum flat and narrow. Premaxillaries well developed dorsally. Lateral pterygoid pits small. – *Distribution:* Ranging from tropical Mexico through Central America to South America (west of the Andes barely to Peru) east to Tobago and eastern Brazil. – Four species with two additional subspecies.

1. *S. bilineata* (TEMMINCK 1838). – Size relatively large (forearm length, 41–51 mm). Free edge of plagiopatagium attached to tibia. Two longitudinal whitish lines on blackish back. – *Distribution:* Same as for genus. – Two subspecies are currently recognized.

S. b. bilineata (distribution except for northern Venezuela, Trinidad, and Tobago). *S. b. perspicillifer* (northern Venezuela, Trinidad and Tobago).

2. *S. leptura* (SCHREBER 1774). – Size medium (forearm length, 37–43 mm; maxillary toothrow length, 5.8–6.2 mm). Free edge of plagiopatagium attached to tibia. Two longitudinal whitish lines on brown back. – *Distribution:* Ranging from tropical Mexico through Central America to South America (west of the Andes not south of Ecuador) east to Tobago and eastern Brazil. – No subspecies.

3. *S. canescens* THOMAS 1901. – Size relatively small (forearm length, 35–41 mm; maxillary toothrow length, 5.0–5.5 mm). Free edge of plagiopatagium attached to tibia. Ear relatively narrow. Indistinct longitudinal pale lines on a more or less grizzled back. – *Distribution:* In tropical South America from central Colombia and eastern Peru to the mouth of the Amazon. – Two subspecies here recognized:

S. c. canescens (range except eastern Venezuela). *S. c. pumila* (eastern Venezuela).

4. *S. gymnura* THOMAS 1901. – Size small (forearm length, 33–35 mm). Free edge of plagiopatagium attached to metatarsus. Blackish back without markings. – *Distribution:* Known only from a small area along the lower Amazon in Brazil. – No subspecies.

Genus **Balantiopteryx** PETERS 1867 (Fig. 71)

Dental formula i 1/3, c 1/1, p 2/2, m 3/3 x 2 = 32. Postorbital processes long, curved, and slender, not obscured by supraorbital ridges. A short wing sac in the middle of the propatagium, opening medially. Clavicle not expanded. Tibia without a longitudinal groove. Premaxillaries greatly reduced dorsally. Anterior upper premolar a structureless spicule. Lateral pterygoid pits small. Rostrum greatly inflated. – *Distribution:* From the northern edge of the tropical zone in northern (Baja California, Sonora) and eastern (San Luis Potosi) Mexico south to Costa Rica. Ecuador. – Three currently recognized species, one additional subspecies.

1. *B. infusca* (THOMAS 1897). – Size medium (forearm length, 37–41 mm; total length of skull, 12.6.–13.1). Wing and leg bones not particularly slender. Rostrum inflated posteriorly. Inter-

pterygoid fossa broad anteriorly. Basial pit relatively short. No white line on edge of plagiopatagium. – *Distribution:* This poorly known species is only recorded from a single locality in northwestern Ecuador.

2. *B. io* THOMAS 1904. – Size small (forearm length, 35–39 mm; total length of skull, 12.3–12.9 mm). Wing and leg bones unusually slender. Rostrum not inflated posteriorly, with shallow frontal depression. Interpterygoid fossa broad anteriorly. Basial pit relatively long. No white line on edge of plagiopatagium. – *Distribution:* Confined to tropical Mexico, Guatemala, and Belize. – No subspecies.

3. *B. plicata* PETERS 1867. – Size relatively large (forearm length, 38–47 mm; total length of skull, 13.0–14.8 mm). Rostrum not inflated posteriorly, with deep frontal depression. Interpterygoid fossa narrowed anteriorly. Basial pit relatively long. White line on edge of plagiopatagium present. – *Distribution:* Same as that of genus, excluding Ecuador. – Two subspecies are currently recognized:

B. p. plicata (main part of range), *B. p. pallida* (northern Sinaloa, southwestern Chihuahua, southern Sonora, extreme southern Baja California).

Genus *Cormura* PETERS 1867 (= *Myropteryx* MILLER 1906) (Fig. 70)

Dental formula i 1/3, c 1/1, p 2/2, m 3/3 x 2 = 32. Postorbital processes long, curved, and slender, not obscured by supraorbital ridges. A long wing sac running from edge of propatagium to near elbow, opening laterally. Clavicle not expanded. Tibia without a longitudinal groove. Premaxillaries well developed dorsally. Anterior root of zygoma so expanded that upper molars and premolars are hidden in dorsal view. Interpterygoid fossa broad anteriorly, the palate with an evident median projection. Anterior upper premolar round with anterior and posterior cusps. Rostrum relatively broad but with no angle between it and the forehead. Free edge of plagiopatagium attached to metatarsus. – *Distribution:* Ranging from Nicaragua through southern Central America to South America (east of the Andes only) to Amazonian Peru and Brazil. – A single species.

1. *C. brevirostris* (WAGNER 1843) (= *pullus* MILLER, 1906). – Size medium (forearm length, 43–50 mm). – *Distribution:* same as for genus. – No subspecies.

Genus *Peropteryx* PETERS 1867 (Fig. 69)

Dental formula i 1/3, c 1/1, p 2/2, m 3/3 x 2 = 32. Postorbital processes long, curved, and slender, not obscured by postorbital ridges. A short wing sac present near anterior edge of propatagium, opening laterally. Clavicle not expanded. Tibia without a longitudinal groove. Premaxillaries well developed dorsally. Anterior root of zygoma so expanded that upper molars and premolars are hidden in dorsal view. Basisphenoid pits not separated by a medial septum. Anterior upper premolar a structureless spicule. Interpterygoid fossa strongly narrowed anteriorly, the palate without a median projection. Upper surface of rostrum convex and broad, but not inflated. Free edge of plagiopatagium attached to metatarsus. – *Distribution:* From tropical Mexico through Central America to South America (west of the Andes south to Ecuador) east to southeastern Brazil. – Two subgenera, three species.

Subgenus *Peropteryx* PETERS 1867

Rostrum so much swollen anteriorly that its dorsal profile is nearly parallel with the maxillary tooth row. Lateral pterygoid pits small. Ears separate. – *Distribution:* Same as that of genus. – Two species and three additional subspecies.

1. *P. macrotis* (WAGNER 1843). – Size relatively small (forearm length, 38–49 mm; total length of skull, 12–15 mm). Wing entirely black. – *Distribution:* Same as for subgenus except that west of the Andes it is not known south of Colombia. – Three currently recognized subspecies:

P. m. macrotis (entire range except Trinidad, Grenada, and possibly Margarita island and parts of Venezuela). *P. m. trinitatis* (Trinidad and possibly Margarita island and parts of Venezuela). *P. m. phaea* (Grenada).

2. *P. kappleri* PETERS 1867. Size relatively large (forearm length, 45–54 mm; total length of skull, 16–18 mm). Wing entirely black. – *Distribution:* From tropical Mexico through Central America to South America (west of the Andes south to Ecuador) east to the mouth of the Amazon. Also known from southeastern Peru and southeastern Brazil. – Two subspecies are currently recognized:

P. k. intermedia (southeastern Peru), *P. k. kappleri* (remaining distribution).

Subgenus *Peronymus* PETERS 1868

Rostrum so little swollen anteriorly that its dorsal profile forms a conspicuous angle with the maxillary tooth row. Lateral pterygoid pits large. Ears joined together by their medial edges. – *Distribution:* Confined to tropical South America (chiefly Amazonian) from southeastern Colombia and eastern Peru to the Guianas and northeastern Brazil. – A single species with one additional subspecies.

3. *P. leucopterus* PETERS 1867. – Size relatively small (forearm length, 41–47 mm; total length of skull, 14–17 mm). Chiropatagium white. – *Distribution:* Same as for subgenus. – Two subspecies are currently recognized:

P. l. leucopterus (distribution except for eastern Peru). P. l. cyclops (eastern Peru).

Genus *Cyttarops* THOMAS 1913 (Fig. 74)

No special modifications of the uropatagium. Postorbital processes long and slender, not obscured by the supraorbital ridges. Clavicle expanded. Tibia with a longitudinal groove. Dental formula i 1/3, c 1/1, p 2/2, m 3/3 x 2 = 32. Fur dark in color. A large outer lobe on the tragus. – *Distribution:* Ranging from Nicaragua to the mouth of the Amazon. – A single species.

1. *C. alecto* THOMAS 1913. – Size relatively small (forearm length, 46–48 mm). – *Distribution:* Same as for genus. – No subspecies.

Genus *Diclidurus* WIED-NEUWIED 1819 (Fig. 75)

Uropatagium considerably modified with the tail arching ventral to it and supporting a vertical extension of the uropatagium before projecting on the dorsal side. Posterior to the arched tail, there are a pair of peculiar glandular structures. Postorbital processes largely fused with supraorbital ridges. Clavicle expanded. Tibia with a longitudinal groove. Dental formula i 1/3, c 1/1, p 2/2, m 3/3 x 2 = 32. Fur pale in color. No outer lobe on tragus. – *Distribution:* Ranging from tropical Mexico to eastern Brazil. – Two subgenera, four species.

Subgenus *Depanycteris* THOMAS 1920

Thumb relatively unreduced with a distinct claw. Feet relatively long in relation to tibiae. Posterior border of palate evenly concave and reaching level of last upper molar. Fur light brown in color. – *Distribution:* Known only from southern Venezuela and northern Brazil. – A single species.

1. *D. isabella* THOMAS 1920. – Size fairly small (forearm length, 54 mm). – *Distribution:* Same as for subgenus. – No subspecies.

Subgenus *Diclidurus* WIED-NEUWIED 1819

Thumb greatly reduced with a vestigial claw. Feet relatively short in relation to tibiae. Posterior border of palate with a median cleft. Fur white or very light gray. – *Distribution:* Same as for genus. – Three species, one additional subspecies.

2. *D. scutatus* PETERS 1869. – Size relatively small (forearm length, 51–58 mm). Upper incisor with anterior and posterior cingular secondary cusps. Upper premolars in contact. – *Distribution:* Confined to southern Venezuela, the Guianas, and Amazonian parts of Peru and Brazil. – No subspecies.

3. *D. albus* WIED-NEUWIED 1819. – Size medium (forearm length, 63–69 mm). Upper incisors and upper premolars variable. – *Distribution:* Ranging from tropical western Mexico (Sinaloa) to eastern Brazil, including Trinidad (west of the Andes not south of Colombia). – Two subspecies:

D. a. albus (presumeably northeastern Peru, southern Venezuela, Guianas, Amazonian and eastern Brazil), D. a. virgo (Middle America, Colombia, northern Venezuela, and Trinidad).

4. *D. ingens* HERNANDEZ CAMACHO 1955. – Size relatively large (forearm length, 70–73 mm). Upper incisor with prominent posterior secondary cusp. Upper premolars separate. – *Distribution:* Known only from eastern Colombia, southern and eastern Venezuela, Guyana, and western Amazonian Brazil. – No subspecies.

Superfamily **Rhinolophoidea** GRAY 1825 (Slit faced, Old World Leaf-nosed and Horseshoe Bats).

Seventh cervical vertebra at least partially fused with first thoracic vertebra and often with additional fusions in the anterior thoracic area. Reduction or loss of nasal branch of the premaxillary. Modifications of the muzzle or rhinarium, of some sort, always present. Inguinal false nipples usually present. – *Distribution:* In Africa and Madagascar, temperate and tropical Eurasia and throughout the East Indies to the New Hebrides and to northern and eastern Australia. – Three families, one additional subfamily, 15 genera, 147 species.

Family **Nycteridae** VAN DER HOEVEN 1855 (Slit-faced Bats).

Structure: Anterior thoracic modifications restricted to last cervical and anterior thoracic vertebrae (which are incompletely fused) and to strengthening of the first rib. Loss of nasal branch of premaxillary, but palatal branch well-developed, filling space between maxillaries. No nose-leaf but a prominent slit in the dorsal side of the muzzle, margined by cutaneous outgrowths and leading into a large chamber, partially divided by several partitions, which is supported by a basin-like modification of the rostrum. Ears large, but more or less separate, and with relatively small tragi. Uropatagium broad, supported by a long tail terminating in an odd T-shaped cartilage. Trochiter of humerus small and not articulating with scapula. Inguinal false nipples absent.

Ecology: Mainly insectivorous, but, in at least one species, partially carnivorous. In habitat ranging from rain forests to semi-deserts. Roosting habits equally varied among various species, ranging from tree-holes, overhanging vegetation, porcupine and aard-vark burrows, to caves and similar man-made structures. Largely tropical.

Distribution (Fig. 10): Africa, mainly sub-saharan, but extending to Morocco on the west and Egypt in the east. Western and central portions of the Arabian peninsula. Madagascar. Southeast Asia to the edge the Sunda shelf and possibly to Celebes and Timor.

Systematics: A single genus and 13 species are here recognized.

Genus *Nycteris* CUVIER & GEOFFROY 1795 (Fig. 76)

Dental formula i 2/3, c 1/1, p 1/2, m 3/3 x 2 = 32. – *Distribution:* Same as for family. – Thirteen species and 18 additional subspecies are recognized, largely following CAKENBERGHE and VREE 1985.

1. *N. javanica* E. GEOFFROY 1813 [*javanica* group]. – Posterior lower premolar a functional tooth, its crown rising well above cingula of anterior lower premolar and first lower molar. Upper incisors more or less tricuspid. Tragus more or less falciform with a marked concavity in its anterior margin. Size medium (forearm length, 41–51 mm; condylocanine length, 16–19 mm; maxillary tooth row length, 6.4–7.8 mm). – *Distribution:* Restricted to Java, Bali, and the Kangean islands except for an old and probably erroneous record from Timor. – Two subscpecies:

N. j. javanica (Java, Bali), *N. j. bastiani* (Kangeans).

2. *N. tragata* (ANDERSEN 1912) [*javanica* group]. – Upper incisors more or less bicuspid. Size fairly large (forearm length, 46–53 mm; condylocanine length, 17–20 mm; maxillary toothrow length, 7.1–8.8 mm). – *Distribution:* Ranging from extreme southern Burma and Thailand through Malaya and Sumatra to Borneo and perhaps Celebes. – No subspecies.

3. *N. nana* (ANDERSEN 1912) [*javanica* group]. – Upper incisors more or less bicuspid. Size small (forearm length, 31–37 mm; condylocanine length, 12–14 mm; maxillary tooth row length, 4.4–5.3 mm). – *Distribution:* Confined to the forest zones of central Africa from Ivory Coast to Kenya and south to northern Angola. – No subspecies are currently recognized.

4. *N. intermedia* AELLEN 1959 [*javanica* group]. – Upper incisors more or less bicuspid. Size fairly small (forearm length, 33–38 mm; condylocanine length, 14–16 mm; maxillary tooth row length, 5.0–6.3 mm). – *Distribution:* Confined to the forest zones of central Africa from Liberia to western Tanzania and south to Angola. – No subspecies.

5. *N. arge* THOMAS 1903 [*javanica* group]. – Upper incisors more or less bicuspid. Size medium (forearm length, 33–50 mm; condylocanine length, 15–18 mm; maxillary tooth row length, 5.1–7.7 mm). – *Distribution:* Confined to forest zones of central Africa from Sierra Leone to western Kenya and south to northern Angola and southern Zaire. – No subspecies.

6. *N. major* (ANDERSEN 1912) [*javanica* group]. – Upper incisors more or less bicuspid. Size fairly large (forearm length, 44–50 mm; condylocanine length, 17–20 mm; maxillary tooth row length, 6.6–7.8 mm). – *Distribution:* Confined to tropical central Africa from Ivory Coast to Zaire and south to Zambia. – No subspecies.

7. *N. grandis* PETERS 1871 [*hispida* group]. – Posterior lower premolar reduced, its crown rising scarcely above cingula of anterior lower premolar and first lower molar. Upper incisors definitely tricuspid. Tragus more or less semilunate with no well marked concavity in its anterior margin. Size large (forearm length, 51–66 mm; condylocanine length, 20–25 mm; maxillary toothrow length, 7.9–9.7 mm). – *Distribution:* Confined to forest

and woodland areas of tropical Africa from Senegal to Kenya and south to Zimbabwe and Mozambique, including Pemba and Zanzibar islands off the east coast. – Two subspecies:

N. g. marica (southeastern Kenya south to Zimbabwe and Mozambique, including Pemba and Zanzibar). *N. g. grandis* (= *proxima*) (remaining range).

8. N. hispida (SCHREBER 1775) [*hispida* group]. – Size medium to small (forearm length, 33–45 mm; condylocanine length, 12–16 mm; maxillary tooth row length, 4.6–6.7 mm). – *Distribution*: Widely distributed in forest and savanna areas of Africa from Mauretania to Sudan and Somalia and south to Angola, Botswana, and Natal, including Zanzibar island. – Three subspecies are here recognized:

N. h. aurita (southern Somalia, northern and eastern Kenya, northeastern Tanzania), *N. h. villosa* (southern Mozambique and Natal), *N. h. hispida* (remainder of range). In eastern Kenya, *h. aurita* and *h. hispida* both occur and may act as separate species (KOOPMAN 1975).

9. N. woodi ANDERSEN 1914 [*macrotis* group]. – Posterior lower premolar reduced, its crown rising scarcely, if at all, above cingula of anterior lower premolar and first lower molar. Upper incisors definitely bicuspid. Tragus more or less semilunate with no well-marked concavity in its anterior margin. Size medium to small (forearm length, 36–43 mm; condylocanine length, 14–17 mm; maxillary toothrow length, 5.0–6.1 mm). – *Distribution*: Known from Cameroon and from Ethiopia south to Transvaal. – Four subspecies are recognized here:

N. w. benuensis (Cameron), *N. w. parisii* (Ethiopia, Somalia), *N. w. woodi* (Tanzania, Malawi, Zambia). *N. w. sabiensis* (Mozambique, Zimbabwe, Transvaal).

10. N. macrotis DOBSON 1876 [*macrotis* group]. – Size fairly large (forearm length, 40–55 mm; condylocanine length, 16–20 mm; maxillary tooth row length, 5.6–7.9 mm). – *Distribution*: Widely distributed in forest and savanna regions of sub-Saharan Africa from Senegal to Ethiopia and south to Malawi and northern Zimbabwe including Zanzibar and Madagascar. – Five currently recognized subspecies:

N. m. macrotis (chiefly forested regions from Gambia to Uganda and south to northern Angola and southern Zaire), *N. m. aethiopica* (chiefly savanna regions from Senegal to Ethiopia), *N. m. luteola* (northeastern Zaire and southern Somalia south to Tanzania, including Zanzibar), *N. m. oriana* (Tanzania, Malawi, Zambia, and Zimbabwe), *N. m. madagascariensis* (Madagascar).

11. N. vinsoni DALQUEST 1965 [*thebaica* group]. – Posterior lower premolar reduced, its crown rising scarcely, if at all, above cingula of anterior lower premolar and first lower molar. Upper incisors definitely bicuspid. Tragus pyriform with no well-marked concavity in its anterior margin. Size fairly large (forearm length, 50–52 mm; condylocanine length, 19–20 mm; maxillary tooth row length, 7.8–7.9 mm). – *Distribution*: Known only by the two original specimens from southern Mozambique. – No subspecies.

12. N. thebaica E. GEOFFROY 1818 [*thebaica* group]. – Size medium to fairly large (forearm length, 34–52 mm; condylocanine length, 15–19 mm; maxillary tooth row length, 5.0–7.5 mm). Ear relatively long (28–37 mm). – *Distribution*: Widely distributed in savanna and semi-desert regions of sub-Saharan Africa, south to the Cape Province, and north to Morocco on the west and Egypt on the east, continuing northeast to Northern Israel. Also central and southeastern Arabia. – Nine subspecies here recognized:

N. t. brockmani (northern Somalia), *N. t. media* (eastern Ethiopia), *N. t. adana* (southwestern Arabia), *N. t. najdiya* (central Arabia), *N. t. thebaica* (Israel, Egypt, northern Sudan and probably through the southern fringes of the Sahara to Senegal and north to Morocco), *N. t. labiata* (northern Ethiopia and central Sudan to northern Tanzania, including Pemba and Zanzibar, and Uganda, probably west through savanna woodland as far as Upper Volta), *N. t. capensis* (southern Tanzania, southeastern Zaire, and Zambia, to the Cape Province), *N. t. angolensis* (central Angola), *N. t. damarensis* (southern Angola, Namibia, Botswana).

13. N. gambiensis (ANDERSEN 1912) [*thebaica* group]. – Size small (forearm length, 35–44 mm; condylocanine length, 14–17 mm; maxillary tooth row length, 5.2–6.2 mm). Ear relatively short (25–28 mm). – *Distribution*: Confined to savanna and woodland areas of western Africa from Senegal to Benin. – No subspecies.

Family **Megadermatidae** H. ALLEN 1864
(Old World Leaf nosed Bats)

Structure: Broadened presternum more or less fused with first pair of ribs, first thoracic and last cervical vertebrae fused to form a solid ring. Loss of palatal branch of premaxillary and loss or great reduction of its nasal branch. A large but relatively simple noseleaf present. Ears very large and fused by their inner margins for at least a third of their length, and with very large bifid tragi. Uropatagium broad but tail vestigial or absent. Trochiter of humerus small and not articulating with scapula.

Ecology: Insectivorous or carnivorous. Almost entirely tropical but in habitat ranging from rain forests to semi-deserts. Roosting habits varied including caves, houses, hollow trees and open branches of trees.

Distribution (Fig. 11): Widely distributed in tropical Africa from Sengal to Ethiopia and south to Zambia. Also Afghanistan and Ceylon east through southeastern Asia and the East Indies to the Philippines and Moluccas. Tropical Australia.

Systematics: In the Recent, there are four genera, one additional subgenus, and five species.

Genus *Megaderma* E. GEOFFROY 1810 (Fig. 77)

Dental formula, i 0/2, c 1/1, p 2/2, m 3/3 x 2 = 28. Interorbital region not conspicuously concave; frontal expansion little developed. Mesostyles of upper molars reduced. Noseleaf relatively small. – *Distribution:* Ranging from Afghanistan and Ceylon east through southeastern Asia and the East Indies to the Philippines and Moluccas. – Two subgenera, and two species are recognized.

Subgenus *Megaderma* E. GEOFFROY 1810

Lacrimal width much less than distance from orbit to canine. Basisphenoid pits shallow but distinct. Hamular processes of pterygoids well developed. – *Distribution:* Ranging from peninsular India and Ceylon east through southeastern Asia and the East Indies to the Philippines and Moluccas. – A single species and 15 additional subspecies.

1. *M. spasma* (LINNAEUS 1758). Size relatively small (forearm length, 52–61 mm). – *Distribution:* Same as for subgenus. – There are 16 subspecies currently recognized:

M. s. horsfieldi (peninsular India), *M. s. ceylonense* (Ceylon), *M. s. majus* (northeastern India, northern Burma), *M. s. minus* (Indo-China, Thailand, except peninsular), *M. s. medium* (southern Burma, Malay penisula, northern Sumatra), three subspecies (west Sumatran islands), three subspecies (islands in the South China Sea), *M. s. trifolium* (southern Sumatra, Java except southeast, Borneo except Mt. Kinabalu), *M. s. pangandarana* (southeastern Java), *M. s. kinabalu* (Mt. Kinabalu in northeastern Borneo), *M. s. celebensis* (Celebes), *M. s. spasma* (Philippines, northern Moluccas).

Subgenus *Lyroderma* PETERS 1872

Lacrimal width greater than distance from orbit to canine. Basisphenoid pits virtually absent. Hamular processes of pterygoids small. – *Distribution:* Ranging from Afghanistan to southern China and south to Ceylon and Malaya. – A single species with one additional subspecies.

1. *M. lyra* E. GEOFFROY 1810. – Size medium (forearm length, 64–75 mm). – *Distribution:* Same as for subgenus. – Two subspecies are currently recognized:

M. l. lyra (Afghanistan and Ceylon east to Burma), *M. l. sinensis* (southern China south to Malaya).

Genus *Macroderma* MILLER 1906 (Fig. 78)

Dental formula: i 0/2, c 1/1, p 1/2, m 3/3 x 2 = 26. Interorbital region conspicuously concave. Frontal expansion little developed. Mesostyles of upper molars reduced. Noseleaf relatively small. – *Distribution:* Endemic to Australia; in the Recent virtually confined to the tropics. – A single species.

1. *M. gigas* (DOBSON 1880). – Size very large (forearm length, 102–112 mm). – *Distribution:* same as for genus. – No currently recognized subspecies.

Genus *Cardioderma* PETERS 1873 (Fig. 79)

Dental formula: i 0/2, c 1/1, p 1/2, m 3/3 x 2 = 26. Interorbital region conspicuously concave. Frontal expansion well developed, forming prominent postorbital processes. Mesostyles of upper molars reduced. Noseleaf relatively small. – *Distribution:* Confined to eastern Africa from Ethiopia through Somalia and Kenya to Tanzania, also eastern Sudan, Uganda, and Zanzibar. – A single species.

1. *C. cor* (PETERS 1872). – Size relatively small (forearm length, 51–57 mm). – *Distribution:* Same as for genus. – No subspecies.

Genus *Lavia* GRAY 1838 (Fig. 80)

Dental formula: i 0/2, c 1/1, p 1/2, m 3/3 x 2 = 26. Interorbital region conspicuously concave. Frontal expansion well developed, forming prominent postorbital processes. Mesostyles of upper molars well-developed. Noseleaf greatly enlarged. – *Distribution:* In tropical Africa from Senegal to Ethiopia and south to Zambia. There is also an old dubious record (the basis for the name *megalotis*) from southern Namibia. – A single species with perhaps two additional subspecies.

1. *L. frons* (E. GEOFFROY 1810). – Size relatively small (forearm length, 49–63 mm). – *Distribution:* Same as for genus. – Three rather poorly defined subspecies:

L.f. affinis (Sudan, Chad, northern Uganda, extreme northwestern Zaire), L.f. frons (western portion of range, presumeably from Senegal to western Zaire), L.f. rex (Ethiopia south at least to Tanzania and west to eastern Zaire).

Family **Rhinolophidae** GRAY 1825
(Horseshoe Bats)

Structure: Presternum, first (and at least ventral part of second) rib, first (and sometimes second) thoracic and last cervical vertebrae fused to form a solid ring of bone. Loss of nasal branch of premaxillary with some reduction of palatal branch so that it is free from the maxillary except at its posterior end. A complex noseleaf always present. Ears variable in size and shape, tragi absent. Skull always with some rostral inflation. Uropatagium well-developed but tail with varying degress of development. Trochiter of humerus fairly large and definitely articulating with scapula. Ischium and pubis reduced in size but broadened so that the space between them is reduced.

Ecology: As far as is known entirely insectivorous. Mainly tropical but with a number of species extending into warm temperate regions and a few extending even farther outside the tropics. In habitat ranging from rain forests to deserts. Roosting habits are also varied including caves and rock crevices (or similar man-made structures), but also tree hollows or even branches in the open.

Distribution (Fig. 12): Widely distributed in the eastern hemisphere including the whole of Africa with its surrounding islands, Madagascar, the southern half of Eurasia including the British and Japanese islands, the Indo-Australian archipelago out to the New Hebrides, and northern and eastern Australia.

Systematics: In the Recent, there are two subfamilies, one additional tribe, 10 genera, 129 species.

Subfamily **Rhinolophinae** GRAY 1825

Complex noseleaf present composed of four parts: horseshoe, sella, connecting process, and lancet. First and second ribs only partially fused. No precetabular foramen or lumbar vertebral fusion. Except for hallux, each toe has three phalanges. – *Distribution:* Range same as for family, but absent from Madagascar, the Solomons, Santa Cruz islands, New Hebrides, and in Australia is confined to the eastern margin. – A single genus, 64 species.

Genus ***Rhinolophus*** LACEPEDE 1799 (Fig. 81)

Dental formula i 1/2, c 1/1, p 2/3, m 3/3 x 2 = 32. Size small to fairly large (forearm length, 31–78 mm). – *Distribution:* Same as for subfamily. – Currently, 64 species and 127 additional subspecies are recognized, but there are many outstanding taxonomic problems. (SINHA 1973, for many species).

1. *R. simplex* ANDERSEN 1905 [*ferrumequinum* group]. – Sella small, constricted in middle, rounded on top, without lateral processes. Lancet long, almost cuneate. Median groove of horseshoe simple, no papilla at posterior end. Connecting process rounded. Periotic bones not enlarged, no narrowing of median basioccipital. Metacarpals of digits 3, 4, and 5 subequal; second phalanx of third digit of wing not shortened, more than two thirds length of first phalanx. Supraorbital width of skull greater than width of nasal swellings. Palatal bridge relatively long (one third length of maxillary tooth row). Middle lower premolar not extruded from toothrow. Supraorbital crest behind mid-orbit. Size relatively small (forearm length, 44–45 mm). – *Distribution:* Confined to the western Lesser Sundas (Lombok to Komodo). – No subspecies.

2. *R. megaphyllus* GRAY 1834 [*ferrumequinum* group]. – Sella abruptly constricted in middle, rounded or truncate on top. Size small to medium (forearm length, 41–50 mm). – *Distribution:* Eastern edge of Australia from Victoria to Cape York, eastern New Guinea, East Papuan islands, and Bismarcks. – Four currently recognized subspecies:

R. m. megaphyllus (eastern Australia), R. m. fallax (southeastern New Guinea, D'Entrecasteaux islands), R. m. monachus (Louisiade islands), R. m. vandeuseni (northeastern New Guinea, Bismarcks).

3. *R. keyensis* PETERS 1871 [*ferrumequinum* group]. – Sella more or less straight sided, truncate

on top. Size small to medium (forearm length, 39–47 mm). – *Distribution:* Confined to the Moluccas and Wetar (just north of Timor). – Four subspecies are currently recognized:

R. k. truncatus (Batchian island in the northern Moluccas), R. k. nanus (Ceram and nearby islands), R. k. keyensis (Kcis), R. k. annectens (Wetar).

4. R. *borneensis* PETERS 1861 [*ferrumequium* group]. – Sella rounded on top. Lancet more or less hastate. Supraorbital width of skull less than width of nasal swellings which are not enlarged. Palatal bridge relatively short. Middle lower premolar extruded from toothrow. Supraorbital crests in front of mid-orbit. Braincase relatively inflated. Size relatively small (forearm length, 41–47 mm; total length of skull, 18–20 mm). – *Distribution:* Confined to a small area of southern Indo-China, Borneo and nearby islands, and Java. – There are four currently recognized subspecies (HILL 1983).

R. b. borneensis (Borneo), R. b. spadix (South Natunas and Karimata in the South China Sea), R. b. importunus (Java), R. b. chaseni (Cambodia and extreme southern Vietnam).

5. R. *nereis* ANDERSEN 1905 [ferrumequinum group]. – Like R. *borneensis*, except for greater length of second phalanx of third digit of wing and overall size (forearm length, 45 mm), but skull larger (total length 21–22 mm). – *Distribution:* Confined to Anamba and South Natuna islands in South China Sea. – No subspecies.

6. R. *celebensis* ANDERSEN 1905 [*ferrumequinum* group]. – Supraorbital width of skull subequal to width of nasal swellings. Supraorbital crests in front of the mid-orbit. Braincase relatively uninflated. Size relatively small (forearm length 38–45 mm; total length of skull, 18–19 mm). – *Distribution:* Confined to Java and surrounding islands, Timor, and Celebes. – Four currently recognized subspecies (HILL 1983):

R. c. celebensis (Celebes), R. c. javanicus (Java, Bali), R. c. madurensis (Madura), R. c. parvus (Timor).

7. R. *virgo* ANDERSEN 1905 [*ferrumequinum* group]. – Nasal swellings relatively narrow. Horseshoe of noseleaf relatively narrow. Ears relatively short. Size small (forearm length, 37–39 mm; total length of skull, 16–18 mm). – *Distribution:* Restricted to but widely distributed in the Philippines. – No subspecies.

8. R. *malayanus* BOHOTE 1903 [*ferrumequinum* group]. – Median anterior nasal swellings definitely larger than lateral anterior nasal swellings. Size relatively small (forearm length, 40–43 mm; total length of skull, 18–19 mm. – *Distribution:* Ranging from northern Vietnam and Laos through Thailand to Malaya. – No subspecies.

9. R. *stheno* ANDERSEN 1905 [*ferrumequinum* group]. – Anterior nasal swellings enlarged and posterior nasal swellings reduced. Second phalanx of third digit of wing elongate. Lower leg elongated but tail reduced. Size medium (forearm length, 45–48 mm; total length of skull, 19–21 mm). – *Distribution:* Ranging from Thailand through the Malay Peninsula and Sumatra to Java. – No subspecies.

10. R. *anderseni* CABRERA 1901 [*ferrumequinum* group]. – Anterior nasal swellings enlarged. Second phalanx of third digit of wing elongate and first phalanx of fourth shortened. Tail relatively long. Size medium (forearm length, 45 mm; total length of skull, 20–21 mm). – *Distribution:* Confined to the Philippines. – Two subspecies are recognized:

R. a. aequalis (Palawan group), R. a. anderseni (probably Luzon).

11. R. *simulator* ANDERSEN 1904 [*ferrumequinum* group]. – Anterior nasal swellings enlarged. Anterior upper premolar greatly reduced, though in toothrow. Second phalanx of third digit of wing elongate. Size medium (forearm length, 40–49 mm). Sella relatively broad. Ears relatively long. – *Distribution:* Eastern Africa from Ethiopia to Natal, also Cameroon, Nigeria, and Guinea. – Two subspecies recognized here (often treated as separate species):

R. s. alticolus (Guinea, Nigeria, Cameroon), R. s. simulator (eastern Africa).

12. R. *swinnyi* GOUGH 1908 [*ferrumequinum* group]. – Anterior nasal swellings enlarged. Anterior upper premolar greatly reduced, though in toothrow. Second phalanx of third digit of wing elongate. Size fairly small (forearm length, 40–44 mm). Sella relatively narrow. Ears relatively short. Front edge of connecting process convex. Sides of lancet concave. – *Distribution:* Ranging in eastern Africa from Zanzibar island and southern Zaire to Cape province; west in Zaire to the mouth of the Congo river. – No subspecies.

13. R. *denti* THOMAS 1904 [*ferrumequinum* group]. – Anterior nasal swellings enlarged. Anterior upper premolar greatly reduced though in toothrow. Second phalanx of third digit of wing elongate. Size small (forearm length, 37–43 mm). Sella relatively narrow. Ears relatively short. Front edge

of connecting process and sides of lancet straight. – *Distribution:* Confined to southern Africa from Namibia and Zimbabwe to the Cape Province; also Guinea. – Two subspecies are recognized:

R. d. knorri (Guinea), *R. d. denti* (southern Africa).

14. R. rouxi TEMMINCK 1835 [*ferrumequinum* group]. – Sella pandurate. Lancet definitely hastate. Metacarpals unusually long. Anterior upper premolar greatly reduced, though in toothrow. Size medium (forearm length, 45–53 mm). – *Distribution:* Ranging from India and Ceylon to southern China and Vietnam. – Two subspecies are recognized:

R. r. rouxi (= *petersi*) (India, Ceylon), *R. r. sinicus* (Burma, southern China, Thailand, Vietnam).

15. R. thomasi ANDERSEN 1905 [*ferrumequinum* group]. – Lancet tip extremely short (very hastate). Metacarpals relatively short. Anterior upper premolar greatly reduced, though in toothrow. Size medium to fairly large (forearm length, 42–55 mm). – *Distribution:* Known from Burma, southwestern China, Thailand, and Vietnam. – Three subspecies are recognized:

R. t. thomasi (Burma, Thailand), *R. t. septentrionalis* (Yunnan in China), *R. t. latifolius* (Vietnam).

16. R. capensis LICHTENSTEIN 1823 [*ferrumequinum* group]. – Second phalanx of third digit relatively elongate. Anterior upper premolar greatly reduced, though in toothrow. Size medium (forearm length, 47–52 mm). – *Distribution:* Confined to southern Africa from Zimbabwe and Mozambique to the Cape Province. – No subspecies.

17. R. adami AELLEN & BROSSET 1968 [*ferrumequinum* group]. Noseleaf relatively large. Bony palate relatively long. Size medium (forearm length, 49 mm). – *Distribution:* Known only from Congo (Brazzaville). – No subspecies.

18. R. affinis HORSFIELD 1823 [*ferrumequinum* group]. – Metacarpal of third digit relatively short, metacarpals of fourth and fifth relatively long and subequal to one another; second phalanx of third shortened, less than two thirds length of first phalanx. Sella pandurate. Lancet definitetly cuneate. Anterior upper premolar in toothrow. Size medium to fairly large (forearm length, 46–56 mm). – *Distribution:* Ranging from northern India to southern China south through southeastern Asia, Sumatra, Borneo, and Java, to the Lesser Sundas, including the Andamans and possibly Ceylon. – Nine subspecies are currently recognized:

R. a. himalayanus (northern India across northern Burma to southwestern China). *R. a. macrurus* (southeastern China through Vietnam and Thailand to southeastern Burma), *R. a. hainanus* (Hainan island), *R. a. tener* (southwestern Burma), *R. a. andamanensis* (Andaman islands), *R. a. superans* (Malay peninsula, Sumatra, Mentawai islands), *R. a. nesites* (Anamba and North Natuna islands, Borneo), *R. a. affinis* (Java), *R. a. princeps* (Lombok, Sumbawa, and Sumba in the Lesser Sundas).

19. R. robinsoni ANDERSEN 1908 [*ferrumequinum* group]. – Connecting process unusually low. Outer margin of sella markedly convex. Size medium (forearm length, 40–48 mm). – *Distribution:* Confined to Malaya and Thailand. – Two subspecies:

R. r. robinsoni (peninsular Thailand), *R. r. klossi* (Malaya, including nearby islands). A third (from northern Thailand) has been proposed, but its name is a homonym of *B. macrotis siamensis*.

20. R. clivosus CRETZSCHMAR 1828 [*ferrumequinum* group]. – Metacarpal of third digit relatively short, metacarpals of fourth and fifth relatively long and subequal to one another; second phalanx of third shortened, less than two thirds length of first phalanx. Sella pandurate. Horseshoe relatively narrow. Anterior upper premolar extruded from toothrow, greatly reduced or absent. Anterior nasal swellings fairly small. Periotic bones not enlarged. Size fairly small to fairly large (forearm length, 43–56 mm). – *Distribution:* Widely, if somewhat discontinuously distributed in Afghanistan, Iran, southern Soviet Central Asia, and Azerbaydzhan, also southern Israel through most of Arabia, Egypt west to southeastern Algeria, south through eastern Africa to southern Africa and west to Liberia, western Zaire, and Angola. – There are nine currently recognized subspecies:

R. c. bocharicus (Afghanistan, Iran, Khirghizia to Turkmenia). *R. c. rubiginosus* (Azerbaydzhan), *R. c. clivosus* (Israel to northeastern Sudan), *R. c. brachygnathus* (Egypt to northeastern Libya and northern Sudan), *R. c. schwartzi* (southeastern Algeria), *R. c. acrotis* (central and southwestern Arabia, Ethiopia, most of central and southern Sudan, Somalia, most of Kenya), *R. c. keniensis* (southeastern Sudan, Mount Kenya), *R. c. zuluensis* (Uganda, eastern and southern Zaire, south through eastern southern Africa), *R. c. augur* (western southern Africa).

21. R. ferrumequinum (SCHREBER 1774) [*ferrumequinum* group]. – Metacarpals of third and fourth digits relatively short. Sella pandurate. Horseshoe relatively narrow. Anterior upper premolar extruded from toothrow, greatly reduced or absent. Anterior nasal swellings fairly small. Periotic bones not enlarged. Size fairly large (forearm length, 52–63 mm). – *Distribution:* Widely distributed across temperate Eurasia from Britain to Japan south to northwestern Africa, Palestine and northern India. – Six subspecies are here recognized:

R. f. ferrumequinum (Europe, northwestern Africa), *R. f. creticus* (Crete), *R. f. proximus* (= *irani*) (southwestern Asia east to Kashmir), *R. f. tragatus* (= *regulus*) (northern India, southwestern China), *R. f. nippon* (northern and central China, Korea, Japan).

22. <u>*R. deckeni*</u> PETERS 1868 [*ferrumequinum* group]. – Horseshoe relatively broad. Anterior upper premolar extruded from toothrow and greatly reduced or lost. Anterior nasal swellings enlarged. Periotic bones somewhat enlarged, narrowing median basioccipital. Size fairly large (forearm length, 48–55 mm). – *Distribution:* Restricted to Uganda, Kenya, and Tanzania, including Pemba and Zanzibar islands. – No subspecies.

23. <u>*R. silvestris*</u> AELLEN 1959 [*ferrumequinum* group]. – Horseshoe relatively broad. Anterior upper premolar extruded from toothrow and greatly reduced. Anterior nasal swellings enlarged. Periotic bones somewhat enlarged, narrowing basioccipital. Size fairly large (forearm length, 49–56 mm). Probably only a subspecies of *R. deckeni*. – *Distribution:* Confined to Gabon and Congo (Brazzaville). – No subspecies.

24. <u>*R. darlingi*</u> ANDERSEN 1905 [*ferrumequinum* group]. – Metacarpal of third digit unusually short. Sella markedly pandurate. Horseshoe relatively broad. Anterior upper premolar extruded from toothrow and greatly reduced. Anterior nasal swellings enlarged. Size medium (forearm length, 45–50 mm). – *Distribution:* More or less confined to dryer parts of southern and eastern Africa from Tanzania to Angola and the Cape Province. – Two subspecies are currently recognized:

R. d. darlingi (same range as species except Namibia), *R. d. damarensis* (Namibia).

25. <u>*R. acuminatus*</u> PETERS 1871 [*pusillus* group]. – Sella small, without lateral processes. Median groove of horseshoe simple, no papilla at posterior end. Connecting process pointed, triangular in profile. Periotic bones not enlarged, no narrowing of median basioccipital. Size relatively large (forearm length, 46–53 mm). – *Distribution:* Ranging from Laos through Cambodia, Thailand, Malaya and Sumatra to Borneo and the southwestern Philippines and to Java and Lombok, also the west Sumatran islands. – Five subspecies are currently recognized:

R. a. sumatranus (Sumatra, Borneo), *R. a. circe* (Nias island), *R. a. calypso* (Engano island), *R. a. acuminatus* (Java), *R. a. audax* (Bali, Lombok). Subspecific allocations of mainland and Philippine populations are uncertain.

26. <u>*R. alcyone*</u> TEMMINCK 1852 [*pusillus* group]. – Connecting process triangular in profile. First phalanx of fourth digit of wing shortened. Size relatively large (forearm length, 49–54 mm; total length of skull, 22–24 mm); maxillary toothrow length, 8.6–8.9 mm. – *Distribution:* Confined to the forest zone from Senegal to Uganda, including Fernando Poo. – No subspecies.

27. <u>*R. guineensis*</u> EISENTRAUT 1960 [*pusillus* group]. – Connecting process triangular in profile. First phalanx of fourth digit of wing shortened. Size relatively large (forearm length, 44–49 mm), but skull relatively small (total length, 20–22 mm; maxillary toothrow length, 7.2–7.8 mm). – *Distribution:* Known only from Senegal, Guinea, and Sierra Leone. – No subspecies.

28. <u>*R. landeri*</u> MARTIN 1838 [*pusillus* group]. – Connecting process triangular in profile. First phalanx of fourth digit of wing shortened. Size medium (forearm length, 38–46 mm; total length of skull, 17–20 mm; maxillary tooth row length, 6.3–7.1 mm). – *Distribution:* Widely distributed in tropical Africa from Gambia to Ethiopia and south to Namibia and Transvaal. – Three subspecies are currently recognized:

R. l. landeri (Gambia to Cameroon and south to the mouth of the Congo river), *R. l. lobatus* (Sudan and Ethiopia south to Transvaal, including Zanzibar island), *R. l. angolensis* (western Angola and perhaps Namibia).

29. <u>*R. lepidus*</u> BLYTH 1844 [*pusillus* group]. – Connecting process triangular in profile. Lancet extremely hastate. Size medium (forearm length, 37–43 mm) with relatively large skull (condylocanine length, 14–17 mm; maxillary tooth row length 6.0–7.1 mm; rostral width, 4.4–5.0 mm). Median rostral swellings enlarged. – *Distribution:* Ranging from Afghanistan through northern India and Burma to western China, south through Thailand and Malaya to Sumatra. – Six subspecies are currently recognized:

R. l. monticola (Afghanistan and northwestern India), *R. l. lepidus* (central and northeastern India), *R. l. shortridgei* (northern Burma and perhaps western China), *R. l. feae* (southern Burma, northern Thailand), *R. l. refulgens* (Malay peninsula), *R. l. cuneatus* (Sumatra).

30. <u>*R. osgoodi*</u> SANBORN 1939 [*pusillus* group]. – Connecting process triangular but somewhat rounded. Lancet not extremely hastate. Size medium (forearm length, 41–46 mm) but skull relatively small (greatest length, 16–17 mm; maxillary tooth row length, 5.5–5.8 mm). Rostrum relatively slender. – *Distribution:* Known only from Yunnan. – No subspecies.

31. <u>*R. pusillus*</u> TEMMINCK 1834 [*pusillus* group]. – Connecting process usually triangular in profile

but with some variation. Size relatively small (forearm length, 35–40 mm; total length of skull, 15–16 mm; maxillary tooth row length, 5.7–6.0 mm). Median rostral swellings not enlarged. – *Distribution:* Ranging from India to southern China, south to Malaya, also the Mentawai islands, Anamba islands, Borneo and Java. – Eight subspecies are currently recognized:

R. p. blythi (northwestern India), *R. p. gracilis* (southern India), *R. p. szechwanus* (northeastern India, Burma, southwestern China, Thailand), *R. p. calidus* (eastern China, Vietnam), *R. p. parcus* (Hainan island), *R. p. minutillus* (Malay peninsula, Anamba islands), *R. p. pagi* (Mentawai islands), *R. p. pusillus* (Java, Borneo).

32. <u>*R. cornutus*</u> TEMMINCK 1834 [*pusillus* group]. – Connecting process varying from subtriangular to somewhat horn-like and curved, its anterior margin somewhat concave. Size relatively small (forearm-length, 38–41 mm; total length of skull, 16–17 mm; maxillary tooth row length, 5.7–6.3 mm). Median rostral swellings not enlarged. – *Distribution:* Confined to Japan, the Ryukyu islands (except the extreme southwesten Iriomote group) and possibly in southeastern China. – Five subspecies are currently recognized:

R. c. cornutus (main islands of Japan) and the remainder on various parts of the Ryukyu chain. Probably conspecific with *pusillus*.

33. <u>*R. monoceros*</u> ANDERSEN 1905 [*pusillus* group]. – Connecting process typically forming a slender sharply pointed horn, but with much variability. Lancet triangular with nearly straight sides and rounded tip. Size reatively small (forearm length, 34–39 mm; total length of skull, 14–16 mm; maxillary tooth row length, 5.4–5.7 mm). – *Distribution:* Confined to Taiwan. – No subspecies. Probably a subspecies of *pusillus*.

34. <u>*R. imaizumii*</u> HILL & YOSHIYUKI 1980 [*pusillus* group]. – Connecting process forming an erect narrow horn. Lateral margins of lancet slightly concave, its upper part more or less spatulate. Size medium (forearm length, 40–43 mm; total length of skull, 18–19 mm; maxillary tooth row length, 6.4–6.8 mm). Median anterior narial swellings not enlarged. Posterior palatal emargination equal to or narrower than anterior palatal emargination. Upper canines relatively massive. – *Distribution:* Confined to Iriomote island at the extreme southwestern end of the Ryukyu chain. – No subspecies.

35. <u>*Rhinolophus subbadius*</u> BLYTH 1844 [*pusillus* group]. – Connecting process an erect narrow horn. Lancet triangular with straight sides and a rounded tip. Size unusally small (forearm length, 31–36 mm; total length of skull, 15–16 mm; maxillary tooth row length, 5.1–5.4 mm). Rostrum relatively slender. – *Distribution:* Known only from northeastern India, Burma, and Vietnam. – No subspecies.

36. <u>*R. cognatus*</u> ANDERSEN 1906 [*pusillus* group]. – Connecting process an erect narrow horn. Lateral margins of lancet slightly concave, its upper part more or less spatulate. Size medium (forearm length, 39–40 mm; total length of skull, 17–19 mm; maxillary tooth row length, 6.1–6.9 mm). Median anterior nasal swellings moderately enlarged. Posterior palatal emargination wider than anterior palatal emargination. Upper canines relatively slender. – *Distribution:* Confined to the Andaman islands. – Two subspecies:

R. c. famulus (North Andaman), *R. c. cognatus* (South Andaman).

37. <u>*R. euryale*</u> BLASIUS 1853 [*pusillus* group]. – Connecting process rises to a high narrow horn. First phalanx of fourth finger not notably shortened. Sella parallel sided. Anterior lower premolar reduced. Lancet gradually narrowing. Size relatively large (forearm length, 44–51 mm). – *Distribution:* Ranging from northwestern Africa through southern Europe (including several Mediterranean islands) to southwestern Asia south to Palestine and Iraq and east to Turkmenia and Iran. – Four poorly defined subspecies are here recognized:

R. e. meridionalis and *R. e. barbarus* (northwestern Africa), *R. e. euryale* (= *nordmanni*) (southern Europe east to Turkmenia and Iran), *R. e. judaicus* (Syria and Iraq south).

38. <u>*R. mehelyi*</u> MATSCHIE 1901 [*pusillus* group]. – Connecting process rises to a high narrow horn. First phalanx of fourth finger not notably shortened. Sella parallel sided. Anterior lower premolar reduced. Lancet abruptly narrowing to a linear tip. Size relatively large (forearm length, 43–54 mm). – *Distribution:* Ranging across Africa (north of the Sahara) and southern Europe (including several Mediterranean islands) to Turkey, Transcaucasia and western Iran. – Two subspecies are here recognized:

R. m. mehelyi (European and western Asian parts of range), *R. m. tunetae* (northern Africa).

39. <u>*R. blasii*</u> PETERS 1866 [*pusillus* group]. – Connecting process rises to a high narrow horn. First phalanx of fourth finger not notably shortened. Sella cuneate (wedge-shaped). Anterior lower premolar not reduced. Size relatively large (fore-

arm length, 40–49 mm). – *Distribution:* Occurring in northwestern Africa and from southern Europe (including several Mediterranean islands) east to Afghanistan; south through western Arabia and in eastern Africa from Ethiopia to Transvaal. – Four subspecies are currently recognized:

R. b. blasii (northwestern Africa, southern Europe, and southwestern Asia), *R. b. meyeroehmi* (Iran, Turkmenia, Afghanistan), *R. b. andreinii* (= *brockmani*) (Ethiopia, Somalia), *R. b. empusa* (southeastern Africa from southern Zaire to Transvaal).

40. R. *hipposideros* (BECHSTEIN 1800) [*hipposideros* group]. – Sella small, without lateral processes. Median groove of horseshoe simple, no papilla at posterior end. Connecting process very low and rounded off. Periotic bones enlarged, causing narrowing of the median basioccipital. Size relatively small (forearm length, 34–42 mm). – *Distribution:* Widely distributed in the western Palearctic from Ireland to southeastern Kazakhstan and Kashmir, south to northwestern Africa and through western Arabia to Ethiopia and eastern Sudan. – Seven subspecies are here recognized:

R. h. escalerae and *R. h. vespa* (northeastern Africa), *R. h. minimus* (southern Europe to the eastern end of the Mediterranean, including several islands and south to Ethiopia and the Sudan), *R. h. majori* (Corsica), *R. h. minutus* (Britain and Ireland), *R. h. hipposideros* (continental Europe north of the Alps east to the eastern end of the Black Sea), *R. h. midas* (Transcaucasia and Iraq to Kazakhstan and Kashmir).

41. R. *philippinensis* WATERHOUSE 1843 [*luctus* group]. – Large broad ears with well-developed antitragal lobe. Horseshoe broad. Skull with high projecting median anterior rostral swellings. Anterior upper and middle lower premolars in toothrow. Palatal bridge long, more than 1/3 length of maxillary toothrow. Connecting process extremely low and rounded off. Zygomatic width less than mastoid width. Upper incisors minute and widely separated. Infraorbital canal short. Sella broad, without expanded lappets at base, but internarial lobes forming a large cup which is twice as broad as the sella. Lancet tall and weakly haired, its tip rounded. Size fairly large (forearm length, 47–55 mm). – *Distribution:* Known from Borneo, main Philippines, Celebes, Timor, Keis, and northeastern Queensland in Australia. – Six subspecies are currently recognized:

R. p. sanborni (Borneo), *R. p. alleni* (Mindoro in the Philippines), *R. p. philippinensis* (remaining Philippine islands), *R. p. maros* (Celebes), *R. p. montanus* (Timor), *R. p. achilles* (Keis), *R. p. robertsi* (northeastern Queensland).

42. R. *maclaudi* POUSARGUES 1897 [*luctus* group]. – Anterior upper premolar considerably reduced, though in toothrow. Distal phalanges of third digit of wing lengthened. Mental grooves obliterated. Size relatively large (forearm length, 55–68 mm). – *Distribution:* Known only from Guinea in western Africa and from eastern Zaire, Uganda, and Rwanda in eastern Africa. – Two subspecies are currently recognized:

R. m. ruwenzorii (= *hilli*) (eastern range), *R. m. maclaudi* (Guinea).

43. R. *macrotis* BLYTH 1844 [*luctus* group]. – Sella broad and densely haired with incipient lappets at base, internarial lobes forming a small cup which is scarcely wider than the sella. Connecting process relatively high. Size medium (forearm length, 36–45 mm). – *Distribution:* Known from northern India, southern China, Thailand, Vietnam, Malaya, Sumatra, and the Philippines. – Six subspecies are currently recognized:

R. m. macrotis (northern India), *R. m. episcopus* (southwestern China), *R. m. caldwelli* (southeastern China to Vietnam), *R. m. siamensis* (Thailand to Vietnam), *R. m. dohrni* (Malaya, Sumatra), *R. m. hirsutus* (Philippines).

44. R. *hildebrandti* PETERS 1878 [*luctus* group]. – Anterior upper premolar greatly reduced and extruded from toothrow. Fifth metacarpal of wing somewhat longer than fourth or third. Lateral mental grooves obliterated. Size relatively large (forearm length, 60–67 mm; condylocanine length, 23–25 mm). – *Distribution:* Ranging from Ethiopia and northeastern Zaire to Transvaal. – No subspecies.

45. R. *eloquens* ANDERSEN 1905 [*luctus* group]. – Anterior upper premolar greatly reduced and extruded from toothrow. Fifth metacarpal of wing somewhat longer than fourth or third. Lateral mental grooves obliterated. Size fairly large (forearm length, 54–60 mm; condylocanine length, 21–23 mm). – *Distribution:* Ranging from southern Somalia to eastern Zaire and south to northern Tanzania (including Pemba and Zanzibar islands). – Two subspecies are recognized:

R. e. perauritus (southern Somalia), *R. e. eloquens* (southern Sudan to northern Tanzania).

46. R. *fumigatus* RÜPPELL 1842 [*luctus* group]. – Anterior upper premolar reduced and extruded from toothrow. Fifth metacarpal of wing somewhat longer than fourth or third. Lateral mental grooves obliterated. Size medium to fairly large (forearm length, 47–60 mm; condylocanine length, 18–21 mm). – *Distribution:* Ranging from Gambia to Ethiopia and south to the Cape Province. – Six subspecies are here recognised:

R. f. exsul (central Sudan to Tanzania), *R. f. fumigatus* (Ethiopia), *R. f. abae* (northeastern Zaire), *R. f. foxi* (Central African Republic to Upper Volta), *R. f. diversus* (Guinea, Sierra Leone, Gambia, Senegal), *R. f. aethiops* (Zambia and Angola to the Cape Province). Their boundaries, however, have never been worked out.

47. *R. rex* G. M. ALLEN 1923 [*luctus* group]. – Sella very broad, without lappets, internarial cup very broad, subcircular, and enclosing the base of the sella proper. Lancet very low and rounded. Infraorbital canal long. Size relatively large (forearm length, 59–63 mm). – *Distribution:* Known only from Szechwan and Kweichow provinces of China. – No subspecies.

48. *R. paradoxolophus* (BOURRET 1951) [*luctus* group]. – Sella broad, without lappets, internarial cup very broad, subcircular, and enclosing the base of the sella proper. Lancet very low and rounded. Infraorbital canal long. Size fairly large (forearm length, 54 mm). – *Distribution:* Known only from Thailand and Vietnam. – No subspecies.

49. *R. marshalli* THONGLONGYA 1973 [*luctus* group]. – Sella broad and very short with rudimentary lappets; internarial cup very broad, trapezoid rather than subcircular and not enclosing base of sella proper. Lancet fairly low but triangular. Infraorbital canal short. Median rostral swellings of skull unusually enlarged. Size medium (forearm length, 45–46 mm). – *Distribution:* Known only from Thailand. – No subspecies.

50. *R. trifoliatus* TEMMINCK 1834 [*luctus* group]. – Zygomatic width greater than mastoid width. Upper incisors minute and widely separated. Sella high and cuneate with expanded lappets at its base; internarial lobes at base small, less distinctly cup-shaped. Third metacarpal shortened, first phalanx lengthened. Size medium (forearm length, 46–55 mm). – *Distribution:* Known from northeastern India, Thailand through the Malay peninsula and Sumatra to Java and Borneo. – Four subspecies are here recognized:

R. t. edax (mainland range), *R. t. trifoliatus* (Sumatra, Java, Borneo), *R. t. niasensis* (Nias island, west of Sumatra), *R. t. solitarius* (Banka island, east of Sumatra).

51. *R. sedulus* ANDERSEN 1905 [*luctus* group]. – Zygomatic width greater than mastoid width. Sella high, parallel-sided, with expanded lappets at its base; internarial lobes at base small, less distinctly cup-shaped. Third metacarpal short, its first phalanx lengthened. Pelage woolly. Size fairly small (forearm length, 42–50 mm). – *Distribution:* Known only from Malaya and Borneo. – No subspecies.

52. *R. mitratus* BLYTH 1844 [*luctus* group]. – Sella short and rounded, apparently without expanded lappets; internarial lobes forming a deep cup. Lateral mental grooves absent. Size fairly large (forearm length, 57–58 mm). – *Distribution:* Known only from northeastern India. – No subspecies.

53. *R. pearsoni* HORSFIELD 1851 [*luctus* group]. – Zygomatic width greater than mastoid width. Upper incisors relatively large and approximated. Sella moderately high and broad with poorly developed lappets at its base; internarial lobes at base of sella forming a broad, rather flat saucer. Third metacarpal short. Lateral mental grooves absent. Size medium (forearm length, 48–56 mm). – *Distribution:* Ranging from northern India and southern China south to Malaya. – Two subspecies are recognized:

R. p. pearsoni (northern India to southwestern China), *R. p. chinensis* (southeastern China to Malaya).

54. *R. yunanensis* DOBSON 1872 [*luctus* group]. – Zygomatic width greater than mastoid width. Upper incisors relatively large and approximated. Sella moderately high and broad with poorly developed lappets at its base; internarial lobes at base of sella forming a broad, rather flat, saucer. Third metacarpal short. Lateral mental grooves absent. Size fairly large (forearm length, 58–64 mm). – *Distribution:* Ranging from northeastern India and southwestern China to southern Thailand. – No subspecies.

55. *R. luctus* TEMMINCK 1835 [*luctus* group]. – Zygomatic width greater than mastoid width. Upper incisors relatively large and approximated. Sella moderately high and broad with expanded lappets at its base; internarial lobes at base of sella forming a broad, rather flat, saucer. Pelage woolly. Size relatively large (forearm length, 57–78 mm). – *Distribution:* Ranging from northern India east to Taiwan and south to Ceylon and Malaya, Sumatra to Bali and Borneo. – Nine subspecies are currently recognized:

R. l. sobrinus (Ceylon), *R. l. beddomei* (southern India), *R. l. perniger* (northern India, Burma, perhaps northern Thailand), *R. l. lanosus* (southeastern China), *R. l. formosae* (Taiwan), *R. l. spurcus* (Hainan), *R. l. morio* (Malay peninsula, northern Sumatra), *R. l. luctus* (southern Sumatra, Java, Bali), *R. l. foetidus* (Borneo).

56. *R. euryotis* TEMMINCK 1834 [*euryotis* group]. – Large broad ears with well-developed antitragal lobe. Horseshoe broad. Skull with high projecting median anterior rostral swellings. Anterior upper premolar in toothrow but middle lower extruded. Palatal bridge short, less than 1/3 length of maxil-

lary toothrow. Connecting process high. Anterior edge of horseshoe scarcely emarginated but with broad, parallel swollen longitudinal ridges extending back to the internarial region, enclosing a groove that widens posteriorly to terminate at a low median projection. Size relatively large (forearm length, 52–58 mm). – *Distribution:* Ranging from Celebes through the Moluccas and New Guinea to the Bismarcks. – Six subspecies are currently recognized (though some are of dubious validity, HILL 1983):

R. e. tatar (Celebes), *R. e. timidus* (northern Moluccas through New Guinea to the Bismarcks), *R. e. burius* (Buru), *R. e. euryotis* (Ceram to Timorlaut), *R. e. praestans* (Keis), *R. e. aruensis* (Arus).

57. <u>*R. creaghi*</u> THOMAS 1896 [*euryotis* group]. – Anterior edge of horseshoe narrowly emarginated, the emargination extended posteriorly as a narrow groove reaching less than halfway to internarial region. Upper part of sella thickened posteriorly. Connecting process obsolete, hairs at base forming a dense, bristly subconical tuft. Size fairly large (forearm length, 47–50 mm). – *Distribution:* Confined to Borneo and Madura (off northern coast of Java). – Two subspecies are recognized:

R. c. creaghi (Borneo), *R. c. pilosus* (Madura).

58. <u>*R. canuti*</u> THOMAS & WROUGHTON 1909 [*euryotis* group]. – Anterior edge of horseshoe narrowly emarginated, the emargination extended posteriorly as a narrow groove reaching less than halfway to internarial region. Upper part of sella lacking any posterior thickening. Connecting process low and rounded, hairs at base dense, long and dispersed. Size fairly large (forearm length, 47–52 mm). – *Distribution:* Confined to Java and Timor. – Two subspecies are recognized:

R. c. canuti (Java), *R. c. timoriensis* (Timor).

59. <u>*R. coelophyllus*</u> PETERS 1867 [*euryotis* group]. – Anterior edge of horseshoe narrowly emarginated, the emargination extended posteriorly as a narrow groove reaching less than halfway to internarial region. Base of well-developed connecting process at most sparsely haired; recessed into a vertical fissure at base of lancet. Postnarial rostral depression of skull moderately deep, elongate, enclosed by broad, well-developed supraorbital ridges. Size medium (forearm length, 41–45 mm). – *Distribution:* Known only from Burma, Thailand, and Malaya. – No subspecies.

60. <u>*R. shameli*</u> TATE 1943 [*euryotis* group]. – Anterior edge of horseshoe narrowly emarginated, the emargination extended posteriorly as a narrow groove reaching less than halfway to internarial region. Base of well-developed connecting process at most sparsely haired, recessed into a vertical fissure at base of lancet. Postnarial rostral depression of skull shallow, short, enclosed by narrow supra-orbital ridges. Size medium (forearm length, 41–48 mm). – *Distribution:* Known from Burma, Thailand, Cambodia, and Malaya. – No subspecies.

61. <u>*R. inops*</u> ANDERSEN 1905 [*euryotis* group]. – Anterior edge of horseshoe narrowly emarginated, the emargination extended posteriorly as a narrow groove reaching less than halfway to internarial region. Base of well-developed connecting process at most sparsely haired, not recessed into base of lancet. Upper part of sella forming a small triangular pouch opening downwards. Size relatively large (forearm length, 53–54 mm). – *Distribution:* Known only from Mindanao in the Philippines. – No subspecies.

62. <u>*R. rufus*</u> EYDOUX & GERVAIS 1836 [*euryotis* group]. – Anterior edge of horseshoe narrowly emarginated, the emargination extended posteriorly as a narrow groove reaching less than halfway to internarial region. Base of well-developed connecting process at most sparsely haired, not recessed into base of lancet. Upper part of sella unmodified. Size quite large (forearm length, 66–71 mm). Frontal depression of skull prominent. – *Distribution:* Confined to the main Philippine islands. – No subspecies.

63. <u>*R. subrufus*</u> ANDERSEN 1908 [*euryotis* group]. – Anterior edge of horseshoe narrowly emarginated, the emargination extended posteriorly as a narrow groove reaching less than halfway to internarial region. Base of well-developed connecting process at most sparsely haired, not recessed into base of lancet. Upper part of sella unmodified. Size relatively large (forearm length, 53–57 mm). Internarial cup narrow. Frontal depression of skull prominent. – *Distribution:* Confined to the main Philippine islands. – Two subspecies are recognized:

R. s. subrufus (northern and central Philippines), *R. s. bunkeri* (Mindanao).

64. <u>*R. arcuatus*</u> PETERS 1871 [*euryotis* group]. – Anterior edge of horseshoe narrowly emarginated, the emargination extended posteriorly as a narrow groove reaching less than halfway to internarial region. Base of well-developed connecting process at most sparsely haired, not recessed into base of lancet. Upper part of sella un-

modified. Size medium (forearm length, 42–51). Internarial cup broad. Frontal depression of skull shallow. – *Distribution:* Ranging from Sumatra through Borneo to the Philippines, also Buru in the Moluccas, Wettar island (just north of Timor), and a small area of central New Guinea. – Seven subspecies are currently recognized (HILL & SCHLITTER 1982):

R. a. beccarii (Sumatra), *R. a. proconsularis* (Borneo), *R. a. arcuatus* (northern Philippines), *R. a. exiguus* (southern Philippines), *R. a. toxopeusi* (Buru), *R. a. angustifolius* (Wettar), *R. a. mcintyrei* (New Guinea).

Subfamily **Hipposiderinae** FLOWER & LYDEKKER 1891

Complex noseleaf with both anterior and posterior portions, but only the anterior can be clearly homologized with a part (horseshoe) of the noseleaf in Rhinolophinae. First and second ribs and their corresponding vertebrae fully fused. A preacetabular foramen formed by a bridge of bone uniting the ocicular process with the anterior end of the ilium. Lumbar vertebrae frequently fused. Number of phalanges reduced by fusion to two on each toe. – *Distribution:* Ranging widely in the Old World tropics from western Africa east to the New Hebrides, including Madagascar. Extending into the temperate zones in Africa and southern Asia, but not in Australia. – Two tribes, one additional subtribe, nine genera, and 65 species are here recognized.

Tribe Hipposiderini FLOWER & LYDEKKER 1891

Ears with strengthening "ribs" of cartilage. Horseshoe of noseleaf relatively unmodified. First metacarpal not greatly lengthened. Tail not greatly reduced. Canine bearing portion of maxilla not greatly enlarged. Upper canine without enlarged internal supplemental cusp. Mandibular symphysis V-shaped. – *Distribution:* Same as for subfamily. – Two subtribes, seven genera, 62 species.

Subtribe Hipposiderina FLOWER & LYDEKKER 1891

Noseleaf without an anterior median straplike process extending anteriorly from the internarial region. Posterior leaf without any deep pockets or cells. – *Distribution:* Virtually the same as for tribe. – Four genera, 58 species.

Genus ***Hipposideros*** GRAY 1831 (Fig. 82)

Sagittal crest not developed primarily in the immediate postorbital region. Fusion of extra phalanges of foot complete. No well defined dorsal processes on free edge of posterior noseleaf. Tail well developed. Lumbar vertebrae not fused. Dental formula, i 1/2, c 1/1, p 2/2, m 3/3 x 2 = 30. Size small to very large (forearm length, 32–115 mm). – *Distribution:* Ranging through sub-Saharan Africa, Morocco, Madagascar, southwestern Arabia, southern Asia from Afghanistan to the Riukius, south and east through the Malay archipelago to the New Hebrides and northern Australia. – 53 species and 65 additional subspecies (see HILL 1962a).

1. *H. megalotis* (HEUGLIN 1862) [*megalotis* group]. – Large rounded ears united by a low frontal band. Noseleaf small and simple, without lateral leaflets. Periotics large, the width of each approximately equal to three times their distance apart. Upper incisors widely spaced and weakly bilobed. Crown area of outer lower incisors much larger than that of inner lower incisors. No frontal depression. Rostral eminences moderately inflated. Sphenoidal bridge moderate, not concealing lateral apertures. Size relatively small (forearm length, 34–38 mm). – *Distribution:* Confined to Ethiopia and Kenya. – No subspecies.

2. *H. bicolor* (TEMMINCK 1834) [*bicolor* group]. – Ears fairly large and rounded, but separate. Noseleaf moderate in size and simple, without lateral leaflets. Periotics relatively small, the width of each approximately equal to their distance apart. Upper incisors weak with the outer lobe virtually absent. Crown area of outer lower incisors at most slightly larger than that of inner lower incisors. Anterior upper premolar very small and extruded from toothrow. Anterior lower premolar hardly more than half the length and two thirds to three quarters the height of the posterior lower premolar. Internarial septum not greatly expanded or modified, more or less parallel sided, uninflated. Interorbital region greatly constricted, its width considerably less than that of the rostrum. Anterior noseleaf without a median emargination. Posterior noseleaf with three supporting septa. No glandular ridge on muzzle beneath margin of anterior noseleaf. Anterior half of zygoma massive. Posterior projecting portion of vomer thickened. Size small to medium (forearm length, 39–48 mm). – *Distribution:* Ranging from the Malay peninsula and Sumatra to Java, Borneo, the Philippines, and Timor. – Four subspecies are currently recognized (HILL & al. 1986):

H. b. atrox (Malay peninsula, Sumatra), *H. b. major* (West Sumatran islands), *H. b. bicolor* (Java, Borneo), *H. b. erigens* (Philippines). Timor specimens have not been allocated subspecifically.

3. <u>*H. pomona*</u> ANDERSEN 1918 [*bicolor* group]. – Ears relatively large. Periotics relatively large, the width of each definitely greater than their distance apart. Size fairly small (forearm length, 37–44 mm). – *Distribution:* Ranging from India, east to southern China and south to the Malay peninsula. – Three subspecies are recognized (HILL & al. 1986):

H. p. pomona (southern India), *H. p. gentilis* (northern India to Vietnam and Malaya), *H. p. sinensis* (southern China to northern Thailand).

4. <u>*H. macrobullatus*</u> TATE 1941 [*bicolor* group]. – Ears relatively large. Periotics of medium size, but the width of each definitely greater than their distance apart. Size fairly small (forearm length, 41–42 mm). – *Distribution:* Known only from Celebes, Ceram, and Kangean (northeast of Java). – No subspecies but possibly conspecific with *H. pomona* (HILL & al. 1986).

5. <u>*H. ater*</u> TEMPLETON 1848 [*bicolor* group]. – Internarial septum thickened and bulbous. Anterior half of zygoma slender; a low superior projection on the posterior half. Size relatively small (forearm length, 33–43 mm). – *Distribution:* Known from India and Ceylon, several limited areas in southeastern Asia, the Nicobar islands and Sumatra to the Philippines, Bismarcks, and northern Australia. – Seven subspecies are recognized:

H. a. ater (India, Ceylon), *H. a. nicobarulae* (Nicobars), *H. a. saevus* (southeastern Asia east to the Moluccas), *H. a. antricola* (Philippines), *H. a. aruensis* (= *albaniensis*) (New Guinea, Bismarcks, northern Queensland), *H. a. amboiensis* (Amboina in the Moluccas), *H. a. gilberti* (northern coast of the Northern Territory and Western Australia).

6. <u>*H. fulvus*</u> GRAY 1838 [*bicolor* group]. – Posterior projecting portion of vomer blade-like. Anterior lower premolar much reduced, not more than one third the length nor one half the height of the posterior lower premolar. Size relatively small (forearm length 38–44 mm). – *Distribution:* Ranging from Afghanistan east to Vietnam and south to Ceylon and southern Thailand. – Two subspecies are recognized:

H. f. pallidus (Afghanistan, Pakistan, and the dryer parts of northern India), *H. f. fulvus* (remainder of range).

7. <u>*H. halophyllus*</u> HILL & YENBUTRA 1984 [*bicolor* group]. – Internarial septum expanded into a small disc-like structure. Tympanic bullae elongate and at right angles to transversely elongate periotics. Size small (forearm length, 35–39 mm). – *Distribution:* Known only from central Thailand. – No subspecies.

8. <u>*H. cineraceus*</u> BLYTH 1853 [*bicolor* group]. – Internarial septum thickened and bulbous. Anterior half of zygoma slender, superior projection on posterior half absent or poorly developed. Anterior upper premolar not extruded from toothrow. Size small (forearm length, 32–37 mm). – *Distribution:* Ranging from Pakistan and southern India east to Vietnam, south through the Malay peninsula to Borneo and probably the Philippines. – Three subspecies may be recognized (HILL & FRANCIS 1984):

H. c. durgadasi (southern India), *H. c. wrighti* (Philippines), *H. c. cineraceus* (= *micropus*) (remainder of range).

9. <u>*H. nequam*</u> ANDERSEN 1918 [*bicolor* group]. – Posterior projecting portion of vomer blade-like. Anterior lower premolar half the length and height of posterior lower premolar. Anterior upper premolar not extruded from toothrow. Size medium (forearm length, 45–46 mm). – *Distribution:* Known only from Malaya. – No subspecies.

10. <u>*H. calcaratus*</u> DOBSON 1877 [*bicolor* group]. – Interorbital region relatively broad, its width nearly equal to that of the rostrum. Anterior upper premolar not extruded from toothrow. Sphenoid bridge relatively narrow and emarginated, not concealing optic foramina. Basisphenoid depression shallow and troughlike. Size medium to fairly large (forearm length, 45–56 mm). – *Distribution:* Occurring on New Guinea, Bismarcks, Solomons and nearby islands. – Two subspecies are recognized:

R. c. cupidus (New Guinea), *R. c. calcaratus* (East Papuan islands, Bismarcks, Solomons).

11. <u>*H. maggietaylorae*</u> SMITH & HILL 1981 [*bicolor* group]. – Interorbital region relatively broad, its width nearly equal to that of the rostrum. Anterior upper premolar extruded from toothrow. Sphenoid bridge relatively broad, more or less concealing optic foramina. Basisphenoid depression deep, nearly as wide as long. Size relatively large (forearm length, 50–68 mm). – *Distribution:* Confined to New Guinea and the Bismarcks. – Two subspecies are recognized:

H. m. erroris (New Guinea), *H. m. maggietaylorae* (Bismarcks).

12. <u>*H. coronatus*</u> (PETERS 1871) [*bicolor* group]. – Apparently similar to *H. calcaratus*, but status uncertain. Posterior leaf without supporting septa.

Size medium (forearm length, 46–47 mm). – *Distribution:* Known only from Mindanao in the Philippines. – No subspecies.

13. *H. ridleyi* ROBINSON & KLOSS 1911 [*bicolor* group]. – Internarial septum expanded to form a concave subcircular disc. An incipient lateral leaflet present. Posterior margin of bony palate U-shaped, no post-palatal spicule. Size medium (forearm length, 47–48 mm). – *Distribution:* Known only from Malaya and Borneo. – No subspecies.

14. *H. jonesi* HAYMAN 1947 [*bicolor* group]. – Ears pointed. Internarial septum expanded to form very broad transversely oval disc. Well-developed lateral leaflet present. Posterior margin of bony palate square, with a small post-palatal spicule. Posterior leaf high and triangular. Size medium (forearm length, 44–55 mm). – *Distribution:* Tropical forest and woodland areas of western Africa from Guinea to Nigeria. – No subspecies.

15. *H. dyacorum* THOMAS 1902 [*bicolor* group]. – Anterior upper premolar greatly reduced, completely extruded from toothrow. Anterior lower premolar only one quarter the length and height of the posterior lower premolar. Anterior leaf without a median emargination. Posterior leaf supported by a well-defined median septum and weaker lateral septa. Pterygoids long, sphenoidal bridge wide. Size relatively small (forearm length, 39–42 mm). – *Distribution:* Confined to Borneo. – No subspecies.

16. *H. sabanus* THOMAS 1898 [*bicolor* group]. – Anterior upper premolar absent. Anterior lower premolar only one quarter the length and height of the posterior lower premolar. Anterior noseleaf with a well-defined median emargination. Posterior noseleaf lacking supporting septa. Pterygoids short, sphenoidal bridge narrow. Size relatively small (forearm length, 37–38 mm). – *Distribution:* Known from Malaya, Sumatra, and Borneo. – No subspecies.

17. *H. doriae* (PETERS 1871) [*bicolor* group]. – Anterior upper premolar absent. Anterior lower premolar less than half the size of the posterior lower premolar. Anterior noseleaf without median emargination. Posterior noseleaf lacking supporting septa. Probably a senior synonym of *H. sabanus*. – *Distribution:* Known only from Borneo. – No subspecies.

18. *H. obscurus* (PETERS 1871) [*bicolor* group]. – Anterior noseleaf with a small median emargination. Posterior noseleaf without supporting septa. A low glandular ridge present on muzzle beneath margin of anterior leaf. Size relatively small (forearm length, 41–44 mm). – *Distribution:* Restricted to the main Philippine islands. – No subspecies.

19. *H. marisae* AELLEN 1954 [*bicolor* group]. – Internarial septum expanded to form an ellipsoidal structure. A single, very small lateral leaflet present. Size relatively small (forearm length, 40–41 mm). – *Distribution:* Confined to a small area in western Africa, including parts of Guinea, Liberia, and Ivory Coast. – No subspecies.

20. *H. pygmaeus* (WATERHOUSE 1843) [*bicolor* group]. – Two lateral leaflets present, anterior one extending anteriorly beneath anterior noseleaf to the median line. Intermediate noseleaf without a median eminence. Posterior noseleaf supported by three septa of equal width, not deeply pocketed. Ears pointed. Size relatively small (forearm length, 36–40 mm). – *Distribution:* Restricted to the main Philippine islands. – No subspecies.

21. *H. galeritus* CANTOR 1846 [*bicolor* group]. – Two lateral leaflets present, neither one extending anteriorly beneath anterior noseleaf to the median line. Posterior noseleaf without a well-developed transverse serrated supplementary structure on its posterior face, but supported by three septa. Internarial septum swollen, but not expanded. Pterygoids long, sphenoidal bridge wide, partly concealing lateral apertures. Ears haired for two thirds of their length. Tips of upper incisors not strongly convergent. Intermediate noseleaf as wide or wider than posterior noseleaf. Rostrum short and markedly inflated. Size medium (forearm length, 39–53 mm). – *Distribution:* Known from India, Ceylon, the Malay peninsula, Sumatra, Java, and Borneo. – Four subspecies are currently recognized (JENKINS & HILL 1980):

H. g. brachyotis (India, Ceylon), *H. g. galeritus* (Malay peninsula, Sumatra), *H. g. longicaudus* (Java), *H. g. insolens* (Borneo).

22. *H. cervinus* (GOULD 1854) [*bicolor* group]. – Two lateral leaflets present, neither one extending anteriorly beneath anterior noseleaf to the median line. Posterior noseleaf without a well-developed transverse serrated supplementary structure on its posterior face, but supported by three septa. Internarial septum not swollen or expanded. Pterygoids long, sphenoidal bridge wide, partly concealing lateral apertures. Ears haired for two thirds of their length. Tips of upper incisors not strongly convergent. Intermediate noseleaf nar-

rower than posterior noseleaf. Rostrum relatively long and uninflated. Size medium (forearm length, 40–52 mm). – *Distribution:* Ranging from Malaya and Sumatra to the Philippines, New Hebrides, and northeastern Queensland (Australia). – Four subspecies are currently recognized (JENKINS & HILL 1980):

H. c. labuanensis (= *schneideri*) (Malaya, Sumatra, Borneo, Philippines), *H. c. cervinus* (= *celebensis*) (Celebes east to the New Hebrides and northeastern Queensland, except for Batjan and Biak), *H. c. batchianensis* (Batjan in the northern Moluccas), *H. c. misorensis* (Biak island northeast of New Guinea).

23. *H. crumeniferus* (LESEUR & PETIT 1807). – [*bicolor* group]. Two lateral leaflets present. Posterior noseleaf supported by three septa. Ears apparently not haired. Size medium (forearm length, apparently 53 mm). A poorly known form, probably a senior synonym of *H. galeritus* or *H. cervinus*. – *Distribution:* Known only from Timor. – No subspecies.

24. *H. breviceps* TATE 1941 [*bicolor* group]. – Two lateral leaflets present, neither one extending anteriorly beneath anterior noseleaf to median line. Posterior noseleaf without a well-developed transverse serrated supplementary structure on its posterior face, but supported by three septa. Internarial septum not swollen or expanded. Pterygoids long, sphenoidal bridge wide, partly concealing lateral apertures. Ears haired for one half of their length. Tips of upper incisors strongly convergent. Rostrum markedly inflated. Size medium (forearm length, 44–45 mm). – *Distribution:* Confined to the Mentawai islands. – No subspecies.

25. *H. curtus* G. M. ALLEN 1921 [*bicolor* group]. – Two lateral leaflets present, neither one extending anteriorly beneath anterior noseleaf to median line. Posterior noseleaf without a well-developed transverse serrated supplementary structure on its posterior face, but supported by three septa. Internarial septum expanded to form a small disclike structure. Pterygoids short, sphenoidal bridge narrow, not concealing lateral apertures. Size medium (forearm length, 42–47 mm). – *Distribution:* Confined to southern Cameroon and Fernando Poo in tropical western Africa. – No subspecies.

26. *H. fuliginosus* (TEMMINCK 1853) [*bicolor* group]. – Two lateral leaflets present, neither one extending anteriorly beneath anterior noseleaf to median line. Posterior noseleaf without a well-developed transverse serrated supplementary structure on its posterior face, or well-defined septa. Size relatively large (forearm length, 56–64 mm). – *Distribution:* Forested regions of tropical Africa from Guinea to eastern Zaire. – No subspecies.

27. *H. caffer* (SUNDEVALL 1846) [*bicolor* group]. – Two lateral leaflets present, neither one extending anteriorly beneath anterior noseleaf to median line. Posterior noseleaf with a well-developed transverse serrated supplementary structure on its posterior face, but no well-defined septa. Anterior upper premolar small but not extruded from toothrow. Median posterior narial compartment of rostral swelling relatively narrow, with wide lateral inflations. Size medium (forearm length 42–51 mm). – *Distribution:* Widely distributed in sub-Saharan Africa, south to the Cape Province (including Pemba and Zanzibar islands), but largely avoiding forested areas; also Morocco and southwestern Arabia). – Four subspecies are currently recognized (HAYMAN & HILL 1971):

H. c. caffer (entire range of species except for Morocco, the dry savannas immediately south of the Sahara, northeastern Zaire, and southwestern Africa), *H. c. tephrus* (Morocco, also dry sub-Saharan belt from Mauretania and Guinea to Sudan), *H. c. nanus* (northeastern Zaire), *H. c. angolensis* (Gabon south to Namibia).

28. *H. lamottei* BROSSET 1984 [*bicolor* group]. – Two lateral leaflets present, neither one extending anteriorly beneath anterior noseleaf to median line. Posterior noseleaf without well-defined septa. Anterior upper premolar small but not extruded from toothrow. Size fairly large (forearm length, 54–57 mm), but skull (total length, 18–29 mm) and ear (16 mm) relatively short. – *Distribution:* This poorly known species has been reported only from the vicinity of Mount Nimba in Guinea. – No subspecies.

29. *H. ruber* (NOACK 1893) [*bicolor* group]. – Two lateral leaflets present, neither one extending anteriorly beneath anterior noseleaf to median line. Posterior noseleaf with a well-developed transverse serrated supplementary structure on its posterior face, but no well-defined septa. Anterior upper premolar small but not extruded from toothrow. Median posterior nasal compartment of rostral swelling relatively wide, with narrow lateral inflations. Size fairly large (forearm length, 44–60 mm). – *Distribution:* More or less confined to forest and wet savanna areas from Senegal to Ethiopia and south to Malawi (including the islands in the Gulf of Guinea). – Two subspecies are here recognized (HAYMAN & HILL (1971):

H. r. ruber (= *centralis, niapu*) (eastern Africa from Ethiopia to Malawi and west to Central African Republik and northern Angola), *H. r. guineensis* (western Africa from Senegal to Gabon, including Fernando Poo, Principe, and Sao Tome). The true picture may, however, be more complex.

30. *H. beatus* ANDERSEN 1906 [*bicolor* group]. – Two lateral leflets present, neither one extending anteriorly beneath anterior noseleaf to median line. Posterior noseleaf with a well-developed transverse serrated supplementary structure on its posterior face, but no well-defined septa. Anterior upper premolar greatly reduced, extruded from toothrow. Size fairly small (forearm length, 41–46 mm). – *Distribution:* Confined to forested regions of tropical Africa from Sierra Leone to northeastern Zaire. – Two subspecies are currently recognized:

H. b. beatus (Sierra Leone to Gabon), *H. b. maximus* (Central African Republic, extreme southwestern Sudan, northeastern Zaire).

31. *H. coxi* SHELFORD 1901 [*bicolor* group]. – Two lateral leaflets present, the anterior one extending anteriorly beneath anterior noseleaf to the median line. Intermediate noseleaf with a prominent median eminence. Posterior noseleaf supported by a narrow median septum and two broad lateral septa, deeply pocketed. Ear pointed. Size fairly large (forearm length, 51–55 mm). – *Distribution:* Confined to Borneo. – No subspecies.

32. *H. papua* (THOMAS & DORIA 1886) [*bicolor* group]. – Two lateral leaflets and an incipient third present, the second extending anteriorly beneath the anterior noseleaf. Ear acutely pointed. Size fairly large (forearm length, 50–51 mm). – *Distribution:* Recorded only from Biak island (Geelvink Bay, New Guinea). – No subspecies.

33. *H. cyclops* (TEMMINCK 1853) [*cyclops* group]. – Ears long, narrow, and pointed, but separate. Periotics greatly enlarged, the width of each approximately four times their distance apart. Two lateral leaflets, the posterior one not extending anteriorly beneath the anterior noseleaf. Premaxillae wide, enclosing the anterior palatal foramina, making a broad junction with the maxillae. Antorbital foramen relatively large. Size fairly large (forearm length, 56–73 mm). – *Distribution:* Confined to forested regions of Africa from Senegal to Kenya. – No subspecies.

34. *H. camerunensis* EISENTRAUT 1956 [*cyclops* group]. – Posterior lateral supplementary leaflet not extending anteriorly beneath anterior noseleaf. Premaxillae wide, enclosing the anterior palatal foramina, making a broad junction with the maxillae. Antorbital foramen relatively small. Size relatively large (forearm length, 74–76 mm).

– *Distribution:* Known only from southern Cameroon, eastern Zaire and western Kenya. – No subspecies.

35. *H. muscinus* (THOMAS & DORIA 1886) [*cyclops* group]. – Posterior lateral leaflet extending anteriorly beneath the anterior noseleaf. Premaxillae narrow not enclosing the anterior palatal foramina, making a narrow spatulate junction with the maxillae. Median process of intermediate noseleaf not greatly developed or club-shaped. Frontal depression of skull absent. A shallow sphenoidal depression present. Posterior ridge of last upper molar greatly reduced. Posterior noseleaf with a median process, but without a transverse supplementary structure developed from its posterior face. Periotics greatly enlarged, the width of each four or five times their distance apart. Size relatively small (forearm length, 45–46 mm). – *Distribution:* Apparently confined to eastern New Guinea. – No subspecies.

36. *H. wollastoni* THOMAS 1913 [*cyclops* group]. – Posterior lateral leaflet extending anteriorly beneath the anterior noseleaf. Premaxillae narrow, not enclosing the anterior palatal foramina, making a narrow spatulate junction with the maxillae. Median process of intermediate noseleaf not greatly developed or club-shaped. Frontal depression of skull absent. A shallow sphenoidal depression present. Posterior ridge of last upper molar greatly reduced. Posterior noseleaf without a median process, but with a transverse supplementary structure developed from its posterior face. Periotics tremendously enlarged, the width of each six or more times their distance apart. Size relatively small (forearm length, 42–43 mm). – *Distribution:* Apparently confined to western New Guinea. – No subspecies.

37. *H. corynophyllus* HILL 1985 [*cyclops* group]. – Posterior lateral leaflet extending anteriorly beneath the anterior noseleaf. Premaxillae narrow, not enclosing the junction with the maxillae. Median process of intermediate noseleaf well developed and club-shaped. Skull with a shallow frontal depression. A shallow sphenoidal depression present. Posterior ridge of last upper molar considerably reduced. Width of each cochlea about four times their distance apart. Rostrum relatively elongate. Interorbital region relatively wide. Size medium (forearm length, 48–49 mm). – *Distribution:* Known only from central New Guinea. – No subspecies.

38. *H. semoni* MATSCHIE 1903 [*cyclops* group]. – Posterior lateral leaflet extending anteriorly beneath the anterior noseleaf. Premaxillae narrow, not enclosing the anterior palatal foramina, making a narrow spatulate junction with the maxillae. Median process of intermediate noseleaf well developed and club-shaped. Skull with a shallow frontal depression. Sphenoidal depression absent. Posterior ridge of last upper molar little reduced. Width of each cochlea at least six times their distance apart. Interorbital region relatively wide. Median process of posterior noseleaf well-developed. Rostrum relatively short, its eminences greatly inflated, Antorbital foramen relatively small. Size relatively small (forearm length, 42–49 mm). – *Distribution:* Known only from eastern New Guinea and northeastern Queensland (Australia). – No subspecies.

39. *H. stenotis* THOMAS 1913 [*cyclops* group]. – Posterior lateral leaflet extending anteriorly beneath the anterior noseleaf. Premaxillae narrow, not enclosing the anterior palatal foramina, making a narrow spatulate junction with the maxillae. Median process of intermediate noseleaf well developed and club-shaped. Skull with a deep frontal depression. Sphenoidal depression absent. Posterior ridge of last upper molar little reduced. Width of each cochlea at least six times their distance apart. Interorbital region relatively wide. Median process of posterior noseleaf poorly developed. Rostrum relatively short, its eminences moderately inflated. Antorbital foramen relatively large. Size relatively small (forearm length, 42–45 mm). – *Distribution:* Confined to the dryer portions of tropical Australia from northern Western Australia to northwestern Queensland. – No subspecies.

40. *H. pratti* THOMAS 1891 [*pratti* group]. – Ears bluntly pointed and triangular, but separate. Periotics relatively small, the width of each approximately equal to their distance apart. Upper incisors strong, retaining most of the outer lobe. Crown area of outer lower incisors much greater than that of inner lower incisors. Noseleaf with two lateral leaflets. Frontal depression well-defined. Maxillae elongated. Lateral margins of anterior and posterior noseleaves not continuous. Rostrum without lateral pits. An abrupt step-like discontinuity between roofs of narial and mesopterygoid canals. Vomer projecting posteriorly behind palate. Size relatively large (forearm length, 81–89 mm). – *Distribution:* Confined to southern China and Vietnam. – No subspecies.

41. *H. lylei* THOMAS 1913 [*pratti* group]. – Lateral margins of anterior and posterior noseleaves continuous. Rostrum with lateral pits. Roofs of narial and mesopterygoid canals merging smoothly. Vomer not projecting posteriorly behind palate. Size fairly large (forearm length, 73–84 mm). – *Distribution:* Known from southwestern China south to Malaya. – No subspecies.

42. *H. armiger* (HODGSON 1835) [*armiger* group]. – Ears pointed and triangular, with a small projection at the antitragal fold, but separate. Periotics relatively small, the width of each approximately equal to their distance apart. Upper incisors strong with two subequal lobes. Crown area of outer lower incisors much greater than that of inner lower incicisors. Noseleaf with three large and one very small lateral leaflet. Frontal depression absent. Maxillae not elongated. Posterior noseleaf narrower than anterior noseleaf with a slightly lobate border. Rostrum flattened, its eminences not inflated. A distinct discontinuity between roofs of narial and mesopterygoid canals. Size large (forearm length, 79–100 mm). – *Distribution:* The species ranges from northern India and southern China south to Malaya. – Three subspecies are currently recognized:

H. a. armiger (entire range except Taiwan and Indo-China), *H. a. terasensis* (Taiwan), *H. a. tranninhensis* (Vietnam, Laos).

43. *H. turpis* BANGS 1901 [*armiger* group]. – Noseleaf with three lateral leaflets. Size fairly large (forearm length, 67–80 mm). – *Distribution:* Known only from the Riukiu islands and peninsular Thailand. – Two subspecies are recognized:

H. t. pendleburyi (peninsular Thailand), *H. t. turpis* (Riukius).

44. *H. abae* J. A. ALLEN 1917 [*speoris* group]. – Ears pointed and triangular, but separate. Periotics relatively small, the width of each approximately equal to their distance apart. Upper incisors strong but weakly bilobed. Crown area of outer lower incisors much greater than that of inner lower incisors. Noseleaf with three lateral leaflets. No definite frontal depression. Maxillae not elongated. Posterior noseleaf equal in width to anterior noseleaf, its border more or less semicircular, supporting septa absent. Rostrum flattened with prominent supraorbital ridges, its eminences moderately inflated. Roofs of narial and mesopterygoid canals not sharply discontinuous. Ear with a small projection at the antitragal fold. Pterygoid wings of skull undeveloped. Anterior noseleaf without median emargination. Vomer not projecting behind palate. Upper canine with a shallow antero-internal groove and low posterior cusp. Posterior edge of bony palate nearly square, with prominent posterior palatal

foramina. Sphenoidal bridge constricted, not concealing lateral apertues. Size medium (forearm length, 55–66 mm). – *Distribution:* Occurring in woodlands of northern tropical Africa from Guinea (Bissau) to southern Sudan. – No subspecies.

45. *H. larvatus* (HORSFIELD 1823) [*speoris* group]. – Anterior noseleaf with a definite median emargination. Rostrum rounded, supraorbital ridges barely developed. Vomer projecting behind palate. Upper canines not grooved and lacking a posterior cusp. Size medium (forearm length, 53–67 mm). – *Distribution:* Ranging from northwestern India and southern China through the Malay peninsula and Sumatra to Borneo, Java, and the Lesser Sundas. – Eight subspecies are currently recognized, though there is considerable uncertainty concerning their ranges and validity:

H. l. leptophyllus (northern India), *H. l. grandis* (Burma, Thailand, southern Vietnam). *H. l. alongensis* (northern Vietnam), *H. l. poutensis* (southeastern China), *H. l. neglectus* (Malay peninsula, Sumatra, Borneo, and most neighboring small islands), *H. l. barbensis* (several small islands in the South China Sea), *H. l. larvatus* (Java, Bali), *H. l. sumbae* (Sumba island in the lesser Sundas).

46. *H. speoris* (SCHNEIDER 1800) [*speoris* group]. – Posterior noseleaf with supporting septa. Posterior edge of bony palate U-shaped without posterior palatal foramina. Sphenoidal bridge narrow but unconstricted. Size fairly small (forearm length, 45–45 mm). – *Distribution:* Confined to India and Ceylon. – No currently recognized subspecies.

47. *H. lekaguli* THONGLONGYA & HILL 1974 [*diadema* group]. – Ears pointed and triangular, broad based, but separate, slightly thickened at the antitragus. Periotics relatively small, the width of each approximately equal to their distance apart. Upper incisors strong and irregularly bilobed, not in contact. Crown area of outer lower incisors somewhat greater than that of inner lower incisors. Noseleaf with three lateral leaflets. No frontal depression. Maxillae not elongated. Posterior noseleaf approximately equal in width to anterior noseleaf, its border heavily lobulated, supported by three well-developed septa that isolate four deep pockets. Intermediate noseleaf with a prominent medial projection. Rostrum flattened, its eminences moderately inflated. Roofs of narial and mesopterygoid canals not sharply discontinuous. Sphenoidal bridge wide, partially concealing lateral apertures. Pterygoid wings expanded. Postorbital processes not projecting. Upper canine without an anterior groove or posterior cusp. Nasofrontal region rounded. Posterior border of bony palate more or less U-shaped. Sphenoidal depression shallow. Size fairly large (forearm length, 71–80 mm). – *Distribution:* Known only from southern Thailand and Malaya. – No subspecies.

48. *H. lankadiva* KELAART 1850 [*diadema* group]. – Posterior border of bony palate V-shaped. Sphenoidal depression well-developed. Interorbital region relatively constricted. Rostrum relatively long. Size relatively large (forearm length, 76–93 mm). Skull relatively large (condylocanine length exceeding 26.5 mm; maxillary toothrow length exceeding 12.0 mm). – *Distribution:* Confined to peninsular India and Ceylon. – Four currently recognized subspecies:

H. l. unitus, H. l. indus, H. l. mixtus (all in India), *H. l. lankadiva* (Ceylon).

49. *H. schistaceus* ANDERSEN 1918 [*diadema* group]. – Posterior border of bony plate V-shaped. Sphenoidal depression well-developed. Interorbital region relatively constricted. Rostrum relatively long. Size fairly large (forearm length, 73–82 mm). Skull relatively small (condylocanine length less than 25.5 mm; maxillary toothrow less than 11.8 mm). – *Distribution:* Known only from peninsular India. – No subspecies.

50. *H. diadema* (GEOFFROY 1813) [*diadema* group]. – Frontal region concave with a shallow frontal depression. Interorbital region relatively narrow. Zygomatic arches markedly flared. Size relatively large (forearm length, 65–95 mm; condylocanine length less than 31 mm; maxillary toothrow length less than 13.8 mm). Intermediate noseleaf with a prominent medial ridge. – *Distribution:* An extensive range from Burma and Vietnam south through the Malay peninsula and Sumatra and east to the Philippines, Solomons and tropical Australia. – Eighteen subspecies are currently recognized (though many are of dubious validity):

H. d. masoni (southeastern Asia, Sumatra, Borneo), *H. d. nicobarensis* (Nicobar islands), *H. d. natunensis* (North Natuna islands), *H. d. enganus* (Engano island, west of Sumatra), *H. d. diadema* (Java, Lesser Sundas), *H. d. griseus* (Philippines), *H. d. speculator* (Celebes), *H. d. euotis* (Northern Moluccas), *H. d. ceramicus* (central Moluccas), *H. d. custos* (Keis), *H. d. pullatus* (New Guinea, Bismarcks), *H. d. mirandus* (Admirality islands), *H. d. trobrius* (Trobriand islands), *H. d. oceanitis, H. d. malaitensis, H. d. demissus* (all three in Solomons), *H. d. reginae* (northern Queensland, Australia), *H. d. inornatus* (northern Northern Territory, Australia).

51. *H. dinops* ANDERSEN 1905 [*diadema* group]. – Frontal region concave with a shallow frontal de-

pression. Interorbital region relatively narrow. Zygomatic arches markedly flared. Size large (forearm length, 93–97 mm; condylocanine length more than 32 but less than 33 mm; maxillary toothrow length more than 14.0 mm). Intermediate noseleaf with a prominent medial ridge. – *Distribution:* Known only from Celebes and the Solomons. – Two subspecies are recognized:

H. d. pelingensis (Celebes), *H. d. dinops* (Solomons).

52. *H. inexpectatus* LAURIE & HILL 1954 [*diadema* group]. – Frontal region concave with a shallow frontal depression. Interorbital region relatively narrow. Zygomatic arches markedly flared. Size very large (forearm length, 100–101 mm; condylocanine length more than 34 mm). Intermediate noseleaf swollen but without a prominent median ridge or projection. – *Distribution:* Confined to Celebes. – No subspecies.

53. *H. commersoni* (E. GEOFFROY 1813) [*diadema* group]. – Ears narrow at base. Post-orbital processes projecting. Naso-frontal region pentagonal. Upper incisors widely separated. Upper canine with an anterior groove and a high posterior cusp. Interorbital region relatively narrow. Frontal depression present. Zygomatic arches markedly flared. Size large to very large (forearm length, 77–115 mm). – *Distribution:* Widely distributed in tropical Africa from Senegal to Ethiopia and south to Namibia and Transvaal, including Pemba and Zanzibar in the east and São Tomé island in the Gulf of Guinea; also Madagascar. – Five subspecies are here recognized:

H. c. commersoni (Madagascar), *H. c. marungensis* (eastern Africa from Ethiopia to Transvaal), *H. c. niangarae* (northeastern Zaire), *H. c. gigas* (western Africa from Senegal and Central African Republic to Namibia), *H. c. thomensis* (São Tomé island).

Genus *Anthops* THOMAS 1888 (Fig. 83)

Sagittal crest not developed primarily in the immediate postorbital region. Fusion of extra phalanges of foot complete. Three raised rounded processes on free edge of posterior noseleaf. Tail considerably shortened. Lumbar vertebrae fused. Dental formula, i 1/2, c 1/1, p 2/2, m 3/3 x 2 = 30. – *Distribution:* Confined to the Solomons. – A single species.

1. *A. ornatus* THOMAS 1888. – Size fairly small (forearm length, 48–51 mm). – *Distribution:* Same as for genus. – No subspecies.

Genus *Aselliscus* TATE 1941 (Fig. 84)

Sagittal crest poorly developed. Three pointed processes on free edge of posterior noseleaf. Tail well developed. Premaxillae divergent anteriorly. Rostrum greatly inflated. Dental formula: i 1/2, c 1/1, p 2/2, m 3/3 x 2 = 30. – *Distribution:* Markedly disjunct; Burma and southern China south to Malaya; Moluccas east to the New Hebrides. – Two species, three additional subspecies.

1. *A. tricuspidatus* (TEMMINCK 1835). – Two lateral leaflets. Tail relatively short. Less sloping rostrum. Anterior lower premolar less compressed by canine and posterior premolar. Size small (forearm length, 37–44 mm). – *Distribution:* Ranging from the Moluccas east to the New Hebrides. – Four subspecies are currently recognized (SCHLITTER & al. 1983):

A. t. tricuspidatus (Moluccas), *A. t. novaeguineae* (New Guinea mainland), *A. t. koopmani* (East Papuan islands, Bismarcks, Solomons, Santa Cruz islands), *A. t. novaehebridensis* (New Hebrides).

2. *A. stoliczkanus* (DOBSON 1871) (= *trifida* PETERS, 1871; *wheeleri* OSGOOD 1932). – Three lateral leaflets. Tail relatively long. More sloping rostrum. Anterior lower premolar compressed by canine and posterior premolar. Size small (forearm length, 39–44 mm). – *Distribution:* Confined to southeastern Asia from Burma and southern China south to Malaya. – No subspecies (SANBORN 1952).

Genus *Asellia* GRAY 1838 (Fig. 85)

Sagittal crest developed primarily in the immediate postorbital region. Fusion of extra phalanges of foot not complete. Three pointed processes on free edge of posterior noseleaf. Tail well developed. Lumbar vertebrae fused. Premaxillae fused with one another. Rostrum not greatly inflated. Dental formula: i 1/2, c 1/1, p 1/2, m 3/3 x 2 = 28. – *Distribution:* Widely distributed in northern Africa south to Senegal and Somalia (the Zanzibar record is erroneous); east across southwestern Asia to Pakistan (including Socotra). – Two species and two additional subspecies.

1. *A. tridens* (E. GEOFFROY 1818). – Size relatively large. Forearm length, 43–53 mm; forearm slightly shorter than combined length of third metacarpal and its first phalanx. – *Distribution:* Same as for genus. – Four poorly delimited subspecies are here recognized:

A. t. diluta (Senegambia to Morocco), *A. t. italosomalica* (Somalia, southern Ethiopia), *A. t. tridens* (Algeria to Yemen and Egypt), *A. t. murraiana* (remainder of range).

2. *A. patrizii* DE BEAUX 1931. – Size relatively small. Forearm length (36–41 mm) slightly longer than combined length of third metacarpal and its first phalanx. – *Distribution:* Confined to northern Ethiopia including adjacent Red Sea islands. – No subspecies.

Subtribe Rhinonycterina GRAY 1866

Noseleaf with an anterior strap-like process extending anteriorly from the internarial region. Posterior noseleaf with a number of deep pockets or cells. – *Distribution:* Confined to sub-Saharan Africa (chiefly eastern), Madagascar and nearby islands, southwestern Asia, and tropical Australia. – Three genera, four species (HILL 1982).

Genus *Rhinonycteris* GRAY 1847 (Fig. 86)

Zygoma expanded into a wide plate, braincase much higher than rostrum (which has prominent inflations). Sagittal crest terminating abruptly in front of middle of braincase. Premaxillae unusually thick, a distinct ridge along their line of contact, which terminates anteriorly in a backwardly curved point. Posterior noseleaf without tall pointed projections. Upper canine with large secondary cusp. Dental formula: i 1/2, c 1/1, p 2/2, m 3/3 x 2 = 30. – *Distribution:* Confined to tropical Australia from northwestern Western Australia to northwestern Queensland. – A single species.

1. *R. aurantius* (GRAY 1845). – Size fairly small (forearm length, 45–50 mm). – *Distribution:* Same as for genus. – No subspecies.

Genus *Cloeotis* THOMAS 1901 (Fig. 87)

Zygoma not specially expanded. Rostrum unusually short with inflations poorly developed. Premaxillae unmodified. Posterior noseleaf with three tall pointed projections. Ears unusually short. Upper canine with a well developed secondary cusp. Dental formula: i 1/2, c 1/1, p 2/2, m 3/3 x 2 = 30, but small anterior upper premolar may be lost. – *Distribution:* Confined to southeastern Africa from Kenya to Transvaal. – A single species, one additional subspecies.

1. *C. percivali* THOMAS 1901. – Size small (forearm length, 30–39 mm). – *Distribution:* Same as for genus. – Two subspecies:

C. p. percivali (southeastern Kenya, northeastern Tanzania), *C. p. australis* (southern Zaire to Botswana and Swaziland).

Genus *Triaenops* DOBSON 1871 (Fig. 88)

Zygoma expanded into a wide plate. Braincase scarcely higher than rostrum which has prominent inflations. Sagittal crest low and unmodified. Premaxillae unusually thick, a distinct ridge along their line of contact which terminates anteriorly in an upwardly directed point. Posterior noseleaf with three tall pointed projections. Upper canine with a strong posterior secondary cusp. Dental formula: i 1/2, c 1/1, p 2/2, m 3/3 x 2 = 30. – *Distribution:* Ranging from Iran across southern Arabia to Ethiopia and south through eastern Africa to Mozambique (including Zanzibar); also a small area in southern Congo Republic and northern Angola; Madagascar and nearby islands. – Two species, three additional subspecies.

1. *T. furculus* TROUESSART 1906. – Noseleaf with a thickened supplementary ridge. Median anterior process forked anteriorly. Outer margins of lateral projections of posterior noseleaf basally smoothly convex. Anterior margin of ear evenly rounded. Rostrum with lateral profile of nares nearly vertical. Size fairly small (forearm length, 43–46 mm). – *Distribution:* Confined to Madagascar and the Aldabra group to the northwest. – No subspecies.

2. *T. persicus* DOBSON 1871. – Noseleaf with two lateral supplementary leaflets. Anterior margin of median anterior process linear or slightly concave. Outer margins of lateral projections of posterior noseleaf basally emarginated. Anterior margin of ear with central step-like emargination. Rostrum with lateral profile of nares sloping posteriorly. Size medium (forearm length, 45–61 mm). – *Distribution:* Same as for genus except for absence from the Aldabra group. – Four subspecies are currently recognized:

T. p. persicus (Iran, Oman), *T. p. afer* (Yemen south through eastern Africa to Mozambique). *T. p. majusculus* (Uganda, Congo Republic, Angola), *T. p. rufus* (Madagascar).

Tribe Coelopsini TATE 1941

Ears without strengthening "ribs" of cartilage. Horseshoe of noseleaf divided into two parts by a median notch which extends back to the nasal septum. First metacarpal greatly lengthened. Tail vestigial or absent. Canine-bearing portion of maxilla greatly extended forwards. Basicranial foramina greatly enlarged. Upper canine with an internal supplemental cusp. Mandibular symphysis U-shaped. – *Distribution:* Occurs in northeastern India and southern China south to Malaya, also Java, Borneo, and the Philippines. – Two genera, three species.

Genus *Coelops* BLYTH 1848 (Fig. 89)

Ears relatively small. Uropatagium emarginated. Rostrum relatively less inflated. Dental formula i 1/2, c 1/1, p 2/2, m 3/3 x 2 = 30. – *Distribution:* Same as for tribe. – Two species, five additional subspecies.

1. *C. frithi* BLYTH 1848. – Lappets projecting from the supplementary leaflets flanking the anterior noseleaf narrow and elongate. Narial part of the anterior noseleaf not depressed nor sharply demarcated from the intermediate noseleaf, the intervening ridges low and indefinite. Posterior part of the upper surface of the rostrum sloping anteriorly, rostral inflations low. A definite recess within the maxillary root of the zygoma. Upper toothrows convergent anteriorly. Posterior lower premolar elongate and narrow. Lower molars with prominent horizontal external cingula. Size medium (forearm length, 35–47 mm). – *Distribution:* Range same as for the genus except for Borneo and the Philippines. – Five subspecies are here recognized:

C. f. frithi (northeastern India to Thailand), *C. f. sinicus* (southwestern China), *C. f. inflatus* (southeastern mainland China to Vietnam and Thailand), *C. f. formosanus* (Taiwan), *C. f. bernsteini* (Java, Bali). Malay material has not been allocated to subspecies.

2. *C. robinsoni* BONHOTE 1908. – Lappets projecting from the supplementary leaflets flanking the anterior noseleaf wide and rounded. Narial part of the anterior noseleaf depressed and sharply demarcated from the intermediate noseleaf by prominent ridges. Posterior part of the upper surface of the rostrum nearly horizontal, rostral inflations large and convex. No definite recess within the maxillary root of the zygoma. Upper toothrows nearly parallel. Posterior lower premolar wide. Lower molars with low, upwardly curving cingula. Size relatively small (forearm length, 33–37 mm). – *Distribution:* Known only from the Malay peninsula, Borneo, and the Philippines. – Two subspecies are here recognized:

C. r. robinsoni (Malay peninsula, Borneo), *C. r. hirsuta* (Philippines).

Genus *Paracoelops* DORST 1947

Ears relatively large. Uropatagium not emarginated. Rostrum greatly inflated. Dental formula i 1/2, c 1/1, p 2/2, m 3/3 x 2 = 30. – *Distribution:* Known only from Vietnam. – A single species.

1. *P. megalotis* DORST 1947. – Size medium (forearm length, 42 mm). – *Distribution:* Same as for genus. – No subspecies (known by the type specimen only).

Infraorder **Yangochiroptera** KOOPMAN 1985
Bats with fused premaxillae

Premaxillaries always fused with maxillaries in adults. – *Distribution:* Virtually coextensive with that of the suborder. – Two superfamilies, ten families, 104 genera, 546 species.

Superfamily **Noctilionoidea** GRAY 1821
(New World Leaf-nosed Bats)

Tail, if present, usually shorter than uropatagium. Rhinarium, muzzle, or both, always more or less modified. Never more than two pairs of lower incisors. Always a single phalanx on the second digit of the wing. – *Distribution:* Widely distributed in tropical and subtropical North and South America, including Middle America and the West Indies. – Three families seven additional subfamilies, 51 genera, 149 species.

Family **Noctilionidae** GRAY 1821
(Hare-lipped Bats)

Structure: Rostrum not tilted upward relative to the braincase. Trochiter of humerus small, barely making contact with scapula. Nasal branches of premaxillaries well-developed but palatal branches reduced. No noseleaf, but muzzle pointed with a projecting pad. Full lips forming definite cheek pouches. Orifice of mouth transverse. Nares opening anteriorly and somewhat tubular. Single phalanx of second digit of wing poorly developed. Fibula threadlike and cartilaginous proximally. Foot with greatly developed bony calcar supported by enlarged, distally expanded, flattened calcaneum. Tail well-developed, extending approximately to middle of uropatagium.

Ecology: Either insectivorous or piscivorous, but in general catching prey from the surface of the water either fresh or salt, and therefore never found far from water. Usually roost in holes or crevices, but these may be in caves, trees, or buildings. They are relatively slow flyers, confined to the tropics.

Distribution (Fig. 12): Widely distributed in tropical Middle and South America and the West Indies (except for most of the Bahamas).

Systematics: A single genus and two species.

Genus **Noctilio** LINNAEUS 1766 (Fig. 90)

Dental formula: i 2/1, c 1/1, p 1/2, m 3/3 x 2 = 28. Hind claws more or less enlarged to form gafflike hooks. – *Distribution:* Same as for family. – Two subgenera, two species, and five additional subspecies are here recognized.

Subgenus **Dirias** MILLER 1904

Length of tibia and foot together less than half length of leg. – *Distribution:* Confined to mainland tropical America from southern Mexico to northern Argentina (west of the Andes, not south of Colombia). – A single species and three additional subspecies.

1. *N. albiventris* DESMAREST 1818. – Size relatively small (forearm length, 54–70 mm). – *Distribution:* Same as for subgenus. – Four subspecies:

N. a. minor (southern Mexico to northwestern Venezuela), *N. a. affinis* (Guianas and northern Venezuela through western Amazonia to Bolivia), *N. a. albiventris* (southeastern Venezuela through eastern Amazonia to eastern Brazil), *N. a. cabrerai* (southwestern Brazil through Paraguay to northern Argentina).

Subgenus **Noctilio** LINNAEUS 1766

Length of tibia and foot together more than half length of leg. – *Distribution:* Same as for genus (west of the Andes not south of Ecuador). – A single species and two additional subspecies.

2. *N. leporinus* (LINNAEUS 1758). – Size relatively large (forearm length, 73–92 mm). – *Distribution:* Same as for subgenus. – Three subspecies:

N. l. mastivus (Sinaloa, Veracruz, and extreme southern Bahamas south to western Ecuador and Venezuela), *N. l. leporinus* (Guianas and eastern Ecuador to eastern Brazil), *N. l. rufescens* (Bolivia and southern Brazil to northern Argentina).

Family **Mormoopidae** DE SAUSSURE 1860 (Spectacled Bats)

Structure: Rostrum tilted upward to a greater or lesser degree. Trochiter of humerus well-developed, but not contacting scapula. Palatal branches of premaxillaries well-developed. Noseleaf represented only by a naso-labial plate, lower lip with plate-like outgrowths. Single phalanx of second digit of wing small. Fibula cartilaginous proximally. Foot lacking special modifications. Tail well-developed, extending approximately to middle of uropatagium.

Ecology: Apparently strictly insectivorous and usually roosting in large colonies in caves. Occurrence in tropical and subtropical climates ranging from rain forests to semi-deserts.

Distribution (Fig. 13): Southwestern United States and Baja California south through Middle America to central Brazil (west of the Andes not south of northern Peru) and virtually throughout the West Indies except for the Bahamas (where, however, known fossil).

Systematics: Two genera and eight species.

Genus **Pteronotus** GRAY 1838 (Fig. 91)

Rostral portion of skull only slightly elevated. Dental formula: i 2/2, c 1/1, p 2/3, m 3/3 x 2 = 34, but middle lower premolar greatly reduced and more or less excluded lingually from the toothrow. Ears relatively long and pointed not united by a pronounced band. Tragus relatively simple. – *Distribution:* Same as for family except for absence from Baja California and southwestern United States. – Three subgenera, six species, and 13 additional subspecies are currently recognized.

Subgenus **Phyllodia** GRAY 1843

Inner edges of plagiopatagia attached to sides of body. Rostral breadth equal to or less than maxillary toothrow length. Basioccipital relatively constricted between periotics. Basisphenoid with two narrow deep furrows. – *Distribution:* Ranging from tropical Mexico to northeastern Brazil (not west of the Andes in South America); Greater Antilles. – A single species and eight additional subspecies.

1. *P. parnellii* (GRAY 1843). – Size relatively large (forearm length, 48–66 mm; condylobasal length, 16–22 mm). – *Distribution:* Same as for subgenus. – Nine subspecies are currently recognized:

P. p. parnellii (= *boothi*) (Cuba and Jamaica), *P. p. pusillus* (Hispaniola), *P. p. gonavensis* (Gonave island off east coast of Hispaniola), *P. p. portoricensis* (Puerto Rico), *P. p. mexicanus* (Sonora and Tamaulipas to Oaxaca and Veracruz in Mexico), *P. p. mesoamericanus* (on the western coast from Chiapas to western Panama and on the east coast from Veracruz and Yucatan to Honduras), *P. p. rubiginosus* (Honduras to Panama; also Tobago, Trinidad and southern Venezuela to Surinam, eastern Peru, and northeastern Brazil), *P. p. fuscus* (northeastern Colombia and northern Venezuela except for the Paraguana peninsula), *P. p. paraguensis* (Paraguana peninsula of northern Venezuela).

Subgenus *Chilonycteris* GRAY 1839

Inner edges of plagiopatagia attached to sides of body. Rostral breadth equal to or less than maxillary toothrow length. Basioccipital not constricted between periotics. Basisphenoid with two wide shallow furrows. – *Distribution:* Ranging from northwestern and northeastern tropical Mexico to northeastern Brazil (but not west of the Andes in South America); Greater Antilles. – Three species and three additional subspecies.

2. *P. macleayii* (GRAY 1839). – Naso-labial plate with strongly developed lateral spikes. Plagiopatagium and uropatagium attached to tarsus by a long ligament. Size fairly small (forearm length, 41–46 mm; condylobasal length, 14–16 mm). – *Distribution:* Confined to Cuba and Jamaica. – Two subspecies:

P. m. macleayii (Cuba), *P. m. griseus* (Jamaica).

3. *P. quadridens* (GUNDLACH 1840). Nasolabial plate with fairly well-developed lateral processes. Plagiopatagium and uropatagium attached to tarsus by a long ligament. Size small (forearm length, 35–40 mm; condylobasal length, 12–14 mm). – *Distribution:* Confined to the Greater Antilles. – Two subspecies are currently recognized:

P. q. fuliginosus (= *inflata*) (Jamaica, Hispaniola, Puerto Rico), *P. q. quadridens* (= *torrei*) (Cuba).

4. *P. personatus* (WAGNER 1843). – Nasolabial plate without lateral processes. Plagiopatagium and uropatagium attached to tarsus by a short ligament. Size medium (forearm length, 40–49 mm; condylobasal length, 13–16 mm). – *Distribution:* Includes the entire mainland range of the subgenus. – Two subspecies are recognized:

P. p. psilotis (Sonora and Tamaulipas south to Honduras), *P. p. personatus* (Nicaragua south to Surinam and eastern Peru and from there east to northeastern Brazil).

Subgenus *Pteronotus* GRAY 1838

Inner edges of plagiopatagia attached to middorsal line. Rostral breadth greater than maxillary toothrow length. Basioccipital not constricted between periotics. Basisphenoid with two oval pits. – *Distribution:* Ranging from tropical Mexico to northeastern Brazil (but west of the Andes not south of northern Peru), also Lesser Antilles. – Two species and two additional subspecies.

5. *P. davyi* GRAY 1838. – Size medium (forearm length, 40–50 mm; condylobasal length, 13–16 mm). Plagiopatagium sparsely covered by long irregularly spaced hairs. – *Distribution:* Ranging from tropical Mexico to northern Peru, northeastern Brazil, and north in the Lesser Antilles to Marie Galante. – Three subspecies:

P. d. fulvus (Sonora and Nuevo Leon to Honduras), *P. d. davyi* (Nicaragua to Trinidad and the Lesser Antilles) *P. d. incae* (northewestern Peru). The northeastern Brazilian populations have not been allocated subspecifically.

6. *P. gymnonotus* (WAGNER 1843) (= *suapurensis* J. A. ALLEN 1904). – Size fairly large (forearm length, 49–56 mm; condylobasal length, 15–17 mm). Plagiopatagium covered with many short hairs. – *Distribution:* Ranging from Veracruz in southern Mexico to Guyana and central Brazil (but not west of the Andes in South America). – No subspecies.

Genus *Mormoops* LEACH 1821 (= *Aello* LEACH 1821) (Fig. 92)

Rostral portions of skull markedly elevated. Dental formula: i 2/2, c 1/1, p 2/3, m 3/3 × 2 = 34, with middle lower premolar not greatly reduced and not excluded from the toothrow. Ears very broad and rounded, united by a pronounced band. Tragus relatively complex. – *Distribution:* From Baja California and southwestern United States to Honduras, in northwestern South America from northwestern Peru to eastern Venezuela (including several offshore islands); Greater Antilles. – Two species and four additional subspecies are currently recognized.

1. *M. megalophylla* PETERS 1864. – Rostrum relatively broad. Frontal region of braincase not greatly inflated. Basioccipital and basisphenoid broad with a median septum. Size relatively large (forearm length, 49–61 mm; condylobasal length, 13–16 mm). Infralabial plate only partly divided. – *Distribution:* Occupying the mainland (and

South American offshore island) distribution. – Four currently recognized subspecies:

M. m. megalophylla (Baja California, southern Arizona, and southern Texas to Honduras), *M. m. tumidiceps* (northern Colombia, northern Venezuela, Margarita and Trinidad islands), *R. m. intermedia* (Aruba, Curacao, and Bonaire islands), *M. m. carteri* (coastal Ecuador and northwestern Peru).

2. *M blainvillii* LEACH 1821 (= *cuvieri* LEACH 1821). – Rostrum relatively narrow. Frontal region of braincase greatly inflated. Basioccipital and basisphenoid narrow, without a median septum. Size relatively small (forearm length, 44–49 mm; condylobasal length, 12–14 mm). Infralabial plate completely divided. – *Distribution:* Confined to Cuba, Jamaica, Hispaniola, Puerto Rico, and small nearly islands. – No currently recognized subspecies.

Family **Phyllostomidae** GRAY 1825
(New World Leaf-nosed Bats)

Structure: Rostrum not tilted upward relative to the braincase. Trochiter of humerus well-developed and making contact with scapula. Palatal branches of premaxillaries well-developed. A true noseleaf almost always present, at least as represented by a more or less circular nosepad and usually with an upright portion as well. Single phalanx of second digit of wing small. Fibula variously developed. Foot usually lacking special modifications. Virtually all known combinations of tail and uropatagium relationships, both of which can vary from long to absent.

Ecology: Exhibiting virtually every type of food habit known for bats with the exception of piscivory. Roosting habits also highly varied though sually hanging free and avoiding crevices. Confined to the tropics and sub-tropics, but within this zone, ranging from rain forests to deserts. – *Distribution* (Fig. 14): Ranging from the southwestern United States through Middle and South America to northern Chile and northern Argentina, also throughout the West Indies.

Systematics: Eight subfamilies, one additional tribe, 48 genera, 139 species.

Subfamily **Phyllostominae** GRAY 1825

Upper incisors and canine not bladelike. Noseleaf well developed with a prominent upright portion. Tongue not elongate, lacking conspicuous bristle-like papillae; musculature exhibits a transverse pattern in section, supplied by paired longitudinal arteries. Anterior upper premolar in contact with canine and with posterior upper premolar. Zygomatic arch complete. Upper molars dilambdodont with a distinct W-shaped ectoloph. – *Distribution:* Ranging from the southwestern United States through Middle America to northernmost Argentina, but west of the Andes not south of northernmost Peru; widely distributed in the Greater Antilles, but not known from the Lesser Antilles north of Grenada. – Eleven genera, 33 species.

Genus ***Micronycteris*** GRAY 1866 (Fig. 93)

Two lower incisors and three lower premolars, middle lower premolar subequal in size to anterior and posterior lower premolars. Rostrum shorter than braincase. Tympanic bullae relatively small, the greatest diameter less than the distance between them. Tail does not extend to edge of uropatagium. – *Distribution:* Ranging from tropical Mexico to northwestern Peru and southern Brazil, north to Grenada. – Seven subgenera, 10 species, and three additional subspecies.

Subgenus ***Micronycteris*** GRAY 1866

Third metacarpal shorter than either fourth or fifth, fifth longest. Ears connected by a high notched band. Skull with a high braincase and a high coronoid process. Dental formula: i 2/2, c 1/1, p 2/3, m 3/3 × 2 = 34. Inner upper incisor not chisel-shaped, its front face convex, crown longer than wide. Anterior upper premolar with the main cusp at the middle of the crown, its apex straight. Lower incisor row concave posteriorly, the individual teeth bifid with expanded crowns. – *Distribution:* Same as for genus. – Three species and three additional subspecies.

1. *M. megalotis* (GRAY 1842). – Interauricular band slightly notched medially. Uropatagium relatively broad. Calcar longer than foot. Middle lower premolar not reduced. Size relatively small (forearm length, 31–38 mm; condylobasal length, 15–18 mm). – *Distribution:* Virtually the same as for subgenus. – Four currently recognized subspecies:

M. m. mexicana (Jalisco and Tamaulipas to western Costa Rica and also the Corn islands off eastern Nicaragua), *M. m. microtis* (eastern Nicaragua to at least part of Venezuela), *M. m. homezi* (Northeastern Venezuela), *M. m. megalotis* (remainder of South American range north to Grenada).

2. *M. schmidtorum* SANBORN 1935. – Interauricular band fairly deeply notched medially. Calcar longer than foot. Middle lower premolar not reduced. Size relatively small (forearm length, 33–38 mm; condylobasal length, 16–18 mm). – *Distribution:* Ranging from southwestern Mexico to southern Venezuela. – No subspecies.

3. *M. minuta* (GERVAIS 1856). – Interauricular band deeply notched medially. Uropatagium relatively narrow. Calcar shorter than foot. Middle lower premolar reduced. Anterior upper premolar shorter than posterior. Size relatively small (forearm length, 31–37 mm; condylobasal length, 15–17 mm). – *Distribution:* Ranging from Nicaragua to southern Brazil, but absent from South America west of the Andes. – No subspecies.

Subgenus *Trinycteris* SANBORN 1949

Fourth metacarpal shorter than either third or fifth, third longest. Ears not connected by a band. Skull with a relatively low braincase but a high coronoid process. Dental formula: i 2/2, c 1/1, p 2/3, m 3/3 × 2 = 34. Inner upper incisors not chisel-shaped, outer upper incisors reduced. Anterior upper premolar reduced and low with a small anterior cusp. Lower incisors small, not crowded, and faintly trifid. – *Distribution:* Ranging from Belize to eastern Peru, Trinidad, and northeastern Brazil but not west of the Andes in South America. – A single species.

4. *M. nicefori* SANBORN 1949. – Size fairly small (forearm length, 35–41mm; condylobasal length, 18–20 mm). – *Distribution:* Same as for subgenus. – No subspecies.

Subgenus *Neonycteris* SANBORN 1949

Fourth metacarpal shorter than either third or fifth, fifth longest. Ears not connected by a band. Skull with a relatively low braincase and a low coronoid process. Dental formula i 2/2, c 1/1, p 2/3, m 3/3 x 2 = 34. Inner upper incisors not chisel-shaped. Posterior, but not anterior, upper premolar reduced and low. Lower incisor row concave posteriorly. – *Distribution:* Known only from a small area in southeastern Colombia and northwestern Brazil. – A single species.

5. *M. pusilla* SANBORN 1949. – Size relatively small (forearm length, 33–35 mm; condylobasal length, 15–16 mm). – *Distribution:* Same as for subgenus. – No subspecies.

Subgenus *Xenoctenes* MILLER 1907

Third metacarpal shorter than either fourth or fifth, fifth longest. Ears connected by a fairly low unnotched band. Skull with a fairly low braincase but a high coronoid process. Dental formula i 2/2, c 1/1, p 2/3, m 3/3 × 2 = 34. Inner upper incisors directed inward at the tip, outer upper incisors reduced. Anterior upper premolar somewhat elongate with the cusp near the anterior end. Lower incisors bifid, and wedged between canines. Upper incisors awl-shaped. Not certainly distinct from *M. (Micronycteris)*. – *Distribution:* Ranging from Honduras to Trinidad, Surinam, and eastern Peru but not west of the Andes in South America. – A single species.

6. *M. hirsuta* (PETERS 1869). – Size medium (forearm length, 39–46mm; condylobasal length, 19–21 mm). – *Distribution:* Same as for Subgenus. – No subspecies.

Subgenus *Lampronycteris* SANBORN 1949

Fifth metacarpal shorter than either third or fourth, third longest. Ears not connected by a band. Skull with a relatively low braincase but a high coronoid process. Dental formula i 2/2, c 1/1, p 2/3, m 3/3 × 2 = 34. Inner upper incisors chisel-shaped, but outer upper incisors well-developed. Lower incisors trifid. Anterior upper premolar long, its cusp interior. Posterior upper premolar straight, not recurved. – *Distribution:* Ranging from southern Mexico to Trinidad and Amazonian Brazil, but not west of the Andes in South America. – A single species.

7. *M. brachyotis* (DOBSON 1878) (= *platyceps* SANBORN 1949). Size fairly small (forearm length, 38–43 mm; condylobasal length, 18–20 mm). – *Distribution:* Same as for subgenus. – No subspecies.

Subgenus *Glyphonycteris* THOMAS 1896

Fourth metacarpal shorter than either third or fifth, fifth longest. Ears not connected by a band. Skull with a high braincase but a low coronoid process. Dental formula i 2/2, c 1/1, p 2/3, m 3/3 x 2 = 34. Inner upper incisors chisel-shaped and outer upper incisors reduced. Upper canine unusually short, but simple. Anterior upper premolar larger than posterior, long, its cusp anterior. Posterior upper premolar recurved. – *Distribution:* Ranging from tropical Mexico to Trinidad, eastern Peru, and southern Brazil. – Two species.

8. *M. sylvestris* (THOMAS 1896). – Size medium (forearm length, 37–44 mm; condylobasal length, 17–19 mm). Grooving on upper incisors prominent. – *Distribution:* Same as for subgenus except for southwestern Brazil. – No subspecies.

9. *M. behni* (PETERS 1865). – Size fairly large (forearm length, 45–47 mm). Grooving on upper incisors somewhat less prominent. – *Distribution:* Known only from southeastern Peru and southwestern Brazil. – No subspecies. Probably a senior synonym of *M. sylvestris*.

Subgenus **Barticonycteris** HILL 1964

Fourth metacarpal shorter than either third or fifth, fifth longest. Ears not connected by a band. Skull with a high braincase, but a low coronoid process. Dental formula i 1/2, c 1/1, p 2/3, m 3/3 × 2 = 32. Single upper incisor chisel-shaped. Upper canine unusually short and premolariform. Anterior upper premolar slightly smaller than posterior. Lower incisors scoop-like and lower premolars caniniform. – *Distribution:* Ranging from Costa Rica to eastern Peru and northeastern Brazil but not west of the Andes in South America. – A single species. A very distinct subgenus often separated as a distinct genus.

10. *M. daviesi* (HILL 1964). – Size relatively large (forearm length, 53–58 mm; condlyobasal length, 22–24 mm). – *Distribution:* Same as for subgenus. – No subspecies.

Genus **Macrotus** GRAY 1843 (Fig. 94)

Dental formula i 2/2, c 1/1, p 2/3, m 3/3 x 2 = 34. Middle lower premolar subequal to anterior and posterior lower premolars. Rostrum shorter than braincase. Tympanic bullae relatively large, their greatest diameters greater than the distance between them. Tail extends beyond edge of uropatagium. – *Distribution:* Ranging from the southwestern United States, south to Guatemala and throughout the Greater Antilles including the Bahamas. – Two species and five additional subspecies.

1. *M. californicus* BAIRD 1858. – Postorbital constriction relatively narrow (3.3–3.8 mm). Size fairly large (forearm length, 48–52 mm; condylobasal length, 19–21 mm). – *Distribution:* Confined to the southwestern United States, northwestern and northeastern Mexico. – No subspecies.

2. *M. waterhousii* GRAY 1843. – Postorbital constriction relatively broad (3.9–4.4 mm). Size fairly large (forearm length, 47–55 mm; condylobasal length, 19–22 mm). – *Distribution:* Ranging from western Mexico, including the Tres Marias islands, to Guatemala and throughout the Greatern Antilles (known only fossil from Puerto Rico). – Six subspecies are currently recognized:

M. w. bulleri (western and central Mexico), *M. w. mexicanus* (southern Mexico and Guatemala), *M. w. minor* (Cuba, the Cayman islands, and Navassa, which is between Jamaica and Hispaniola), *M. w. compressus* (central Bahamas), *M. w. jamaicensis* (Jamaica), *M. w. waterhousii* (Hispaniola and the southeastern Bahamas).

Genus **Lonchorhina** TOMES 1863 (Fig. 95)

Dental formula i 2/2, c 1/1, p 2/3, m 3/3 × 2 = 34. Middle lower premolar greatly reduced and may be displaced lingually from the toothrow. Rostrum shorter than braincase. Dorsal profile of rostrum extremely convex, depression present between the orbits. Tail reaches edge of broad uropatagium. Noseleaf greatly elongated. – *Distribution:* Ranging from southern Mexico to southern Brazil but not west of the Andes south of Ecuador. – Four species and one additional subspecies.

1. *L. aurita* TOMES 1863. – Size medium (forearm length, 47–54 mm; condylobasal length, 17–20 mm). In the noseleaf, the tuber of the „sella" is abruptly dilated and the median ridge encloses a sulcus. – *Distribution:* Same as for genus. – Two subspecies (perhaps of dubious validity):

L. a. occidentalis (western Ecuador and perhaps western Colombia and Middle America), *L. a. aurita* (remainder of distribution).

2. *L. orinocensis* LINARES & OJASTI 1971. – Size relatively small (forearm length, 41–45 mm; condlylobasal length, 16–18 mm). In the noseleaf the tuber of the „sella" is not abruptly dilated and the median ridge is without a sulcus. – *Distribution:* Known only from southwestern Venezuela and southeastern Colombia. – No subspecies.

3. *L. fernandezi* OCHOA & IBAÑEZ 1982. – Size relatively small (forearm length, 41–44 mm; condylobasal length, 15–16 mm). In the noseleaf, the

"sella" is filiform without a tuber and the median ridge does not enclose a sulcus. – *Distribution:* Known only from southwestern Venezuela. – No subspecies.

4. *L. marinkellei* HERNÁNDEZ-CAMACHO & CANA-G. 1978. – Size relatively large (forearm length, 59–60 mm; condylobasal length, 23–24 mm). In the noseleaf, the tuber of the "sella" is abruptly dilated and the median ridge encloses a sulcus. – *Distribution:* Known only from southeastern Colombia. – No subspecies.

Genus *Macrophyllum* GRAY 1838 (Fig. 96)

Dental formula i 2/2, c 1/1, p 2/3, m 3/3 x 2 = 34, but middle lower premolar greatly reduced and displaced lingually from the toothrow. Rostrum less than either length or breadth of braincase. Tail reaches edge of broad uropatagium. Tibia and foot unusually elongated. – *Distribution:* Southern Mexico south to eastern Peru and northeastern Argentina, but west of the Andes not known south of Colombia. – A single species.

1. *M. macrophyllum* (SCHINZ 1821). – Size relatively small (forearm length, 35–38 mm; condylobasal length, 13–15 mm). – *Distribution:* Same as for genus. – No subspecies.

Genus *Tonatia* GRAY 1827 (Fig. 97)

Dental formula i 2/1, c 1/1, p 2/3, m 3/3 x 2 = 32. Middle lower premolar reduced but not displaced lingually from the toothrow. Rostrum shorter than braincase. Tail does not extend to edge of uropatagium. – *Distribution:* Ranging from southern Mexico to Trinidad and northern Argentina, but west of the Andes not south of northern Peru, known fossil from Jamaica in the West Indies. – Six species and three additional living subspecies.

1. *T. bidens* (SPIX 1823). – Size relatively large (forearm length, 51–59 mm; condylobasal length, 22–25 mm). Ear pinna relatively short, without a connecting band; tragus relatively short. Postorbital constriction and sagittal crest poorly developed. – *Distribution:* Range at present from southern Mexico to Trinidad, northern Argentina, and southern Brazil, but not west of the Andes in South America; known fossil from Jamaica. – A single living subspecies (*T. b. bidens*).

2. *T. brasiliense* (PETERS 1866) (= *venezuelae* LYON 1902; *minuta* GOODWIN 1942; *nicaraguae* GOODWIN 1942). – Size relatively small (forearm length, 33–40 mm; condylobasal length, 15–18 mm). Ear pinna relatively long. Postorbital constriction and sagittal crest well developed. – *Distribution:* Ranging from southern Mexico to Trinidad, eastern Peru, and northeastern Brazil, but not west of the Andes in South America. – No currently recognized subspecies.

3. *T. schulzi* GENOWAYS & WILLIAMS 1980. – Size fairly small (forearm length, 42–44 mm; condylobasal length, 18–20 mm). Ear pinna relatively long. Postorbital constriction and sagittal crest well developed. Numerous warts on dorsal surfaces of wing and leg bones. Underpartes grayish. – *Distribution:* Guianas and extreme northern Brazil. – No subspecies.

4. *T. carrikeri* (J. A. ALLEN 1910). – Size medium (forearm length, 43–50 mm; condylobasal length, 18–22 mm). Ear pinna relatively long. Postorbital constriction and sagittal crest well developed. No warts on wing and leg bones. Underparts white. – *Distribution:* Confined to tropical South America east of the Andes from Colombia to Surinam and south to Bolivia. – No subspecies.

5. *T. evotis* DAVIS & CARTER 1978. – Size fairly large (forearm length, 47–53 mm). Ear pinna relatively long. Postorbital constriction and sagittal crest well developed. Underparts not white. – *Distribution:* Confined to northern Central America from southern Mexico to Honduras. – No subspecies.

6. *T. silvicola* (D'ORBIGNY 1836) (= *amblyotis* WAGNER 1843). – Size relatively large (forearm length, 49–60 mm; condylobasal length, 22–25 mm). Ear pinna relatively long, with a low connecting band; tragus relatively long, with a low connecting band. Postorbital constriction and sagittal crest well developed. – *Distribution:* Ranging from Honduras to northern Argentina but west of the Andes not south of northern Peru. – Four currently recognized subspecies:

T. s. centralis (Honduras to Costa Rica), *T. s. silvicola* (Panama through western South America, but not west of the Andes, to northeastern Argentina), *T. s. occidentalis* (western Ecuador, northwestern Peru), *T. s. laephotis* (eastern South America from the Guianas to northeastern Argentina).

Genus *Mimon* GRAY 1847 (Fig. 98)

Dental formula i 2/1, c 1/1, p 2/2, m 3/3 x 2 = 30. Rostrum shorter than braincase. Tail does not extend to edge of uroptagium. – *Distribution:* Ranging from southern Mexico to northern Bolivia and southeastern Brazil but west of the Andes not south of Ecuador. – Two subgenera, two species, and five additional subspecies.

Subgenus *Mimon* GRAY 1847

Tympanic bullae not enlarged, the diameter of each less than the distance between periotics. Border of plagiopatagium attached to tarsus. – *Distribution:* Ranging from southern Mexico to Colombia, also the Guianas and southeastern Brazil. – A single species with one additional subspecies.

1. *M. bennettii* (GRAY 1838). – Size relatively large (forearm length, 53–59 mm; condylobasal length, 21–24 mm). No white line on back. – *Distribution:* Same as for subgenus. – Two subspecies:

M. b. cozumelae (southern Mexico to northwestern Colombia), *M. b. bennettii* (Guianas and southeastern Brazil).

Subgenus *Anthorhina* LYDEKKER 1891

Tympanic bullae enlarged, the diameter of each greater than the distance between periotics. Border of plagiopatagium attached to base of outer toe. – *Distribution:* Ranging from southern Mexico to Trinidad, eastern Peru, and northeastern Brazil. – A single species with four additional subspecies.

2. *M. crenulatum* (E. GEOFFROY 1810). – Size relatively small (forearm length, 45–51 mm; condylobasal length, 18–20 mm). White line on back. – *Distribution:* Same as for subgenus. – Five currently recognized subspecies:

M. c. keenani (southern Mexico south to western Ecuador), *M. c. longifolium* (western Venezuela south to Bolivia), *M. c. crenulatum* (Trinidad and eastern Venezuela to southern Amazonian Brazil), *M. c. picatum* (northeastern Brazil), *M. c. koepckeae* (highlands of central Peru).

Genus *Phyllostomus* LACÉPÈDE 1799 (Fig. 99)

Dental formula i 2/2, c 1/1, p 2/2, m 3/3 × 2 = 32. Rostrum shorter than braincase. Tail does not extend to edge of uropatagium. – *Distribution:* Ranging from southern Mexico to Trinidad, northeastern Argentina and eastern Brazil, but west of the Andes not south of northern Peru. – Four species and two additional subspecies.

1. *P. discolor* (WAGNER 1843). –Size fairly small (forearm length, 56–69 mm; condylobasal length, 25–34 mm). Calcar shorter than hind foot. Sagittal crest weak or absent. – *Distribution:* Ranging from southern Mexico to Trinidad, northwestern Argentina and northeastern Brazil, but west of the Andes not south of northern Peru. – Two subspecies:

P. d. verrucosus (Veracruz to northwestern Peru), *P. d. discolor* (South America east of the Andes south to northwestern Argentina, also Margarita and Trinidad islands off the northern coast of Venezuela).

2. *P. latifolius* (THOMAS 1901). – Size relatively small (forearm length, 56–60 mm; condylobasal length, 23–26 mm). Calcar longer than hind foot. Sagittal crest well developed. – *Distribution:* Known only from southeastern Colombia, the Guianas, and Amazonian Brazil. – No subspecies.

3. *P. elongatus* (E. GEOFFROY 1810). – Size fairly large (forearm length, 61–71 mm; condylobasal length, 24–28 mm). Calcar longer than hind foot. Sagittal crest well developed. – *Distribution:* Ranging across tropical mainland South America from Colombia to eastern Brazil, but west of the Andes not south of Ecuador. – No subspecies.

4. *P. hastatus* (PALLAS 1767). – Size very large (forearm length, 80–97 mm; condylobasal length, 31–38 mm). Calcar longer than hind foot. Sagittal crest well developed. – *Distribution:* Ranging from Honduras to eastern Brazil, but west of the Andes not south of northern Peru, also Margarita, Trinidad, and Tobago islands off northern coast of Venezuela. – Two currently recognized subspecies:

P. h. panamensis (Honduras to Peru and east to western Venezuela), *P. h. hastatus* (remainder of range).

Genus *Phylloderma* PETERS 1865 (Fig. 100)

Dental formula i 2/2, c 1/1, p 2/3, m 3/3 × 2 = 34. Middle lower premolar greatly reduced but not displaced lingually from the toothrow. Rostrum shorter than braincase. Dorsal profile of rostrum not convex, no depression between orbits. Tail does not extended to edge of uropatagium. – *Distribution:* Ranging from southern Mexico to Trinidad, Bolivia, and eastern Brazil, but west of the Andes not south of northern Peru. – A single species with two additional subspecies.

1. *P. stenops* PETERS 1865. – Size relatively large (forearm length, 65–83 mm; condylobasal length,

25–31 mm). – *Distribution:* Same as for genus. – Three currently recognized subspecies:

P. s. septentrionalis (Chiapas to Costa Rica), *P. s. stenops* (Panama and South American range except Bolivia), *P. s. boliviensis* (central Bolivia).

Genus *Trachops* GRAY 1847 (Fig. 101)

Dental formula i 2/2, c 1/1, p 2/3, m 3/3 × 2 = 34, but middle lower premolar greatly reduced and displaced lingually from the toothrow. Rostrum shorter than braincase. Tail shortened, not reaching edge of uropatagium. Lips and chin with numerous small excrescences. – *Distribution:* Ranging from southern Mexico to Trinidad and eastern Brazil, but west of the Andes not south of Ecuador. – A single species with two additional subspecies.

1. *T. cirrhosus* (SPIX 1823). – Size fairly small (forearm length, 57–62 mm; condylobasal length, 24–27 mm). – *Distribution:* Same as for genus. – Three subspecies are recognized:

T. c. coffini (southern Mexico to Nicaragua), *T. c. cirrhosus* (Costa Rica to Amazonian Brazil), *T. c. ehrhardti* (eastern Brazil and perhaps Bolivia).

Genus *Chrotopterus* PETERS 1865 (Fig. 102)

Dental formula i 2/1, c 1/1, p 2/3, m 3/3 x 2 = 32, but middle lower premolar greatly reduced and displaced lingually from the toothrow. Rostrum shorter than braincase. Uropatagium broad, but tail greatly reduced. – *Distribution:* Ranging from southern Mexico to northern Argentina, but not west of the Andes in South America. – A single species.

1. *C. auritus* (PETERS 1856). – Size relatively large (forearm length, 78–84 mm; condylobasal length, 30–32 mm). – *Distribution:* Same as for genus. – No subspecies recognized here.

Genus *Vampyrum* RAFINESQUE 1815 (Fig. 103)

Dental formula i 2/2, c 1/1, p 2/3, m 3/3 x 2 = 34. Molars narrowed with considerable distortion of cusp pattern. Rostrum as long as braincase. Uropatagium broad but tail absent. – *Distribution:* Ranging from southern Mexico to Trinidad, Surinam, and southwestern Brazil, but west of the Andes not south of Ecuador. – A single species.

1. *V. spectrum* (LINNAEUS 1758). – Size extremely large (forearm length, 102–111 mm; condylobasal length, 42–45 mm), the largest of Microchiroptera. – *Distribution:* Same as for genus. – No currently recognized subspecies.

Subfamily **Lonchophyllinae** GRIFFITHS 1982

Upper incisors and canine not sharply bladelike. Noseleaf small but well developed with prominent upright portion. Tongue elongate with deep lateral longitudinal grooves, with lines of bristlelike papillae associated with them. The complex tongue masculature exhibits a crisscross pattern in section, and is supplied by paired longitudinal arteries. Anterior upper premolar separated from canine. Zygomatic arch incomplete. Incisors and molars never reduced in number, nor premolars increased. Anterior upper incisor enlarged and spatulate. – *Distribution:* Ranging from Nicaragua to southern Peru and southeastern Brazil. – Three genera, nine species.

Genus *Lionycteris* THOMAS 1913 (Fig. 104)

Rostrum shorter than braincase. Postcanine maxillary teeth not reduced in size and with last premolar in contact with molars. Dental formula i 2/2, c 1/1, p 2/3, m 3/3 × 2 = 34. Anterior upper premolar unreduced and not laterally compressed. First two upper molars with W-shaped ectolophs. Lower premolars with middle cusps not reduced. – *Distribution:* Ranging from eastern Panama to eastern Peru and Amazonian Brazil, but west of the Andes not south of Colombia. – A single species.

1. *L. spurrelli* THOMAS 1913. – Size relatively small (forearm length, 32–38 mm; condylobasal length, 17–19 mm). – *Distribution:* Same as for genus. – No subspecies.

Genus *Lonchophylla* THOMAS 1903 (Fig. 105)

Rostrum roughly equal to braincase in length. Postcanine maxillary teeth not reduced in size and with last premolar in contact with molars. Dental formula i 2/2, c 1/1, p 2/3, m 3/3 × 2 = 34. Anterior upper premolar reduced and laterally compressed. First two upper molars with W-shaped ectolophs. Lower premolars with middle cusps reduced. – *Distribution:* Range same as for subfamily except for absence from southwestern Peru. – Seven species and one additional subspecies.

1. *L. thomasi* J. A. ALLEN 1904. – Size relatively small (forearm length, 31–34 mm; condylobasal length, 18–21 mm; maxillary toothrow length, 6.2–7.0 mm). Lingual cusp of last upper premolar well-developed, the tooth thus T-shaped. Posterior cusp of anterior lower premolar well developed and hooklike. Posterior palatal emargination V-shaped. – *Distribution:* Ranging from eastern Panama to eastern Peru and Amazonian Brazil but west of the Andes not south of Ecuador. – No subspecies.

2. *L. dekeyseri* TADDEI, VIZOTTO & SAZIMA 1983. – Size fairly small (forearm length, 34–38 mm; condylobasal length, 20–22 mm, maxillary toothrow length, 7.5–7.6 mm). Posterior cusp of anterior lower premolar absent. – *Distribution:* Confined to eastern Brazil. – No subspecies.

3. *L. mordax* THOMAS 1903. – Size fairly small (forearm length, 32–37 mm; condylobasal length, 20–23 mm; maxillary toothrow length, 7.4–8.3 mm). Lingual cusp of last upper premolar poorly developed or absent. Posterior cusp of anterior lower premolar absent or poorly developed. Posterior palatal emargination U-shaped. – *Distribution:* Ranging from Costa Rica to Western Ecuador and also in eastern Brazil. – Two subspecies are recognized, often treated as separate species:

L. m. mordax (eastern Brazil), *L. m. concava* (remainder of range).

4. *L. robusta* MILLER 1912. – Size fairly large (forearm length, 39–46 mm; condylobasal length, 24–26 mm; maxillary toothrow length, 9.1–10.0 mm). Rostrum relatively broad (width across upper molars, 6.1–7.0 mm). Lingual cusp of last upper premolar well developed, the tooth thus triangular. Posterior cusp of anterior lower premolar well developed and hooklike. Posterior palatal emargination U-shaped. – *Distribution:* Ranging from Nicaragua to northern Peru and east to western Venezuela. – No subspecies.

5. *L. handleyi* HILL 1980. – Size relatively large (forearm length, 44–48 mm; condylobasal length, 25–29 mm; maxillary toothrow length, 9.9–11.0 mm). Rostrum relatively broad (width across upper molars, 6.1–7.0 mm). Lingual cusp of last upper molar poorly developed. Posterior cusp of anterior lower premolar well-developed and hook-like. Posterior palatal emargination U-shaped. – *Distribution:* Confined to Ecuador and Peru east of the Andes. – No subspecies.

6. *L. bokermanni* SAZIMA, VIZOTTO & TADDEI 1978. – Size fairly large (forearm length, 38–42 mm; condylobasal length, 23–25 mm; maxillary toothrow length, 7.8–8.6 mm). Rostrum relatively narrow (width across upper molars, 5.2–5.7 mm). Lingual cusp of last upper premolar poorly developed. Posterior cusp of anterior lower premolar poorly developed. – *Distribution:* Known only from southeastern Brazil. – No subspecies.

7. *L. hesperia* G. M. ALLEN 1908. – Size medium (forearm length, 36–41 mm; condylobasal length, 24–27 mm; maxillary toothrow length, 8.3–9.00 mm). Rostrum relatively narrow and elongate (width across upper molars, 5.4–5.8 mm). – *Distribution:* Confined to arid regions of southwestern Ecuador and northwestern Peru. – No subspecies.

Genus **Platalina** THOMAS 1928 (Fig. 106)

Rostrum longer than braincase. Postcanine maxillary teeth reduced in size with evident gaps between them. Dental formula i 2/2, c 1/1, p 2/3, m 3/3 x 2 = 34. Both anterior and posterior upper premolars laterally compressed. Upper molars without W-shaped ectolophs, the teeth triangular with a simplified three cusp pattern. Lower premolars with middle cusps reduced. – *Distribution:* Confined to arid regions of western Peru. – A single species.

1. *P. genovensium* THOMAS 1928. – Size relatively large (forearm length, 46–50 mm; condylobasal length, 29–31 mm). – *Distribution:* Same as for genus. – No subspecies.

Subfamily **Brachyphyllinae** GRAY 1866

Upper incisors and canine not bladelike. Noseleaf reduced, upright portion absent. Tongue not elongate, lacking conspicuous bristlelike papillae; its musculature exhibiting a transverse pattern in section, supplied by paired longitudinal arteries. Anterior upper premolar in contact with canine and with posterior upper premolar. Zygomatic arch complete. Upper molars broad and multi-cusped, no W-shaped ectoloph. Tail greatly reduced. – *Distribution:* Confined to the West Indies. – A single genus with two species.

Genus *Brachyphylla* GRAY 1834 (Fig. 107)

Dental formula i 2/2, c 1/1, p 2/2, m 3/3 x 2 = 32. *Distribution:* Same as for subfamily. – Two species and two additional subspecies.

1. *B. nana* MILLER 1902 (= *pumila* MILLER 1918). – Size relatively small (forearm length, 56–60 mm; condylobasal length, 24–26 mm). – *Distribution:* Confined to Cuba, Hispaniola, Cayman islands, and southern Bahamas; also known as a fossil from Jamaica. – No subspecies.

2. *B. cavernarum* GRAY 1934. – Size relatively large (forearm length, 63–69 mm; condylobasal length, 27–29 mm). – *Distribution:* Confined to Puerto Rico, the Virgin islands, and the Lesser Antilles as far south as St. Vincent. – Three subspecies are currently recognized:

B. c. intermedia (Puerto Rico and the Virgin islands except St. Croix), *B. c. cavernarum* (St. Croix and Anguilla south to St. Vincent), *B. c. minor* (Barbados).

Subfamily **Phyllonycterinae** MILLER 1907

Upper incisors and canine not bladelike. Noseleaf reduced, upright portion absent or poorly developed. Tongue fairly elongate, without lateral longitudinal grooves, but with a brush tip of bristlelike papillae. The tongue musculature exhibits a transverse pattern in section (encircling lingual veins) and is supplied by a single longitudinal artery. Anterior upper premolar not separated from canine. Zygomatic arch may or may not be complete. Molar cuspidation greatly reduced, no W-shaped ectoloph. Tail extending beyond narrow uropatagium. – *Distribution:* Confined to the Greater Antilles, including the Bahamas. – Two genera, one additional subgenus, and four living species.

Genus *Erophylla* MILLER 1906 (Fig. 108)

Zygomatic arch complete. Middle and posterior lower molars distinctly cuspidate. Calcar short but distinct. Upright portion of noseleaf not completely absent. Dental formula i 2/2, c 1/1, p 2/2, m 3/3 × 2 = 32. – *Distribution:* Same as for subfamily.

1. *E. sezekorni* (GUNDLACH 1861). – Braincase relatively low, not rising sharply from rostrum. Size relatively small (forearm length, 45–50 mm; condylobasal length, 22–24 mm). – *Distribution:* Confined to the Bahamas, Cuba, Cayman islands, and Jamaica. – Four subspecies:

E. s. planifrons (northern and central Bahamas), *E. s. mariguanensis* (southeastern Bahamas), *E. s. sezekorni* (Cuba, Caymans), *E. s. syops* (Jamaica).

2. *E. bombifrons* (MILLER 1899). – Braincase relatively high, rising sharply from rostrum. Size relatively small (forearm length, 45–50 mm; condylobasal length, 21–23 mm). – *Distribution:* Confined to Hispaniola and Puerto Rico. – Two subspecies:

E. b. santacristobalensis (Hispaniola), *E. b. bombifrons* (Puerto Rico).

Genus *Phyllonycteris* GUNDLACH 1861 (Fig. 109)

Zygomatic arch incomplete. Middle and posterior lower molars without clear cuspidation. Calcar absent. Upright portion of noseleaf completely absent. Dental formula i 2/2, c 1/1, p 2/2, m 3/3 x 2 = 32. – *Distribution:* Living representatives confined to Cuba, Hispaniola, and Jamaica, a fossil species known from Puerto Rico. – Two subgenera, two living species, and one additional subspecies.

Subgenus *Phyllonycteris* GUNDLACH 1861

No deep groove in the basioccipital region. – *Distribution:* Same as for genus except for Jamaica. – A single living species with one additional subspecies.

1. *P. poeyi* GUNDLACH 1861. – Size relatively small (forearm length, 46–50 mm; condylobasal length, 21–24 mm). – *Distribution:* Confined to Cuba and Hispaniola. – Two subspecies:

P. p. poeyi (Cuba), *P. p. obtusa* (Hispaniola).

Subgenus *Reithronycteris* MILLER 1898

A deep groove in the basioccipital region. – *Distribution:* Confined to Jamaica. – A single species.

2. *P. aphylla* (MILLER 1898). – Size relatively small (forearm length, 44–49 mm; condylobasal length, 21–24 mm). – *Distribution:* Same as for subgenus. – No subspecies.

Subfamily **Glossophaginae** BONAPARTE 1845

Upper incisors and canine not bladelike. Noseleaf small but well developed with prominent upright portion. Tongue elongate, without lateral grooves; but with a brush tip of bristlelike papillae. The tongue musculature exhibits a transverse pattern in section (encircling lingual veins) and is supplied by a single longitudinal artery. Anterior upper premolar separated from canine, posterior premolar, or both. Zygomatic arch may or may not be complete. – *Distribution:* Ranging from southwestern United States south to northern Argentina, but west of the Andes not south of Peru, also islands off the northern coast of South America and throughout most of the West indies. – Nine genera, one additional subgenus, 23 species.

Genus ***Glossophaga*** E. GEOFFROY 1818 (Fig. 110)

Dental formula i 2/2, c 1/1, p 2/3, m 3/3 x 2 = 34. Zygomatic arch complete. Upper premolars usually in contact, filling space between canine and anterior molar. Tail not extending to edge of uropatagium. Stylohyoideus muscle retained. Median circumvallate papillae of tongue retained. – *Distribution:* Ranging from tropical Mexico to western Peru and northern Argentina, also islands off the northern coast of South America, southern Lesser Antilles, and Jamaica, a dubious Bahamian record. – Five species with nine additional subspecies.

1. *G. soricina* (PALLAS 1766). – Upper incisors procumbent, anterior larger than posterior. Lower incisors large, in contact, and of equal size. Winglike lateral processes of pterygoids present. Presphenoid ridge well developed. Basisphenoid pits shallow. Anterior border of premaxillae elongate. Rostrum about equal to braincase in length. Mandibular symphyseal ridge prominent. Anterior upper molar with well-developed parastyle. Size fairly small (forearm length, 32–39 mm; condylobasal length, 19–22 mm). – *Distribution:* Same as for genus except for Lesser Antilles and some islands off the northern coast of South America. – Five subspecies are currently recognized:

G. s. mutica (Tres Marias islands off western Mexico), *G. s. handleyi* (Middle American mainland and western Colombia), *G. s. antillarum* (Jamaica), *G. s. soricina* (South American range east of the Andes, including Trinidad and Grenada), *G. s. valens* (drier areas of western Ecuador and Peru).

2. *G. mexicana* WEBSTER & JONES 1980. – Upper incisors procumbent, anterior larger than posterior. Lower incisors reduced and separate, anterior smaller than posterior. Lateral processes of pterygoids absent or poorly developed. Presphenoid ridge poorly developed. Basisphenoid pits moderately deep. Anterior border of premaxillae elongate. Rostrum shorter than braincase. Size fairly small (forearm length, 32–37 mm; condylobasal length, 19–22 mm). – *Distribution:* Confined to southwestern Mexico. – Two subspecies:

G. m. brevirostris (Michoacan to western Oaxaca), *G. m. mexicana* (eastern Oaxaca and western Chiapas). The correct name for this species may be *G. morenoi* MARTINEZ & VILLA 1938.

3. *G. leachii* (GRAY 1844) (= *morenoi* MARTINEZ & VILLA 1938; *alticola* DAVIS 1944). – Upper incisors not procumbent, subequal to one another in size. Lower incisors moderately developed but separate, equal to one another in size. Winglike lateral processes of pterygoids present. Presphenoid ridge well developed. Basisphenoid pits deep. Anterior border of premaxillae rounded. Rostrum shorter than braincase. Size fairly small (forearm length, 34–39 mm; condylobasal length, 17–20 mm). – *Distribution:* Confined largely to the Pacific side of Middle America from southern Mexico to Costa Rica. – No subspecies.

4. *G. commissarisi* GARDNER 1962. – Upper incisors not procumbent, subequal to one another. Lower incisors greatly reduced and separate, anterior smaller than posterior. Lateral processes of pterygoids absent. Presphenoid ridge poorly developed. Basisphenoid pits deep. Anterior border of premaxillae rounded. Rostrum shorter than braincase. Size relatively small (forearm length, 31–37 mm; condylobasal length, 17–20 mm). – *Distribution:* Occurring in western Mexico, Central America and in a small area of South America including eastern Ecuador, northeastern Peru, southeastern Colombia, and northwestern Brazil. – Three subspecies are recognized:

G. c. hespera (Sinaloa to Colima), *G. c. commissarisi* (Veracruz to Panama), *G. c. bakeri* (South American range).

5. *G. longirostris* MILLER 1898. – Upper incisors procumbent, subequal to one another. Lower incisors not reduced but more or less separate. Lateral processes of pterygoids absent. Presphenoid ridge variably developed. Rostrum equal to or longer than braincase. Mandibular symphyseal ridge reduced. Anterior upper molar with poorly-developed parastyle. Size medium (forearm length, 35–42 mm; condylobasal length, 21–23 mm). – *Distribution:* Confined to northern

South America south to northwestern Ecuador and extreme northern Brazil, also several islands off the northern coast and north in the Lesser Antilles to St. Vincent. – Six subspecies:

G. l. longirostris (northwestern Ecuador, northern Colombia, and northwestern Venezuela), *G. l. major* (eastern Colombia to Margarita island and Trinidad), *G. l. campestris* (southern Venezuela and Guyana), *G. l. reclusa* (central Colombia), *G. l. rostrata* (Tobago and Grenada to St. Vincent), *G. l. elongata* (Aruba, Curacao, Bonaire).

Genus *Monophyllus* LEACH 1821 (Fig. 111)

Dental formula i2/2, c1/1, p2/3, m3/3 x 2 = 34. Zygomatic arch more or less complete. Upper premolars separate from one another and from canines. Tail extending well beyond edge of narrow uropatagium. Stylohyoideus muscle retained. Median circumvallate papillae of tongue retained. – *Distribution:* Confined to the West Indies. – Two species, four additional living subspecies.

1. *M. plethodon* MILLER 1900. – Upper premolars separated by a diastema less than half the length of the anterior upper premolar. Size relatively large (forearm length, 38–46 mm; condylobasal length, 19–23 mm). – *Distribution:* Living populations are confined to the Lesser Antilles, but the species is known fossil from Puerto Rico. – Two living subspecies:

M. p. plethodon (Barbados), *M. p. luciae* (remaining Lesser Antilles from St. Vincent to Anguilla).

2. *M. redmani* LEACH 1821. – Upper premolars separated by a diastema more than half length of anterior upper premolar. Size relatively small to medium (forearm length, 35–43 mm; condylobasal length, 17–23 mm). – *Distribution:* Confined to the Greater Antilles, including the southeastern Bahamas. – Three subspecies:

M. r. redmani (Jamaica), *M. r. clinedaphus* (Cuba, Hispaniola, southeastern Bahamas), *M. r. portoricensis* (Puerto Rico).

Genus *Lichonycteris* THOMAS 1895 (Fig. 112)

Dental formula i2/0, c1/1, p2/3, m2/2 x 2 = 26. Zygomatic arch more or less complete. Upper premolars separated from canine and may be from one another. Tail not extending to edge of uropatagium. Stylohyoideus muscle retained. Median circumvallate papillae of tongue retained. Upper molar pattern variable, but W-shaped ectoloph always considerably modified. – *Distribution:* Ranging from Belize to Bolivia but west of the Andes not south of Ecuador. – A single species is currently recognized.

1. *L. obscura* THOMAS 1895 (= *degener* MILLER 1931). – Size relatively small (forearm length, 30–36 mm; condylobasal length, 16–19 mm). – *Distribution:* Same as for genus. – No subspecies.

Genus *Leptonycteris* LYDEKKER 1891 (Fig. 113)

Dental formula i2/2, c1/1, p2/3, m2/2 x 2 = 30. Zygomatic arch complete. Upper premolars separated from one another and from canines. Tail absent and uropatagium reduced. Stylohyoideus muscle absent or poorly developed. Median circumvallate papillae usually retained. – *Distribution:* Ranging from southwestern United States to Salvador, also dryer parts of extreme northern South America and several adjacent islands. – Three species, one additional subspecies.

1. *L. yerbabuenae* MARTINEZ & VILLA-R 1940 (= *sanborni* HOFFMEISTER 1957). – Size medium (forearm length, 50–56 mm; condylobasal length, 25–27 mm). Dentition relatively light. Fur short and dense. Uropatagium nearly naked. Upper incisors with a median gap. – *Distribution:* Range from Arizona and northeastern Mexico to Salvador. – No subspecies.

2. *L. nivalis* (SAUSSURE 1860). – Size relatively large (forearm length, 50–60 mm; condylobasal length, 26–29 mm). Dentition relatively light. Fur long and lax. Uropatagium moderately haired with a conspicuous fringe. Upper incisors with a median gap. – *Distribution:* Ranging from southwestern Texas to Guatemala. – No subspecies.

3. *L. curasoae* MILLER 1900. – Size medium (forearm length, 50–56 mm; condylobasal length, 26–28 mm). Dentition relatively heavy. Fur short and dense. Uropatagium sparsely haired with a slight fringe. Upper incisors without a median gap. – *Distribution:* Confined to arid areas of northern Venezuela and northwestern Colombia and on nearby islands. – Two subspecies are here recognized:

L. c. tarlosti (mainland, Margarita and Aruba islands), *L. c. carasoae* (Curacao and Bonaire islands).

Genus *Anoura* GRAY 1838 (Fig. 114)

Dental formula i2/0, c1/1, p3/3, m3/3 x 2 = 32. Zygomatic arch complete or incomplete. Upper premolars separated from one another and from canines. Tail greatly reduced or absent. Stylohyoideus muscle absent. Median circumvallate papillae absent. – *Distribution:* Ranging from tropical Mexico south to western Peru, northwestern Argentina, and southeastern Brazil but absent from most of the Amazon-basin, also Trinidad and Grenada north of South America. – Four species, two additional subspecies.

1. *A. caudifera* (GEOFFROY 1818). – Size relatively small (forearm length, 34–39 mm; condylobasal length, 21–24 mm). Tail very short but present. Anterior lower premolar subequal to middle and posterior lower premolars. Upper canine not enlarged, its anterior face flat. Posterior upper premolar without a median internal cusp. Anterior upper molar without an anteroexternal cusp or crest. – *Distribution:* Confined to tropical South America from Colombia to the mouth of the Amazon and to northwestern Argentina and southeastern Brazil, but absent from coastal Ecuador and Peru. – No subspecies are here recognized.

2. *A. cultrata* HANDLEY 1960 (= *brevirostrum* CARTER 1968; *werckleae* STARRETT 1969). – Size fairly large (forearm length, 38–44 mm; condylobasal length, 22–26 mm). Tail vestigal but present. Anterior lower premolar greatly enlarged (in relation to the middle and posterior lower premolars) and bladelike. – *Distribution:* Ranging from Costa Rica to northern Venezuela and Bolivia but not west of the Andes in South America. – No subspecies are here recognized.

3. *A. geoffroyi* GRAY 1838. – Size relatively large (forearm length, 39–47 mm; condylobasal length, 24–26 mm). Tail absent. Anterior lower premolar subequal to middle and posterior premolars. Upper canine not enlarged, its anterior face flat. Posterior upper premolar with a median internal cusp that protrudes from the narrow base of the tooth, remaining premolars narrow. Anterior upper molar with an anteroexternal cusp and crest. – *Distribution:* Virtually same as for genus, but records from Argentina apparently represent *A. caudifera*. – Three subspecies are recognized:

A. g. lasiopyga (tropical Mexico to western Colombia),
A. g. peruana (central Colombia to central Bolivia),
A. g. geoffroyi (Venezuela, the Guianas, Trinidad, and Grenada, also, apparently as a separate area, eastern Bolivia to eastern Brazil).

4. *A. latidens* HANDLEY 1984. – Size relatively large (forearm length, 40–46 mm; condylobasal length, 23–24 mm). Tail absent. Anterior lower premolar subequal to middle and posterior premolars. Upper canine not enlarged, its anterior face flat. Posterior upper premolar with a median internal cusp enclosed in the broad triangular base of the tooth, remaining premolars thick. Anterior upper molar with an anteroexternal cusp and crest. – *Distribution:* Ranging from northern Venezuela to eastern Peru. – No subspecies.

Genus **Hylonycteris** THOMAS 1903 (Fig. 115)

Dental formula i 2/0, c 1/1, p 2/3, m 3/3 x 2 = 30. Zygomatic arch incomplete. Upper premolars separated from one another and from canines. Pterygoids not expanded at base or inflated, pterygoid wings short and not in contact with auditory bullae. Mesostyles on upper molars absent, lower molars long and narrow. Tail present, but not extending beyond margin of uropatagium. Stylohyoideus muscle absent. Median circumvallate papillae absent. – *Distribution:* Confined to tropical Middle America, ranging from western Mexico to western Panama. – A single species with one additional subspecies.

1. *H. underwoodi* THOMAS 1903. – Size relatively small (forearm length, 31–37 mm; condylobasal length, 19–22 mm). – *Distribution:* Same as for genus. – Two subspecies:

H. u. minor (western Mexico from Jalisco to Oaxaca),
H. u. underwoodi (Veracruz to western Panama).

Genus **Scleronycteris** THOMAS 1912 (Fig. 116)

Dental formula i 2/0, c 1/1, p 2/3, m 3/3 x 2 = 30. Zygomatic arch incomplete. Upper premolars separated from one another and from canines. Pterygoids not expanded at base or inflated, pterygoid wings short and not in contact with auditory bullae. Mesostyles on upper molars present, lower molars only moderately compressed. Tail present but not extending beyond margin of uropatagium. – *Distribution:* Known only from southern Venezuela and northwestern Brazil. – A single poorly known species.

1. *S. ega* THOMAS 1912. – Size relatively small (forearm length, 34–35 mm; condylobasal length, 21–22 mm). – *Distribution:* Same as for genus. – No subspecies.

Genus **Choeroniscus** THOMAS 1928 (Fig. 117)

Dental formula i 2/0, c 1/1, p 2/3, m 3/3 x = 30. Zygomatic arch incomplete. A distinct gap between anterior and posterior incisors. Upper premolars separated from one another and from canine, hardly exceeding molars in height. Pterygoids expanded at base and inflated, pterygoid wings long and more or less in contact with auditory bullae. Tail present but not extending beyond margin of uropatagium. Stylohyoideus muscle absent. Median circumvallate papillae absent. – *Distribution:* Ranging from western Mexico to Trinidad and Bolivia, but not south of northern Peru west of the Andes. – Four species are here recognized.

1. *C. godmani* (THOMAS 1903). – Size relatively small (forearm length, 32–36 mm; condylobasal length, 18–21 mm). Postero-lateral margin of palate notched. Cranium relatively high. Rostrum relatively short and slender. Pterygoid inflation moderate. – *Distribution:* Ranging from western Mexico to extreme northern South America, east to Suriname. – No subspecies.

2. *C. intermedius* (J. A. ALLEN & CHAPMAN 1893). – Size fairly small (forearm length, 34–36 mm; condylobasal length, 20–23 mm). Postero-lateral margin of palate unnotched. Cranium relatively low. Rostrum fairly short and slender. Pterygoid inflation moderate. – *Distribution:* Known only from Trinidad, Guyana, Suriname, northern Brazil, and eastern Peru. – No subspecies.

3. *C. minor* (PETERS 1869) (= *inca* THOMAS 1912). – Size fairly large (forearm length, 33–38 mm; condylobasal length, 21–25 mm). Postero-lateral margin of palate unnotched. Cranium relatively low. Rostrum of medium length and slender. Pterygoid inflation moderate. – *Distribution:* Confined to tropical South America from coastal Ecuador to the mouth of the Amazon, north to eastern Venezuela and south to northwestern Bolivia. – No subspecies.

4. *C. periosus* HANDLEY 1966. – Size relatively large (forearm length, 40–42 mm; condylobasal length, 29–30 mm). Postero-lateral margin of palate unnotched. Cranium relatively low. Rostrum relatively long and robust. Pterygoid inflation pronounced. – *Distribution:* Known only from western Colombia and northwestern Venezuela. Two subspecies are here recognized:

C. p. periosus (western Colombia), *C. p. ponsi* (northwestern Venezuela). *C. p. ponsi* was originally described as a subspecies of *Choeronycteris* but is best placed here.

Genus ***Choeronycteris*** TSCHUDI 1844 (Fig. 118)

Dental formula i2/0, c1/1, p2/3, m3/3 x 2 = 30. Zygomatic arch incomplete. No gap between anterior and posterior incisors. Upper premolars separated from one another and from canine, distinctly higher than molars. Pterygoids expanded at base and inflated. Pterygoid wings long and in contact with auditory bullae. Tail present but not extending beyond margin of uropatagium. Stylohyoideus muscle absent. Median circumvallate papillae absent. – *Distribution:* Ranging from southwestern United States to Honduras, including the Tres Marias islands. – Two subgenera and two species.

Subgenus ***Choeronycteris*** TSCHUDI 1844

Rostrum subequal to braincase in length. Last upper molar somewhat reduced in size and lacking a distinct metastyle. – *Distribution:* Same as for genus. – A single species.

1. *C. mexicana* TSCHUDI 1844. – Size relatively large (forearm length, 42–47 mm; condylobasal length, 28–30 mm). – *Distribution:* Same as for subgenus. – No subspecies here recognized.

Subgenus ***Musonycteris*** SCHALDACH & MCLAUGHLIN 1960

Rostrum much longer than braincase. Last upper molar not reduced in size, mesostyle distinct. – *Distribution:* Confined to southwestern Mexico. – A single species.

2. *C. harrisoni* (SCHALDACH & MCLAUGHLIN 1960). – Size fairly large (forearm length, 40–43 mm; condylobasal length, 30–34 mm). – *Distribution:* Same as for subgenus. – No subspecies.

Subfamiliy **Carolliinae** MILLER 1924

Upper incisors and canine not bladelike. Noseleaf well-developed with a prominent upright portion. Tongue not elongate, lacking conspicuous bristle-like papillae; its musculature exhibiting a transverse pattern in section, supplied by paired longitudinal arteries. Dental formula i2/2, c1/1, p2/2, m3/3 x 2 = 32. Anterior upper premolar in contact with canine and with posterior premolar. Zygomatic arch incomplete. Upper molars more or less simplified, without a W-shaped ectoloph.

– *Distribution:* Ranging from tropical Mexico through Middle America to southeastern Brazil, but west of the Andes, not south of northernmost Peru; also Margarita, Trinidad, Tobago, and Grenada islands off the northern coast of South America. West Indian records from Jamaica and Redonda are dubious. – Two genera, seven species.

Genus **Carollia** GRAY 1838 (Fig. 119)

Tail present. Upper premolars approximately equal in size. Upper molars still showing a tritubercular pattern. Lower molars clearly distinct from lower premolars. – *Distribution:* Same as for subfamily. – Four species, two additional subspecies.

1. *C. perspicillata* (LINNAEUS 1758). – Size relatively large (forearm length, 39–47 mm; condylobasal length, 19–23 mm, maxillary toothrow length, 7.2–8.0 mm). Labial margin of maxillary toothrow without a distinct notch and relatively straight. Crown of anterior lower molar not extremely low. Outer lower incisors more or less reduced, tending to be concealed dorsally by cingula of canines. Mandible more or less V-shaped. Pelage usually short and sparse. – *Distribution:* Same as for genus but not known north of southern Mexico. – Three subspecies are currently recognized:

C. p. azteca (Middle America and probably northwestern South America), *C. p. tricolor* (Paraguay and probably southeastern Brazil), *C. p. perspicillata* (remainder of range), but subspecies boundaries are not clear.

2. *C. brevicauda* (SCHINZ 1821). – Size medium (forearm length, 37–43 mm; condylobasal length, 18–22 mm; maxillary toothrow length, 6.7–7.7 mm). Labial margin of maxillary toothrow without a distinct notch but slightly concave. Crown of anterior lower molar not extremely low. Outer lower incisors not reduced, not concealed dorsally by cingula of canines. Mandible somewhat V-shaped. Pelage long and thick. – *Distribution:* Ranging from northeastern Mexico through Middle America (but absent from the Pacific slope north of Nicaragua) and tropical South America to Trinidad, Bolivia and eastern Brazil (but west of the Andes not south of Ecuador). – No subspecies.

3. *C. subrufa* (HAHN 1905). – Size medium (forearm length, 37–43 mm; condylobasal length, 18–21 mm; maxillary toothrow length, 6.5–7.0 mm). Labial margin of maxillary toothrow without a distinct notch but definetely concave. Crown of anterior lower molar not extremely low. Outer lower incisors not reduced, not concealed dorsally by cingula of canines. Mandible somewhat V-shaped. Pelage short and sparse. – *Distribution:* Ranging from southwestern Mexico south (chiefly along the Pacific slope) to Nicaragua; an isolated record from Guyana. – No subspecies.

4. *C. castanea* H. ALLEN 1890. – Size relatively small (forearm length, 33–39 mm; condylobasal length, 16–19 mm). Labial margin of maxillary toothrow with a distinct notch. Crown of anterior lower molar extremely low. Outer lower incisors somewhat reduced but not concealed dorsally by cingula of canines. Mandible somewhat U-shaped. – *Distribution:* Ranging from Honduras through Central and tropical South America to French Guiana and Bolivia but west of the Andes not south of Ecuador. – No subspecies.

Genus **Rhinophylla** PETERS 1865 (Fig. 120).

Tail absent. Anterior upper premolar much smaller than posterior. Upper molars greatly reduced medially. Lower molars so reduced as to be similar to lower premolars. – *Distribution:* Confined to tropical South America from Colombia to Bolivia and eastern Brazil but west of the Andes not south of Ecuador. – Three species.

1. *R. pumilio* PETERS 1865. – First upper incisor relatively broad and with three or four well-defined lobes; no gap between upper incisors and canine. Margin of uropatagium virtually naked. Size medium (forearm length, 32–35 mm; condylobasal length, 16–18 mm). – *Distribution:* Virtually same as for genus but absent from western and northern Colombia and northern Venezuela. – No subspecies.

2. *R. alethina* HANDLEY 1966. – First upper incisor with less than three well-defined lobes, leaving a gap between upper incisors and canine. Margin of uropatagium with a conspicuous fringe of hair. Size relatively large (forearm length, 33–38 mm; condylobasal length, 16–19 mm). – *Distribution:* Confined to the Pacific slope of western Colombia and Ecuador. – No subspecies.

3. *R. fischerae* CARTER 1966. – First upper incisor with less than three well defined lobes, leaving a gap between upper incisors and canine. Margin of uropatagium with a conspicuous fringe of hair. Size relatively small (forearm length, 29–31 mm;

condylobasal length, 14–15 mm). – *Distribution:* Confined to eastern Ecuador and Peru through Amazonian Brazil. – No subspecies.

Subfamily **Stenodermatinae** GERVAIS 1855

Upper incisors and canine not bladelike. Noseleaf almost always well-developed with a prominent upright portion. Tongue not elongate, lacking conspicuous bristle-like papillae, its musculature exhibiting a transverse pattern in section, supplied by paired longitudinal arteries. Anterior upper premolar in contact with canine and almost always with posterior upper premolar. Zygomatic arch almost always complete. Upper molar cusp pattern greatly modified, no W-shaped ectoloph. – *Distribution:* Ranging from tropical Mexico through Central America and tropical South America to Peru and Uruguay, also throughout the West Indies. – Two tribes, 17 genera, 59 species.

Tribe Sturnirini MILLER 1907

Crowns of molars with distinct longitudinal grooves, the cusps strictly lateral. Uropatagium virtually absent. – *Distribution:* Ranging from tropical Mexico to Uruguay and northern Argentina, but west of the Andes not south of northern Peru; north in the Lesser Antilles to Guadeloupe, a dubious record from Jamaica. – One genus, two subgenera, 12 species.

Genus ***Sturnira*** GRAY 1842 (Fig. 121)

Dental formula usually i 2/2, c 1/1, p 2/2, m 3/3 x 2 = 32, but in one species i 2/1, c 1/1, p 2/2, m 3/3 x 2 = 30. – *Distribution:* Same as for tribe. – Two subgenera, 12 species, 8 additional subspecies.

Subgenus ***Sturnira*** GRAY 1842

Outer lower incisors well-developed and functional. Zygomatic arch always complete and well-developed. – *Distribution:* Same as for genus. – Ten species, eight additional subspecies.

1. *S. lilium* (GEOFFROY 1810). – Lingual cusps of anterior and middle lower molars (metaconid and entoconid) well defined by vertical notches. Size fairly small (forearm length, 37–44 mm; condylobasal length, 19–21 mm). Zygomatic arch and maxillary toothrow curved laterally. – *Distribution:* Virtually coextensive with that of genus. – Six subspecies are currently recognized:

S. l. parvidens (northern tropical Mexico south to northern Venezuela and probably northwestern Peru), *S. l. lilium* (South America east of the Andes, including Trinidad and Tobago), four subspecies in the Lesser Antilles from St. Vincent to Dominica.

2. *S. luisi* DAVIS 1980. – Lingual cusps of anterior and middle lower molars (metaconid and entoconid) well defined by vertical notches. Size fairly small (forearm length, 41–45 mm). Zygomatic arch relatively straight, maxillary toothrows parallel. – *Distribution:* Known from Costa Rica and Panama and west of the Andes in Colombia, Ecuador, and northern Peru. – No subspecies.

3. *S. thomasi* DE LA TORRE & SCHWARTZ 1966. – Lingual cusps of anterior and middle lower molars (metaconid and entoconid) well defined by vertical notches. Size medium (forearm length, 45–49 mm; condylobasal length, 22–25 mm). Inner upper incisors bilobed, but pointed, the outer lobes reduced. Zygomatic breadth of skull narrow (11.9–12.7 mm). – *Distribution:* Confined to the island of Guadeloupe in the Lesser Antilles. – No subspecies.

4. *S. tildae* DE LA TORRE 1959. – Lingual cusps of anterior and middle lower molars (metaconid and entoconid) fairly well defined by shallow vertical notches. Size medium (forearm length, 43–48 mm; condylobasal length, 20–23 mm). Inner upper incisors bilobed, the lobes of equal size with a broad cutting edge. Zygomatic breadth of skull wide (13.7–14.7 mm). – *Distribution:* Widely in South America east of the Andes (including Trinidad) from Colombia to Bolivia and southeastern Brazil. – No subspecies.

5. *S. aratathomasi* PETERSON & TAMSITT 1968. – Lingual cusps of anterior and middle lower molars (metaconid and entoconid) well defined by vertical notches. Size relatively large (forearm length, 56–61 mm; condylobasal length, 25–27 mm). – *Distribution:* Restricted to the northern Andean region from western Venezuela to northern Ecuador. – No subspecies.

6. *S. erythromos* (TSCHUDI 1844). – Lingual cusps of anterior and middle molars (metaconid and entoconid) poorly defined, no vertical notches. Size relatively small (forearm length, 38–46 mm; condylobasal length, 18–21 mm). Inner upper incisors pointed, their tips not in contact. – *Distribution:* Ranging through the Andean region from northern Venezuela to northwestern Argentina. – No subspecies.

7. *S. oporaphilum* (TSCHUDI 1844). – Lingual cusps of anterior and middle molars (metaconid and entoconid) poorly defined, no vertical notches. Size medium (forearm length, 44–48 mm; condylobasal length, 21–22 mm). Inner upper incisors pointed, their tips not in contact. – *Distribution:* Ranging through the Andean region from western Venezuela to western Bolivia. – Two subspecies are here recognized:

S. o. bogotensis (Venezuela, Colombia), *S. o. oporaphilum* (Peru, Bolivia).

8. *S. ludovici* ANTHONY 1924. – Lingual cusps of anterior and middle molars (metaconid and entoconid) poorly defined, no vertical notches. Size medium (forearm length, 40–47 mm; condylobasal length, 20–22 mm). Inner upper incisors pointed, their tips not in contact. – *Distribution:* Ranging from tropical Mexico through Central America to Guyana and Ecuador, chiefly in the highlands. – Three subspecies are recognized here:

S. l. occidentalis (Sinaloa to Jalisco in western Mexico), *S. l. hondurensis* (tropical Mexico from Colima and Tamaulipas through Central America to Panama), *S. l. ludovici* (Ecuador, western Colombia, northern Venezuela, Guyana).

9. *S. mordax* (GOODWIN 1938). – Lingual cusps of anterior and middle molars (metaconid and entoconid) poorly defined, no vertical notches. Size fairly large (forearm length, 44–49 mm; condylobasal length, 22–24 mm). Skull narrow. Inner upper incisors spatulate, blunt or bifid, and more or less in contact near their broad cutting edges. – *Distribution:* Known only from Costa Rica. – No subspecies.

10. *S. magna* DE LA TORRE 1966. – Lingual cusps of anterior and middle premolars (metaconid and entoconid) poorly defined, no vertical notches. Size relatively large (forearm length, 55–60 mm; condylobasal length, 24–26 mm). – *Distribution:* Largely restricted to the Andean region from southern Colombia to central Bolivia. – No subspecies.

Subgenus ***Corvira*** THOMAS 1915

Outer lower incisors vestigial or absent. Zygomatic arch weak or incomplete. – *Distribution:* Ranging in the Andean highlands from western Venezuela to southern Peru; a doubtful record from northern Brazil. – Two species.

11. *S. nana* GARDNER & O'NEILL 1971. – Size very small (forearm length, 34–36 mm; condylobasal length, 16–18 mm). Vestigial outer lower incisors frequently present. – *Distribution:* Known only from a small area in the mountains of southern Peru. – No subspecies.

12. *S. bidens* (THOMAS 1915). – Size fairly small (forearm length, 39–44 mm; condylobasal length, 16–18 mm). Outer lower incisors absent. – *Distribution:* Same as for subgenus. – No subspecies.

Tribe Stenodermatini GERVAIS 1855

Crowns of molars with well developed cusps rising from a flattened crushing surface. Uropatagium narrow but definitely present. – *Distribution:* Virtually same as for subfamily. – Sixteen genera, three additional subgenera, 47 species.

Genus ***Uroderma*** PETERS 1866 (Fig. 122)

Dental formula i2/2, c1/1, p2/2, m3/3x2 = 32. Rostrum approximately three quarters as long as braincase, not inflated, and without a deep depression or long nasal emargination. Interpterygoid space not extended by a deep palatal emargination. Inner upper incisors not elongate, less than twice length of outer upper incisors and deeply bifid. Forehead not markedly concave. – *Distribution:* Ranging from southern Mexico through Central America and tropical South America to Trinidad and southeastern Brazil but west of the Andes not south of northern Peru. – Two species, five additional subspecies.

1. *U. bilobatum* PETERS 1866. – Ear pinna with a yellowish edging. Facial stripes well developed. Rostrum relatively shallow. Mesethmoid strap-shaped in frontal view and rodlike in cross section. Forearm length, 38–46 mm; condylobasal length, 19–23 mm. – *Distribution:* Same as for genus. – Six subspecies are currently recognized:

U. b. molaris (Atlantic side of Middle America from Veracruz to western Panama), *U. b. davisi* (Pacific side of Middle America from Oaxaca to Honduras), *U. b. convexum* (western Nicaragua to western Venezuela), *U. b. bilobatum* (most of tropical South America east of the Andes from Venezuela to southeastern Brazil), *U. b. trinitatum* (Trinidad), *U. b. thomasi* (Andean regions from Ecuador to central Bolivia).

2. *U. magnirostrum* DAVIS 1968. – Ear pinna without yellowish edging. Facial stripes poorly developed or absent. Rostrum relatively deep. Mesethmoid shield-like in frontal view and cross-shaped in cross section. Forearm length, 36–47 mm; condylobasal length, 20–22 mm. –

Distribution: Ranging along the Pacific side of Middle America from Guerrero to Panama and in tropical South America east of the Andes to northern Bolivia and northwestern Brazil. – No subspecies.

Genus ***Vampyrops*** PETERS 1865 (= *Platyrrhinus* SAUSSURE 1860) (Fig. 123)

Dental formula i2/2, c1/1, p2/2, m3/3 x 2 = 32. Rostrum approximately three quarters as long as braincase, not inflated, and without a deep depression or long nasal emargination. Interpterygoid space not extended by a deep palatal emargination. Inner upper incisors elongate, more than twice length of outer upper incisors, not deeply bifid. Anterior lower molar with a postero-internal cusp. – *Distribution:* Ranging from southern Mexico south to Uruguay, including Trinidad, but west of the Andes not south of northern Peru. – Nine species, four additional subspecies.

1. *V. vittatus* (PETERS 1860). – Size relatively large (forearm length, 57–65 mm; condylobasal length, 28–32 mm). Dorsal pelage blackish brown with prominent dorsal and facial stripes. Edge of uropatagium conspicuously fringed with hair. – *Distribution:* Ranging from Costa Rica to western Venezuela and south along the Andes to central Bolivia. – No subspecies.

2. *V. infuscus* PETERS 1881. – Size relatively large (forearm length, 53–60 mm; condylobasal length, 26–29 mm). Dorsal pelage brown with poorly developed dorsal and facial stripes. Edge of uropatagium with short hair. – *Distribution:* Ranging from northern Colombia to central Bolivia, including extreme western Brazil. – No subspecies.

3. *V. aurarius* HANDLEY & FERRIS 1972. – Size fairly large (forearm length, 49–54 mm; condylobasal length, 25–27 mm). Dorsal pelage blackish brown with prominent dorsal and facial stripes. Edge of uropatagium conspicuously fringed with hair. Skull broad. Inner upper incisors relatively long and narrow. – *Distribution:* Known only from mountainous areas in southern Colombia, southern Venezuela, and Surinam. – No subspecies.

4. *V. dorsalis* THOMAS 1900. – Size medium (forearm length, 45–50 mm; condylobasal length, 24–27 mm). Dorsal pelage blackish brown with prominent dorsal and facial stripes. Edge of uropatagium with a moderate fringe of hair. Inner upper incisors large and strongly convergent. Rostrum relatively broad (width across upper molars, 10.2–12.3 mm). – *Distribution:* Ranging from eastern Panama to eastern Bolivia, chiefly in Andean regions. – No subspecies.

5. *V. umbratus* LYON 1902. – Size medium (forearm length, 42–49 mm; condylobasal length, 23–26 mm). Dorsal pelage blackish brown with prominent dorsal and facial stripes. Edge of uropatagium conspicuously fringed with hair. Inner upper incisors relatively straight. – *Distribution:* Known only from eastern Panama, western and northern Colombia, and northern Venezuela. – Three subspecies are recognized here:

V. u. aquilus (Panama), *V. u. umbratus* (Colombia), *V. u. oratus* (Venezuela).

6. *V. lineatus* (E. GEOFFROY 1810). – Size medium (forearm length, 38–49 mm; condylobasal length, 21–24 mm). Dorsal pelage dark brown with prominent dorsal stripe but variably developed facial stripes. Edge of uropatagium fringed with hair. Inner upper incisors robust. – *Distribution:* Ranging from northern Colombia south in the Andean region and extending across southern and eastern tropical South America to Uruguay and Suriname. – Two subspecies are here recognized:

V. l. nigellus (Colombia to western Bolivia), *V. l. lineatus* (eastern Bolivia to Uruguay and Suriname).

7. *V. recifinus* THOMAS 1901. – Size medium (forearm length, 40–43 mm; condylobasal length, 20–22 mm). Dorsal pelage brown with a prominent dorsal stripe. Inner upper incisors medium. Postpalatal extension relatively long. Auditory bullae relatively large. – *Distribution:* This poorly known species is apparently confined to eastern Brazil. – No subspecies.

8. *V. brachycephalus* ROCK & CARTER 1972 (= *latus* HANDLEY & FERRIS 1972). – Size relatively small (forearm length, 36–42 mm; condylobasal length, 18–21 mm). Dorsal pelage brown with a fairly prominent dorsal stripe, but variably developed facial stripes. Edge of uropatagium fringed with short sparse hair. Rostrum and zygoma relatively broad and forehead relatively concave. Post-palatal extension relatively short. Posterior lower premolar with well-developed accessory cusps. – *Distribution:* Occurring in northwestern tropical South America (but chiefly east of the Andes) from Ecuador and Peru to Suriname. – No currently recognized subspecies.

9. _V. helleri_ PETERS 1866. – Size relatively small (forearm length, 35–41 mm; condylobasal length, 18–21 mm). Dorsal pelage light to medium brown with prominent dorsal and facial stripes. Edge of uropatagium densely fringed with hair. Rostrum and zygoma relatively slender and forehead relatively flat. Post-palatal extension relatively long. Posterior lower premolar with accessory cusps usually poorly developed. – _Distribution:_ Ranging from southern Mexico through Central America; widely distributed in tropical South America south to Bolivia including Trinidad but west of the Andes not south of northern Peru. The Paraguay record was based on a misidentified specimen of _Pygoderma_. – Two subspecies are here recognized:

V. h. helleri (= _zarhinus_) (Mexico to northwestern Peru and Trinidad), _V. h. incarum_ (east of the Andes from southern Colombia, southern Venezuela, and the Guianas to Bolivia and Amazonian Brazil).

Genus _**Vampyrodes**_ THOMAS 1900 (Fig. 124)

Dental formula i 2/2, c 1/1, p 2/2, m 2/3 x 2 = 30. Rostrum approximately two thirds as long as braincase, not inflated, and without a deep depression or long nasal emargination. Interpterygoid space not extended by a deep palatal emargination. Inner upper incisors elongate, more than twice length of outer upper incisors, not deeply bifid. Posterior lower premolar tricuspidate. Anterior lower molar with a postero-internal cusp. – _Distribution:_ Ranging from southern Mexico to northern Bolivia and Amazonian Brazil, including Trinidad and Tobago, but west of the Andes, not south of Colombia. – A single species, one additional subspecies.

1. _V. caraccioli_ (THOMAS 1889). – Size fairly large (forearm length, 46–54 mm; condylobasal length, 22–25 mm). – _Distribution:_ Same as for genus. – Two subspecies are recognized:

V. c. caraccioli (Venezuela, Trinidad, and Tobago), _V. c. major_ (= _ornatus_) (remainder of range).

Genus _**Vampyressa**_ THOMAS 1900 (Fig. 125)

Dental formula i 2/1–2, c 1/1, p 2/2, m 2/2–3 x 2 = 26–30. Rostrum approximately two thirds as long as braincase, not inflated, and without a deep depression or long nasal emargination. Interpterygoid space not extended by a deep palatal emargination. Inner upper incisors elongate, usually more than twice length of outer upper incisors, not deeply bifid. Posterior lower premolar simple. Anterior lower molar without a posterointernal cusp. Lingual cusps of middle lower molar very large. – _Distribution:_ Ranging from southern Mexico through Central America and much of tropical South America to Bolivia and southeastern Brazil, but west of the Andes not south of Ecuador. – Three subgenera, five species, one additional subspecies.

Subgenus _**Vampyressa**_ THOMAS 1900.

Two pairs of lower incisors. Anterior lower premolar at least as high as long. No dorsal stripe. – _Distribution:_ Ranging from southern Mexico through Central America and in South America to Guyana and southeastern Brazil, but west of the Andes not south of Ecuador. – Two species, one additional subspecies.

1. _V. melissa_ THOMAS 1926. – Last lower molar present. Size relatively large (forearm length, 36–40 mm; condylobasal length, 19–22 mm). – _Distribution:_ Known only from the Andean highlands of Peru and southern Colombia. – No subspecies.

2. _V. pusilla_ (WAGNER 1843). – Last lower molar absent. Size relatively small (forearm length, 29–32 mm; condylobasal length, 15–17 mm). – _Distribution:_ Range virtually the same as for subgenus. – Two subspecies are currently recognized:

V. p. thyone (southern Mexico to Guyana and Peru), _V. p. pusilla_ (Paraguay and southeastern Brazil).

Subgenus _**Metavampyressa**_ PETERSON 1968

Two pairs of lower incisors. Anterior lower premolar longer than high. Last lower molar absent. A dorsal stripe present. – _Distribution:_ Ranging from Nicaragua to western Colombia and from southeastern Colombia to the Guianas; also apparently southeastern Peru. – Two species.

3. _V. nymphea_ THOMAS 1909. – Size fairly large (forearm length, 34–39 mm; condylobasal length, 17–20 mm). Forehead markedly concave. Posterior lower premolar relatively tall. – _Distribution:_ From Nicaragua to western Colombia (a southeastern Peruvian record may be referable to _V. brocki_). – No subspecies.

4. _V. brocki_ PETERSON 1968. – Size fairly small (forearm length, 31–36 mm; condylobasal length, 15–17 mm). Forehead only slightly concave. Posterior lower premolar relatively short. – _Distribu-_

tion: Known range from southeastern Colombia to Suriname. – No subspecies.

Subgenus *Vampyriscus* THOMAS 1900

A single pair of lower incisors. Anterior lower premolars longer than high. Last lower molar present. A dorsal stripe present. – *Distribution:* Confined to tropical South America almost entirely east of the Andes from Colombia and Peru to Suriname and the mouth of the Amazon. – A single species.

5. *V. bidens* (DOBSON 1878). – Size fairly large (forearm length, 35–40 mm; condylobasal length, 16–18 mm). – *Distribution:* Same as for subgenus. – No subspecies.

Genus *Chiroderma* PETERS 1860 (Fig. 126)

Dental formula i2/2, c1/1, p2/2, m2/2 × 2 = 28. Rostrum approximately two thirds as long as braincase, not inflated, and without a deep depression, but with a long nasal emargination. Interpterygoid space not extended by a deep palatal emargination. Inner upper incisors elongate, more than twice length of outer upper incisors, and more or less pointed. – *Distribution:* Ranging from tropical Mexico through Central America and tropical South America to Bolivia and southeastern Brazil, but west of the Andes not south of Colombia. Also on Guadeloupe and Montserrat in the Lesser Antilles. – Five species, three additional subspecies.

1. *C. trinitatum* GOODWIN 1958. – Size relatively small (forearm length, 37–42 mm; condylobasal length, 19–20 mm). Anterior lower premolar relatively large with anterior cusp well developed. Inner upper incisors bluntly pointed, in contact except at tips. Dorsal and facial stripes prominent. – *Distribution:* Occurring in Panama and in South America east of the Andes to Bolivia and the mouth of the Amazon, including Trinidad. – No subspecies.

2. *C. villosum* PETERS 1860. – Size medium (forearm length, 42–51 mm; condylobasal length, 22–24 mm). Anterior lower premolar relatively small with anterior cusp poorly developed. Inner upper incisors sharply pointed, not in contact, and parallel to one another. Dorsal and facial stripes poorly developed. – *Distribution:* Ranging from southern Mexico through Central America and tropical South America (east of the Andes) to southeastern Brazil, including Trinidad and Tobago. – Two subspecies are currently recognized:

C. v. jesupi (Oaxaca and Veracruz to northern Colombia), *C. v. villosum* (remainder of range).

3. *C. salvini* DOBSON 1878. – Size medium (forearm length, 43–52 mm; condylobasal length, 21–25 mm). Anterior lower premolar relatively small with anterior cusp poorly developed. Inner upper incisors strongly converging with tips in contact. Dorsal and facial stripes prominent. – *Distribution:* Ranging form northwestern tropical Mexico through Central America and in the Andean region to northern Venezuela and northwestern Bolivia. – Two subspecies are recognized:

C. s. scopaeum (Pacific side of Mexico from Chihuahua to Oaxaca and perhaps farther), *C. s. salvini* (southeastern Mexico through Central America and South American range).

4. *C. doriae* THOMAS 1891. – Size fairly large (forearm length, 49–56 mm; condylobasal length, 25–27 mm). Anterior lower premolar relatively large with anterior cusp well developed. Inner upper incisors bluntly pointed in contact except at tips. Dorsal and facial stripes generally prominent. – *Distribution:* Confined to southeastern Brazil. – No subspecies.

5. *C. improvisum* BAKER & GENOWAYS 1976. – Size relatively large (forearm length, 57–58 mm; condylobasal length, 27–28 mm). Anterior lower premolar relatively small with anterior cusp poorly developed. Inner upper incisors bluntly pointed, strongly convergent but not in contact. Dorsal stripe prominent but facial stripes poorly developed. – *Distribution:* Known only from Guadeloupe and Montserrat islands in the Lesser Antilles. – No subspecies.

Genus *Mesophylla* THOMAS 1901 (Fig. 127)

Dental formula i2/2, c2/2, p2/2, m2/3 x 2 = 30. Rostrum approximately three quarters as long as braincase, not inflated and without a deep depression or long nasal emargination. Interpterygoid space not extended by a deep palatal emargination. Inner upper incisors elongate, usually more than twice length of outer upper incisors, not deeply bifid. Anterior lower molar without a postero-internal cusp. Lingual cusps of middle lower molar vestigial or absent. Leaflet behind noseleaf present. Ear pinna with small round basal lappet. Posterior upper molar without a median ridge. Middle lower molar scarcely wider than horizontal ramus of mandible, with low cusps and no median ridge. – *Distribution:* Ranging along the Atlantic side of Costa Rica and Panama and through much of tropical South America to Bolivia and Amazonian Brazil (including Trinidad), but west of the Andes not south of Ecuador. – A single species, one additional subspecies.

1. *M. macconnelli* THOMAS 1901. – Size fairly small (forearm length, 29–34 mm; condylobasal length, 15–17 mm). Color light brownish. – *Distribution:* Same as for genus. – Two subspecies are here recognized:

M. m. flavescens (known from Trinidad and the Peruvian highlands), *M. m. macconnelli* (most of the remaining range)

Genus *Ectophylla* H. ALLEN 1892

Dental formula i2/2, c1/1, p2/2, m2/2x2 = 28. Rostrum approximately three quarters as long as braincase, not inflated and without a deep depression or long nasal emargination. Interpterygoid space not extended by a deep palatal emargination. Inner upper incisors elongate, usually more than twice length of outer upper incisors, not deeply bifid. Anterior lower molar without a postero-internal cusp. Lingual cusps of middle lower molar vestigial or absent. No leaflet behind noseleaf. Ear pinna without basal lappet. Posterior upper molar with a median ridge. Posterior lower molar much wider than horizontal ramus of mandible, with a median ridge. – *Distribution:* Confined to the Caribbean side of Central America from eastern Honduras to western Panama, also Colombia. – A single species.

1. *E. alba* H. ALLEN 1892. Size very small (forearm length, 26–30 mm; condylobasal length, 14–16 mm). Color light grayish. – *Distribution:* Same as for genus. – No subspecies.

Genus *Artibeus* LEACH 1821 (Fig. 128)

Dental formula i2/2, c1/1, p2/2, m2–3/2–3x2 = 28–32. Rostrum one half to three quarters as long as braincase, not inflated, and without a deep depression or long nasal emargination. Interpterygoid space not extended by a deep palatal emargination. Inner upper incisors not elongate, less than twice length of outer upper incisors, and usually bifid. Forehead usually markedly concave. – *Distribution:* Ranging from tropical Mexico through Central America and throughout tropical South America, also throughout most of the West Indies. – Three subgenera, 15 species, and 24 additional subspecies are here recognized.

Subgenus *Enchisthenes* ANDERSEN 1906

Inner upper incisors not bifid. Last upper and lower molars well developed. Size relatively small. – *Distribution:* Ranging from tropical Mexico through Central America to Trinidad and Bolivia, mostly in or near mountains. – A single species.

1. *A. harti* THOMAS 1892. – Size fairly small (forearm length, 36–41 mm; condylobasal length, 18–20 mm). – *Distribution:* Same as for subgenus. – No subspecies.

Subgenus *Artibeus* LEACH 1821

Inner upper incisors bifid. Last upper molar vestigial or absent, last lower molar small but normally present. Size relatively large. – *Distribution:* Same as for genus. – Nine species and 12 additional subspecies are here recognized.

2. *A. lituratus* (OLFERS 1818). – Size relatively large (forearm length, 66–75 mm; greatest length of skull, 30–33 mm). Preorbital and postorbital processes well developed. Last upper molar normally absent, but last lower normally present. Both medial and lateral facial stripes well developed. Skull relatively broad with a flattened rostrum. – *Distribution:* Ranging from tropical Mexico south through Central America and tropical South America to northern Argentina, but west of the Andes not south of northern Peru, also Margarita and Trinidad islands north to St. Vincent. – Two subspecies:

A. l. palmarum (tropical Mexico to northwestern Peru and Tobago and north to St. Vincent), *A. l. lituratus* (remainder of range).

3. *A. intermedius* J. A. ALLEN 1897. – Size relatively large (forearm length, 61–69 mm; greatest length of skull, 28–31 mm). Preorbital and postorbital processes well developed. Last upper molar normally absent, but last lower normally present. Medial facial stripe well developed, but lateral absent or poorly developed. Skull relatively broad with a flattened rostrum. – *Distribution:* Ranging from tropical Mexico through Central America at least to northern Colombia, including the Tres Marias islands. – No subspecies

4. *A. amplus* HANDLEY 1987. – Size relatively large (forearm length, 65–76 mm; greatest length

of skull, 30–33 mm). Preorbital and postorbital processes poorly developed. Last upper and lower molars present. Facial stripes present but poorly developed. Skull relatively narrow with a flattened rostum. Uropatagium somewhat haired. – *Distribution:* Known only from Venezuela and northern Colombia. – No subspecies.

5. A. planirostris (SPIX 1823). – Size relatively large (forearm length, 62–71 mm; greatest length of skull, 30–33 mm). Preorbital and postorbital processes poorly developed. Last upper molar present or absent, last lower molar normally present. Medial facial stripe variable, lateral absent or poorly developed. Skull of medium width, rostrum not flattened. Uropatagium virtually hairless. – *Distribution:* Confined to tropical South America east of the Andes from Venezuela to northern Argentina. – Three subspecies are here recognized:

A. p. hercules (eastern Ecuador and Peru), *A. p. planirostris* (eastern Brazil and Paraguay), *A. p. fallax* (most of the remaining range).

6. A. jamaicensis LEACH 1821. – Size fairly large (forearm length, 52–68 mm; greatest length of skull, 26–32 mm). Preorbital and postorbital processes poorly developed. Last upper molar present or absent, last lower normally present. Facial stripes variably developed. Skull of medium width, rostrum not flattened. Uropatagium virtually hairless. – *Distribution:* Ranging from tropical Mexico through Central America and tropical South America to Paraguay, but west of the Andes not south of northern Peru, also virtually throughout the West Indies and with a record from the Florida Keys. – Ten subspecies are here recognized:

A. j. schwartzi (St. Vincent in the Lesser Antilles), *A. j. jamaicensis* (Barbados and St. Lucia north through the Lesser Antilles, east through the Greater Antilles to Jamaica and south to San Andreas in the southwestern Caribbean), *A. j. parvipes* (Cuba and the southern Bahamas), *A. j. triomylus* (western Mexico from Sinaloa to Oaxaca), *A. j. paulus* (Pacific side of Central America from Chiapas to Cost Rica), *A. j. yucatanicus* (Tamaulipas to northern Guatemala, including Cozumel and the Bay islands in the Caribbean), *A. j. richardsoni* (Chiapas through the remainder of Central America to Panama), *A. j. aequatorialis* (Pacific slope from southern Colombia to northern Peru), *A. j. trinitatis* (Central Colombia and northern Venezuela, including several islands off its coast north to Grenada), *A. j. fuliginosus* (South America east of the Andes from southern Venezuela and the Guianas to Bolivia and eastern Brazil). The subspecific status of Paraguayan specimens is uncertain.

7. A. hirsutus ANDERSEN 1906. – Size fairly large (forearm length, 52–60 mm; greatest length of skull, 26–28 mm). Preorbital and postorbital processes absent. Last upper and lower molar present. Facial stripes poorly developed. Uropatagium well haired. Posterior border of palate pointed in midline. – *Distribution:* Confined to western Mexico from Sonora to Guerrero. – No subspecies.

8. A. inopinatus DAVIS & CARTER 1964. – Size medium (forearm length, 48–53 mm; greatest length of skull, 25–27 mm). Preorbital and postorbital processes absent. Last upper and lower molar present. Facial stripes poorly developed. Uropatagium well haired. Posterior border of palate evenly concave. – *Distribution:* Known only from the Pacific side of Central America from Salvador to Nicaragua. – No subspecies.

9. A. fraterculus ANTHONY 1924. – Size fairly large (forearm length, 51–61 mm; greatest length of skull, 26–29 mm). Preorbital and postorbital processes poorly developed. Last upper molar absent, last lower normally present. Medial facial stripe fairly well developed but lateral absent or poorly developed. Uropatagium virtually hairless. – *Distribution:* Confined to arid areas in western Ecuador and western Peru. – No subspecies.

Subgenus **Dermanura** GERVAIS 1856

Inner upper incisors bifid. Last upper and lower molars present or absent. Size relatively small. – *Distribution:* Ranging from tropical Mexico to Bolivia and northeastern Brazil, but west of the Andes not south of Ecuador, also Trinidad, Tobago, and Grenada. – Six species and 12 additional subspecies are here recognized.

10. Artibeus concolor PETERS 1865. – Size fairly small (forearm length, 45–50 mm; greatest length of skull, 20–23 mm). Preorbital and postorbital processes poorly developed. Last upper and lower molars present. Median facial stripe well developed, lateral poorly developed, but present. Uropatagium well haired. – *Distribution:* Confined to tropical South America east of the Andes from southern Venezuela to Peru and northeastern Brazil. – No subspecies.

11. A. aztecus ANDERSEN 1906. – Size fairly small (forearm length, 40–49 mm; greatest length of skull, 21–24 mm). Preorbital and postorbital processes absent. Last upper and lower molars absent. Uropatagium very narrow and well haired. – *Distribution:* Ranging (mainly in mountains) from tropical Mexico to western Panama. – Three subspecies are recognized:

A. a. aztecus (Sinaloa and Tamaulipas to Oaxaca), *A. a. minor* (Chiapas to Honduras), *A. a. major* (Costa Rica and western Panama).

12. *A. toltecus* (SAUSSURE 1860). – Size relatively small (forearm length, 36–43 mm; greatest length of skull, 19–22 mm). Preorbital and postorbital processes absent. Last upper and lower molars absent. Uropatagium very narrow and well haired. – *Distribution:* Ranging from tropical Mexico to Panama. (Alleged South American records are apparently the result of confusion with *A. phaeotis*). – Two subspecies are recognized:

A. t. hesperus (Pacific side of Middle America from Sinaloa to Nicaragua), *A. t. toltecus* (chiefly on the Atlantic side of Middle America from Nuevo Leon to Panama).

13. *A. cinereus* (Gervais 1856). – Size at least fairly small (forearm length, 36–42 mm; greatest length, of skull, 19–22 mm). Preorbital and postorbital processes absent. Last upper molar absent, last lower present or absent. Forehead concave but transition to rostrum relatively gradual. Upper molars relatively narrow. Uropatagium virtually hairless. – *Distribution:* Ranging from southern Mexico through Central America and tropical South America to Bolivia and northeastern Brazil, but west of the Andes not south of Ecuador, also Trinidad, Tobago and Grenada. – Seven subspecies are here recognized:

A. c. watsoni (southern Mexico to western Colombia), *A. c. rosenbergi* (western Ecuador), *A. c. bogotensis* (central Colombia across northern Venezuela to Grenada and extreme northern Brazil), *A. c. pumilio* (western Amazonia from Guyana to northern Bolivia), *A. c. cinereus* (north-central Brazil), *A. c. quadrivittatus* southern Venezuela to northeastern Brazil). Boundaries, however, are not clear and more than one species is probably represented in this complex.

14. *A. phaeotis* (MILLER 1902). – Size relatively small (forearm length, 34–42 mm; greatest length of skull, 17–21 mm). Preorbital and postorbital processes absent. Last upper and lower molars absent. Forehead concave but transition to rostrum varying from gradual to abrupt. Upper molars relatively narrow. Uropatagium virtually hairless. – *Distribution:* Ranging from tropical Mexico to northwestern Ecuador, southeastern Colombia, and Guyana. A Peruvian record probably refers to *A. anderseni*. – Four subspecies are here recognized:

A. p. nanus (western Mexico from Sinaloa to Guerrero), *A. p. palatinus* (Pacific side of Middle America from Oaxaca to Costa Rica), *A. p. phaeotis* (Atlantic side of Middle America from Veracruz to Costa Rica), *A. p. ravus* (Panama and South American range).

15. *A. anderseni* OSGOOD 1916. – Size relatively small (forearm length, 34–37 mm; greatest length of skull, 16–19 mm). Preorbital and postorbital processes absent. Last upper and lower molars absent. Transition between braincase and rostrum abrupt. Upper molars relatively narrow. Uropatagium virtually hairless. – *Distribution:* Largely confined to central South America east of the Andes from Ecuador to Bolivia and French Guiana, but also known from northern Colombia. – No subspecies.

Genus *Ardops* MILLER 1906 (Fig. 129)

Dental formula i 2/2, c 1/1, p 2/2, m 3/3 × 2 = 32. Rostrum approximately two thirds length of braincase, not inflated, and without a deep depression or long nasal emargination. Interpterygoid space extended by a deep parallel-sided palatal emargination. Inner upper incisors scarcely higher than long and less than twice length of outer upper incisors. Forehead almost flat. – *Distribution:* Confined to the Lesser Antilles. – A single species with four additional subspecies.

1. *A. nichollsi* (THOMAS 1891). – Size fairly large (forearm length, 42–55 mm; condylobasal length, 18–22 mm). – *Distribution:* Same as for genus. – Five subspecies:

A. n. montserratensis (St. Eustatius, Monserrat), *A. n. annectens* (Guadeloupe), *A. n. nichollsi* (Dominica), *A. n. koopmani* (Martinique), *A. n. luciae* (St. Lucia, St. Vincent).

Genus *Phyllops* PETERS 1865 (Fig. 130)

Dental formula i 2/2, c 1/1, p 2/2, m 3/3 × 2 = 32. Rostrum slightly more than half length of braincase, not inflated, and without a deep depression or long nasal emargination. Interpterygoid space extended by a deep, more or less V-shaped palatal emargination. Inner upper incisors clearly higher than long, but less than twice height of outer upper incisors. Forehead only slightly concave. – *Distribution:* Confined to Cuba and Hispaniola. – A single living species is here recognized with one additional subspecies.

1. *P. falcatus* (GRAY 1839). – Size fairly small (forearm length, 39–44 mm; condylobasal length, 19–21 mm). – *Distribution:* Same as for genus. – Two subspecies are here recognized:

P. f. falcatus (Cuba), *P. f. haitensis* (Hispaniola).

Genus *Ariteus* GRAY 1838 (Fig. 131)

Dental formula i 2/2, c 1/1, p 2/2, m 2/3 x 2 = 30. Rostrum slightly more than half length of braincase, not inflated, and without a deep depression or long nasal emargination. Interpterygoid space extended by a deep U-shaped palatal emargination. Inner upper incisors scarcely higher than long and less than twice length of outer upper incisors. Forehead only slightly concave. – *Distribution:* Confined to Jamaica. – A single species.

1. *A. flavescens* (GRAY 1831). – Size relatively small (forearm length, 37–44 mm; condylobasal length, 15–18 mm). – *Distribution:* Same as for genus. – No subspecies.

Genus *Stenoderma* E. GEOFFROY 1818 (Fig. 132)

Dental formula i 2/2, c 1/1, p 2/2, m 3/3 × 2 = 32. Rostrum slightly more than half length of braincase, not inflated, and with a deep depression and a fairly extensive nasal emargination. Interpterygoid space extended by a deep U-shaped palatal emargination. Inner upper incisors clearly higher than long, but less than twice height of outer upper incisors. Forehead markedly concave. – *Distribution:* Confined to Puerto Rico and the Virgin Islands. – A single species with one additional living subspecies.

1. *S. rufum* DESMAREST 1820. – Size fairly large (forearm length, 46–52 mm; condylobasal length, 18–20 mm). – *Distribution:* Same as for genus. – Two living subspecies are recognized:

S. r. darioi (Puerto Rico), *S. r. rufum* (St. Thomas and St. John in the Virgin Islands).

Genus *Pygoderma* PETERS 1863 (Fig. 133)

Dental formula i 2/2, c 1/1, p 2/2, m 2/2 x 2 = 28. Rostrum almost as long as braincase, inflated (almost cuboid), but without a depression or nasal emargination. Interpterygoid space not extended by a palatal emargination. Inner upper incisors large, slightly higher than long, outers greatly reduced. Forehead flat. – *Distribution:* Known only from Bolivia, northern Argentina, Paraguay, and southeastern Brazil, with an old record from Suriname. – A single species with one additional subspecies.

1. *P. bilabiatum* (WAGNER 1843). – Size fairly small (forearm length, 36–44 mm; condylobasal length, 16–20 mm). – *Distribution:* Same as for genus. – Two subspecies are currently recognized:

P. b. magna (southeastern Bolivia, northwestern Argentina, *P. b. bilabiatum* (eastern Paraguay to southeastern Brazil).

Genus *Ametrida* GRAY 1847 (Fig. 134)

Dental formula i 2/2, c 1/1, p 2/2, m 3/3 x 2 = 32. Rostrum less than half length of braincase, greatly modified by retraction of external nares. Interpterygoid space so broadened as to be virtually obliterated and with only a shallow palatal emargination. Inner upper incisors triangular, not more than twice length of small outers. Forehead sharply concave. Anterior margin of orbit not greatly modified. Noseleaf relatively unmodified. – *Distribution:* Confined to northern South America east of the Andes from Venezuela to central Brazil, including Bonaire and Trinidad islands off the northern coast. – A single species.

1. *A. centurio* GRAY 1847 (= *minor* H. ALLEN 1894). – Size relatively small (forearm length, 24–34 mm; condylobasal length, 11–14 mm). Females markedly larger than males. – *Distribution:* Same as for genus. – No subspecies.

Genus *Sphaeronycteris* PETERS 1822 (Fig. 135)

Dental formula i 2/2, c 1/1, p 2/2, m 3/3 x 2 = 32. Rostrum less than half length of braincase, greatly modified by retraction of external nares. Interpterygoid space broad but clearly evident, and extended by a V-shaped palatal emargination. Inner upper incisors elongate, more than twice length of greatly reduced outers. Forehead sharply concave. Anterior margin of orbit extended to form a conspicuous plate. True noseleaf considerably reduced in height, but with a ridge-like outgrowth behind it, greatly enlarged in males. – *Distribution:* Occurring in tropical South America east of the Andes from Venezuela to northern Bolivia, east to the mouth of the Amazon. – A single species.

1. *S. toxophyllum* PETERS 1882. – Size fairly small (forearm length, 36–41 mm; condylobasal length, 13–15 mm). – *Distribution:* Same as for genus. – No subspecies.

Genus *Centurio* GRAY 1842 (Fig. 136)

Dental formula i 2/2, c 1/1, p 2/2, m 2/2 x 2 = 28. Rostrum less than half length of braincase but otherwise not greatly modified. Interpterygoid space broad but clearly evident and extended by a fairly deep V-shaped palatal emargination. Inner upper incisors short and less than twice length of small outers. Forehead steeply descending but almost flat. Anterior margin or orbit not greatly modified. True noseleaf greatly reduced in height and difficult to distinguish from a mass of secondary folds and ridges. – *Distribution:* Ranging from tropical Mexico through Central America and extreme northern South America to Trinidad and Tobago. – A single species.

1. *C. senex* GRAY 1842. – Size medium (forearm length, 41–47 mm; condylobasal length, 14–15 mm). – *Distribution:* Same as for genus. – Two subspecies.

A. s. senex (Middle America and Colombia), *A. s. greenhalli* (Trinidad and Tobago). The subspecific identity of Venezuelan specimens is not clear.

Subfamily **Desmodontinae** BONAPARTE 1845.

Anterior upper incisor and canine bladelike. Noseleaf reduced, upright portion absent or poorly developed. Tongue not elongate, lacking conspicuous bristle-like papillae, its musculature exhibiting a transverse pattern in section, supplied by paired longitudinal arteries. Single upper premolar in contact with canine and with anterior upper molar (which may be the only one). Zygomatic arch complete. Upper molars highly modified, reduced and blade-like or absent, no trace of a W-shaped ectoloph. All desmodontines are highly specialized for feeding on vertebrate blood. – *Distribution:* Ranging from tropical Mexico through Central and South America to central Chile and Uruguay, including Margarita and Trinidad islands off the north coast; a single record from southwestern United States (Texas) and known fossil from Cuba. – Three genera, three species.

Genus *Diphylla* SPIX 1823 (Fig. 137)

Ear pinna relatively short. Pollex relatively short, its metacarpal without pads. Calcar small but well developed. Legs and uropatagium thickly haired. Coronoid process of mandible relatively low. Dental formula i2/2, c1/1, p1/2, m2/2x2 = 26. Lower incisors, large, complex, and forming a continuous cutting edge. Postorbital constriction virtually absent. Preorbital process absent. – *Distribution:* Ranging from Texas through eastern Mexico, Central America, and in South America east of the Andes to Bolivia and eastern Brazil. – A single species.

1. *D. ecaudata* SPIX 1823. – Size fairly large (forearm length, 53–57 mm; condylobasal length, 19–21 mm). – *Distribution:* Same as for genus. – No subspecies here recognized.

Genus *Diaemus* MILLER 1906 (Fig. 138)

Ear pinna relatively long. Pollex relatively short, its metacarpal with a single pad. Calcar absent. Legs and uropatagium sparsely haired. Coronoid process of mandible relatively high. Dental formula i1/2, c1/1, p1/2, m2/1x2 = 22. Lower incisors small, simple, and separate from one another. Postorbital constriction poorly developed. Preorbital process present. – *Distribution:* Ranging from northeastern Mexico through Central America and South America to northern Argentina and eastern Brazil, including Margarita and Trinidad, but west of the Andes not south of Colombia. – A single species.

1. *D. youngi* (JENTINCK 1893). – Size fairly large (forearm length, 49–56 mm; condylobasal length, 20–22 mm). – *Distribution:* Same as for genus. – No subspecies here recognized.

Genus *Desmodus* WIED 1826 (Fig. 139)

Ear pinna relatively long, Pollex relatively long, its metacarpal with two pads. Calcar present but greatly reduced. Legs and uropatagium sparsely haired. Coronoid process relatively high. Dental formula i1/2, c1/1, p1/2, m1/1x2 = 20. Lower incisors small, simple, and separate from one another. Postorbital constriction fairly well developed. Preorbital process absent. – *Distribution:* Same as for subfamily (except for the Texas record). – A single species.

1. *D. rotundus* (E. GEOFFROY 1810). – Size fairly large (forearm length, 55–63 mm; condylobasal length, 20–22 mm). – *Distribution:* Same as for genus. – No subspecies are here recognized.

Superfamily **Vespertilionoidea** GRAY 1821 (Plain-nosed bats)

Tail always present, usually at least as long as the well-developed uropatagium. Rhinarium and muzzle rarely much modified. Most frequently three pairs of lower incisors. Phalanx of second digit of wing frequently reduced or absent. – *Dis-*

tribution: Coextensive with that of the infraorder. – Seven families, four additional subfamilies, 53 genera, 397 species.

Family Natalidae GRAY 1866
(Long-legged Funnel-eared bats)

Structure: Second digit of wing reduced to the metacarpal, phalanx lost. Tail long, reaching margin of extensive uropatagium. Trochiter of humerus large, and making a definite articulation with the scapula. Last cervical and first two thoracic vertebrae not fused with one another. Rostrum not shortened. The premaxillaries retain both nasal and palatal branches. Ears large and funnel-shaped. Pollex unmodified except that its claw has a basal talon.
Ecology: Confined to tropical lowlands from mesic to semi-arid. As far as is known always insectivorous. Usually roosts in the deeper parts of caves or cave-like structures.
Distribution (Fig. 15): Ranging from tropical Mexico (including Baja California Sur) through Central America and around the northern and eastern coasts of South America to central Brazil; also throughout most of the West Indies.
– *Systematics:* A single genus, five species.

Genus *Natalus* GRAY 1838 (Fig. 140)

Dental formula i 2/3, c 1/1, p 3/3, m 3/3 x 2 = 38. – *Distribution:* Same as for family. – Three subgenera and five species are here recognized.

Subgenus *Natalus* GRAY 1838

Braincase relatively inflated. Upper canine much longer than last upper premolar. Legs and tail relatively elongate. Muzzle and chin without dermal outgrowths, but lower lip with a shallow cleft. – *Distribution:* Virtually the same as for the genus, except for the Bahamas and (at present) Cuba. – Two species and seven additional living subspecies are here recognized.

1. *N. stramineus* GRAY 1838. – Sides of rostrum not greatly inflated. Posterior border of palate not or only slightly emarginated. Size medium to relatively large (forearm length, 35–46 mm). – *Distribution:* Ranging from northwestern and northeastern Mexico to Panama; also northern Venezuela, eastern and central Brazil, Lesser Antilles, Hispaniola, Jamaica, San Andreas in the southwestern Caribbean, and known fossil from Cuba. – Seven living subspecies are here recognized:

N. s. mexicanus (northwestern and northeastern Mexico to Panama, San Andreas island), *N. s. tronchonii* (northern Venezuela), *N. s. natalensis* (northeastern Brazil), *N. s. esperitosantensis*, (southeastern and probably central Brazil), *N. s. stramineus* (Lesser Antilles), *N. s. major* (Hispaniola) *N. s. jamaicensis* (Jamaica).

2. *N. tumidirostris* MILLER 1900. – Sides of rostrum markedly inflated. Posterior border of palate greatly emarginated. Size medium (forearm length, 35–42 mm). – *Distribution:* Confined to extreme northern South America from Colombia to Suriname including the islands of Curacao, Bonaire, and Trinidad. – Two subspecies are here recognized:

N. t. continentis (mainland range and Trinidad), *N. t. tumidirostris* (Curacao, Bonaire).

Subgenus *Chilonatalus* MILLER 1898

Braincase relatively inflated. Upper canine much longer than last upper premolar. Legs and tail fairly elongate. Muzzle and chin with ridge-like dermal outgrowths and lower lip with a deep groove. – *Distribution:* Confined to the western part of the Antilles from the Bahamas to Providencia island in the southwestern Caribbean. – Two species and one additional subspecies are here recognized.

3. *N. micropus* DOBSON 1880. – Size relatively small (forearm length, 30–37 mm; condylobasal length, 12–13 mm). – *Distribution:* Confined to Cuba, Hispaniola, Jamaica, and Providencia. – Two subspecies are currently recognized:

N. m. macer (Cuba), *N. m. micropus* (remainder of range).

4. *N. tumidifrons* (MILLER 1903). – Size medium (forearm length, 31–35 mm; condylobasal length, 13–15 mm). – *Distribution:* Presently confined to the northern and eastern Bahamas. – No subspecies.

Subgenus *Nyctiellus* GERVAIS 1855

Braincase relatively uninflated. Upper canine subequal to last upper premolar. Legs and tail relatively short. Muzzle and chin without dermal outgrowths, lower lip unmodified. – *Distribution:* Confined to Cuba and the central Bahamas. – A single species.

5. *N. lepidus* (Gervais 1837). – Size relatively small (forearm length, 27–31 mm). – *Distribution:* Same as for subgenus. – No subspecies.

Family **Furipteridae** Gray 1866

Structure: Second digit of wing reduced to the metacarpal. Tail does not quite reach margin of extensive uropatagium. Trochiter of humerus making, at most, only a slight articulation with the scapula. Last cervical and first two thoracic vertebrae not fused with one another. Rostrum not shortened. Ears large and funnel-shaped. Pollex greatly reduced.
Ecology: The two members of this family are poorly known ecologically, but are almost certainly insectivorous. They have been recorded as roosting in caves and man-made structures, but probably use hollow trees as well. Of the two species, one occurs in tropical forests but the other in arid areas.
Distribution (Fig 16): Ranging from Costa Rica to northern Chile and southeastern Brazil, including Trinidad.
Systematics: Two genera, two species.

Genus *Furipterus* Bonaparte 1837 (Fig. 141)

Braincase longer than high. Muzzle and lips virtually without warty outgrowths. Palatal emargination well developed. Dental formula i 2/3, c 1/1, p 2/3, m 3/3 × 2 = 36. – *Distribution:* Ranging from Costa Rica to southeastern Brazil, including Trinidad, but west of the Andes not south of Colombia. – A single species.

1. *F. horrens* (F. Cuvier 1828). – Size fairly small (forearm length, 35–36 mm). – *Distribution:* Same as for genus. – No subspecies.

Genus *Amorphochilus* Peters 1877 (Fig. 142)

Braincase variable but tending to be almost as high as long. Muzzle and lips with conspicuous warty outgrowths. Palatal emargination poorly developed. Dental formula i 2/3, c 1/1, p 2/3, m 3/3 × 2 = 36. – *Distribution:* Confined to arid areas (chiefly west of the Andes) from Ecuador to northern Chile. – A single species.

1. *A. schnablii* Peters 1877. – Size fairly small (forearm length, 34–38 mm). – *Distribution:* Same as for genus. – No subspecies are here recognized.

Familiy **Thyropteridae** Miller 1907

Structure: Second digit of wing reduced to an incomplete metacarpal. Tail long, reaching slightly beyond extensive uropatagium. Trochiter large, articulating with scapula. First two thoracic vertebrae fused with one another but not with the last cervical. Rostrum not shortened. Ears large and funnel-shaped. Pollex with a large pedicillate suction disc (also a suction disc on the pes).
Ecology: Confined to moist tropical forests and as far as is known strictly insectivorous. The two species apparently roost only on smooth leaves, usually before they unroll. This is made possible by its suction discs, the claws being used only to release suction.
Distribution (Fig. 17): Ranging from southern Mexico to northern Bolivia and southeastern Brazil, including Trinidad, but west of the Andes not south of Ecuador.
Systematics: A single genus, two species.

Genus *Thyroptera* Spix 1823 (Fig. 143)

Dental formula i 2/3, c 1/1, p 3/3, m 3/3 × 2 = 38. – *Distribution:* Same as for family. – Two species and three additional subspecies.

1. *T. discifera* (Lichtenstein & Peters 1855). – Underparts brown. Calcar with a single cartilaginous projection extending into the uropatagium. Size relatively small (forearm length, 31–36 mm). – *Distribution:* Known only from eastern Nicaragua and from northern South America south to central Brazil, but west of the Andes not south of Colombia. – Two subspecies:

T. d. abdita (eastern Nicaragua), *T. d. discifera* (South American range).

2. *T. tricolor* Spix 1823. – Underparts whitish. Calcar with two cartilaginous projections extending into the uropatagium. Size relatively large (forearm length, 33–38 mm). – *Distribution:* Virtually same as for genus. – Three subspecies:

T. t. albiventer (southern Mexico through Central and northwestern South America to southeastern Peru, but not west of the Andes south of Ecuador), *T. t. tricolor* (Trinidad and southern Venezuela to northern Bolivia and northeastern Brazil), *T. t. juquiaensis* (southeastern Brazil).

Family **Myzopodidae** Thomas 1904

Structure: Second digit of wing reduced to the metacarpal. Tail long and extending well beyond extensive uropatagium. Trochiter large, articulat-

ing with scapula. Last cervical and first two thoracic vertebrae not fused with one another. Rostrum somewhat shortened. Ears large and somewhat funnel-shaped, but tragus fused to ear pinna and a peculiar mushroom-shaped, process developed at its base. Pollex and pes with large non-pedicillate suction discs.
Ecology: Probably confined to moist tropical forest where it roosts within smooth uncoiled leaves, presumably made possible by the suction discs. Assumed to be insectivorous.
Distribution (Fig. 17): Confined to Madagascar.
Systematics: A single genus and species.

Genus *Myzopoda* MILNE-EDWARDS & GRANDIDIER 1878 (Fig. 144)

Dental formula i 2/3, c 1/1, p 3/3, m 3/3 x 2 = 38. – *Distribution:* Same as for family. A single species.

1. *M. aurita* MILNE-EDWARDS & GRANDIDIER 1878. – Size relatively large (forearm length, 46–50 mm). – *Distribution:* Same as for genus. – No subspecies.

Family **Vespertilionidae** GRAY 1821

Structure: Second digit of wing reduced to the metacarpal and a single small phalanx. Tail long, reaching edge of extensive uropatagium. Trochiter large, making an extensive articulation with the scapula. Last cervical vertebra may be fused with first thoracic, but vertebrae otherwise unmodified. The rostrum may be quite long or show varying degrees of shortening. The premaxillaries lack palatal branches and are usually widely separated from one another. The ears vary from being very long to very short and may or may not be funnel-shaped. Pollex unmodified.
Ecology: Highly diverse, the macrohabitat varying from tropical to cold temperate and from forests to deserts. Roosting habits also highly diverse including virtually all those known for bats. Both migration and hibernation widespread in temperate species. Usually insectivorous, but, in several cases, piscivorous, and in a few cases suspected of being partially carnivorous.
Distribution (Fig. 18): Coextensive with that of the superfamily.
Systematics: Five subfamilies, 35 genera, 308 species.

Subfamily **Kerivoulinae** MILLER 1907

Anterior and middle upper premolars relatively small and conspicuously simpler than posterior upper premolar. Nostrils not elongated as tubes. Sternum short and broad, its length less than twice as great as the breadth of the presternum, which has a relatively small median lobe. Only four or five ribs connecting with sternum. Coracoid process of scapula curved outwards. Ears more or less funnel-shaped. Second phalanx of third digit of wing not greatly elongated. – *Distribution:* Widely distributed in sub-Saharan Africa from Liberia to Ethiopia and south to the Cape Province, but largely avoiding desert areas; also distributed in the Indo-Australian region from India and southern China to the Bismarcks and northeastern Australia. – A single genus with two subgenera, 21 species.

Genus *Kerivoula* GRAY 1842 (Fig. 145)

Dental formula i 2/3, c 1/1, p 3/3, m 3/3 x 2 = 38. – *Distribution:* Same as for subfamily. – Two subgenera, 21 species.

Subgenus *Kerivoula* GRAY 1842

Posterior margin of tragus without a deep notch near its base. Postorbital constriction of skull well developed. Narial emargination usually long and narrow. Anterior palatal emargination relatively long and narrow. Lateral upper incisor well developed. Upper canine at most faintly grooved. – *Distribution:* Virtually coextensive with that of genus, but absent from Australia. – Seventeen species, 15 additional subspecies.

1. *K. whiteheadi* THOMAS 1894. – Rostrum relatively long, slender, and narrowed anteriorly. Anterior and middle upper and lower premolars narrow and elongate. Lingual margins of anterior and middle upper molars rounded with narrow shelves. Size relatively small (forearm length, 27–33 mm; condylobasal length, 11–12 mm). Inner upper incisor bicuspid. Fringe of hair on posterior margin of uropatagium poorly developed. – *Distribution:* Known only from southern Thailand, Borneo, and the Philippines. – Three subspecies are currently recognized:

K. w. bicolor (southern Thailand), *K. w. pusilla* (Borneo, southern Philippines), *K. w. whiteheadi* (northern Philippines).

2. *K. lanosa* (A. SMITH 1847). – Rostrum of medium proportions. Size fairly small (forearm

length, 26–34 mm; condylobasal length, 11–13 mm). Inner upper incisor bicuspid or unicuspid. Fringe of hair on posterior margin of uropatagium fairly well developed. – *Distribution:* Ranging from Liberia to Ethiopia and south through eastern Africa to the Cape Province. – Four subspecies are currently recognized:

K. l. muscilla (Liberia to northern Zaire), *K. l. harrisoni* (Ethiopia to Tanzania), *K. l. lucia* (southern Zaire to Natal), *K. l. lanosa* (Cape Province).

3. *K. picta* (PALLAS 1767). – Rostrum fairly long, but not narrowed anteriorly. Anterior upper premolar more or less circular in section but middle upper, anterior and middle lower premolars elongate. Lingual margin of anterior upper molar nearly square with a fairly broad shelf, but lingual margin of middle upper molar rounded with a narrow shelf. Size medium (forearm length, 32–39 mm; condylobasal length, 12–14 mm). Inner upper incisor bicuspid, but secondary cusp poorly developed. Fringe of hair on posterior margin of uropatagium well developed. Fur (and to some extent membranes) bright orange. – *Distribution:* Ranging from India and Ceylon to southern China, then south through southeastern Asia and Malaysia to Borneo, the Lesser Sundas, and apparently the Moluccas. – Two subspecies.

K. p. bellissima (Hainan island), *K. p. picta* (remainder of range).

4. *K. smithi* THOMAS 1880. – Rostrum fairly short and not narrowed anteriorly. Anterior and middle upper premolars more or less circular in section but anterior and middle lowers elongate. Lingual margins of anterior and middle upper molars more or less rounded, but with broad shelves. Size medium (forearm length, 32–35 mm; condylobasal length, 12–14 mm). Inner upper incisor bicuspid and relatively long. Fringe of hair on posterior margin of uropatagium poorly developed. Outer lower incisors unicuspid. – *Distribution:* Known only from forested regions of Nigeria, Cameroon, northern Zaire, and Kenya. – No subspecies.

5. *K. cuprosa* THOMAS 1912. – Rostrum of medium proportions. Size medium (forearm length, 32–34 mm; condylobasal length, 12–13 mm). Inner upper incisor bicuspid but relatively short. Fringe of hair on posterior margin of uropatagium poorly developed. Outer lower incisor bicuspid. – *Distribution:* Known only from forested regions of Ghana, Cameroon, northern Zaire, and Kenya. – No subspecies.

6. *K. phalaena* THOMAS 1912. – Rostrum relatively short, not narrowed. Anterior and middle premolars elongate. Lingual margins of anterior and middle molars more or less rounded but with a fairly wide shelf. Size relatively small (forearm length, 27–29 mm; condylobasal length, 10–12 mm). Inner upper incisors unicuspid. Fringe of hair on posterior margin of uropatagium poorly developed. – *Distribution:* Known only from forested regions of Liberia, Ghana, Cameroon, and northwestern Zaire. – No subspecies.

7. *K. africana* DOBSON 1878. – Rostrum probably of medium proportions. Size relatively small (forearm length, 27–29 mm; condylobasal length, 10–11 mm). Inner upper incisor bicuspid. Fringe of hair on posterior margin of uropatagium virtually absent. Outer lower incisor tricuspid. – *Distribution:* A poorly known species known only from Tanzania. – No subspecies.

8. *K. eriophora* (HEUGLIN 1877). – Size relatively small (forearm length, *ca.* 28 mm). – *Distribution:* This very poorly known species is only known from northern Ethiopia, but may be an earlier name for *africana*. – No subspecies.

9. *K. muscina* TATE 1941. – Rostrum of medium proportions, not narrowed anteriorly. Anterior upper and lower premolars nearly circular in section, but middle upper and lower premolars somewhat elongate. Lingual margins of anterior and middle upper molars somewhat rounded, their shelves of moderate width. Size medium (forearm length, 32–33 mm; condylobasal length, 12–13 mm). Inner upper incisor bicuspid, but secondary cusp small. Fringe of hair on posterior margin of uropatagium poorly developed. – *Distribution:* Confined to central New Guinea. – No subspecies.

10. *K. agnella* THOMAS 1908. – Rostrum of medium proportions, slightly widened anteriorly. Anterior upper premolar more or less circular in section, but anterior lower and middle upper ovate, and middle lower somewhat elongate. Lingual margins of anterior and middle upper molars rounded, their shelves relatively narrow. Size medium (forearm length, 34–38 mm; condylobasal length, 13–14 mm). Inner upper incisor bicuspid, but secondary cusp small. Fringe of hair on posterior margin of uropatagium poorly developed. – *Distribution:* Confined to the Louisiade and D'Entrecasteaux archipelagos off the eastern end of New Guinea. – No subspecies.

11. *K. minuta* MILLER 1898. – Rostrum relatively short, but somewhat narrowed anteriorly. Anterior upper and lower and middle upper premolars more or less circular in section, but middle lower more ovate. Lingual margins of anterior and middle upper molars rounded, their shelves narrow. Size smallest of Yangochiroptera (forearm length, 24–30 mm; condylobasal length, 10–12 mm). Inner upper incisor bicuspid. No fringe of hair on posterior margin of uropatagium. – *Distribution*: Known only from the Malay peninsula and Borneo. – No subspecies.

12. *K. intermedia* HILL & FRANCIS 1984. – Rostrum fairly long, not narrowed anteriorly. Anterior premolars, above and below, more or less circular in section, middle premolars somewhat elongate. Lingual margins of anterior and middle upper molars rounded, their shelves narrow. Size fairly small (forearm length, 26–31 mm; condylobasal length, 11–12 mm). Inner upper incisor bicuspid. Narial emargination relatively deep. No fringe of hair on posterior margin of uropatagium. – *Distribution:* Known only from Malaya and Borneo. – No subspecies.

13. *K. pellucida* (WATERHOUSE 1845). – Rostrum of medium proportions, slightly narrowed anteriorly. Anterior upper and lower premolars and middle upper premolar nearly circular in section but middle lower slightly ovate. Lingual margins of anterior and middle upper molars more or less square, their shelves broad. Size fairly small (forearm length, 28–35 mm; condylobasal length, 12–14 mm). Inner upper incisor bicuspid, but secondary cusp small. Fringe of hair on posterior margin of uropatagium poorly developed. – *Distribution:* Known from Malaya, Sumatra, Java, Borneo, and the southern Philippines. – No subspecies.

14. *K. hardwickei* (HORSFIELD 1824). – Rostrum short, not narrowed anteriorly. Anterior and middle premolars nearly circular in section. Lingual margins of anterior and middle upper molars more or less square, their shelves broad. Size medium (forearm length, 30–39 mm; condylobasal length, 11–15 mm). Inner upper incisor bicuspid, but secondary cusp small. Fringe of hair on posterior margin of uropatagium poorly developed. – *Distribution:* Ranging from India and Ceylon to southern China and Indo-China, south and east to the Philippines, Celebes, and the Lesser Sundas. – Six subspecies are recognized:

K. h. malpasi (Ceylon), *K. h. crypta* (southern India), *K. h. depressa* (northern India to southern China and Indo-China), *K. h. engana* (Mentawei islands off western Sumatra), *K. h. flora* (Lesser Sundas), *K. h. hardwickei* (remainder of range).

15. *K. myrella* THOMAS 1914. – Rostrum short, not narrowed anteriorly. Anterior and middle premolars nearly circular in section. Lingual margins of anterior and middle upper molars more or less square, their shelves broad. Size medium (forearm length, 33–39 mm; condylobasal length, 12–14 mm). Inner upper incisor bicuspid, but secondary cusp small. Fringe of hair on posterior margin of uropatagium poorly developed. Upper canine somewhat enlarged. – *Distribution:* Confined to the Bismarcks and small nearby islands. – No subspecies.

16. *K. papillosa* (TEMMINCK 1840).– Rostrum relatively short and broad, widened anteriorly. Anterior and middle premolars slightly elongate. Lingual margins of anterior and middle upper molars more or less square, their shelves broad. Size relatively large (forearm length, 38–44 mm; condylobasal length, 14–17 mm). Inner upper incisor bicuspid, but secondary cusp small. No fringe of hair on posterior margin of uropatagium. Upper canine enlarged. – *Distribution:* Ranging from northeastern India to Vietnam, then south and east to Java and Celebes. – Three subspecies are recognized:

K. p. lenis (northeastern India, Burma), *K. p. papillosa* (Java), *K. p. malayana* (remainder of range).

17. *K. argentata* TOMES 1861. – Rostrum relatively short and broad, not narrowed anteriorly. Anterior and middle premolars more or less circular in section. Lingual margins of anterior and middle upper molars more or less square, their shelves broad. Size medium (forearm length, 34–39 mm; condylobasal length, 13–15 mm). Inner upper incisor unicuspid. Fringe of hair on posterior margin of uropatagium well developed. – *Distribution:* Ranging from Kenya and Angola south to Natal. – Three subspecies are recognized:

K. a. argentata (Kenya to Namibia), *K. a. nidicola* (Mozambique), *K. a. zuluensis* (Natal).

Subgenus *Phoniscus* MILLER 1905

Posterior margin of tragus with a deep notch near its base. Postorbital constriction of skull poorly developed. Narial and anterior palatal emarginations relatively short and broad. Lateral upper incisor reduced. Upper canine conspicuously grooved. – *Distribution:* Except for a doubtful record from South Africa, confined to the Malay peninsula east to New Guinea and northeastern

Australia. – Four species, two additional subspecies.

18. *K. jagorii* (PETERS 1866). – Angular concavity on posterior margin of ear near apex well developed. Anterior palatal emargination rounded. Basioccipital pits well developed. Size relatively large (forearm length, 37–39 mm; condylobasal length, 15–16 mm). – *Distribution:* Known from Java and Borneo to the Philippines and Celebes. – Three subspecies are recognized:

K. j. javanus (Java, Bali, Borneo), *K. j. jagorii* (Samar in the Philippines), *K. j. rapax* (Celebes).

19. *K. papuensis* DOBSON 1878. – Angular concavity on posterior margin of ear near apex poorly developed. Anterior palatal emargination angular. Basioccipital pits well developed. Size relatively large (forearm length, 35–41 mm; condylobasal length, 14–16 mm). – *Distribution:* Known only from eastern New Guinea and from Queensland in Australia. – No subspecies.

20. *K. atrox* (MILLER 1905). – Angular concavity on posterior margin of ear near apex virtually absent. Anterior palatal emargination rounded. Basioccipital pits poorly developed. Size relatively small (forearm length, 31–35 mm; condylobasal length, 12–14 mm). – *Distribution:* Known only from the Malay peninsula, Sumatra, and Borneo. – No subspecies.

21. *K. aerosa* TOMES 1858. – Size relatively large (forearm length, 37 mm). Premolars less massive than in other species of *K. (Phoniscus)*. – *Distribution:* A very poorly known form, supposedly from Cape Province, South Africa, but more likely from somewhere in the Malay archipelago. – No subspecies.

Subfamily **Vespertilioninae** GRAY 1821

Anterior and middle upper premolars, if present, relatively small and conspicuously simpler than posterior upper premolar. Nostrils not elongated as tubes. Sternum slender, its length more than twice as great as the breadth of the presternum, which has a relatively small median lobe. Six or seven ribs connecting with sternum. Coracoid process of scapula curved outwards. Ears not funnel-shaped, without a keel, but with a well developed anterior basal lobe. Second phalanx of third digit of wing not greatly elongated. Seventh cervical vertebra free from first thoracic. – *Distribution:* Same as for subfamily. – Seven tribes, 30 genera, 262 species.

Tribe Myotini TATE 1942

Rostrum relatively elongate. Dental formula never less than i2/3, c1/1, p2/2, m3/3x2 = 34. Ears not united by a band. Nostrils not opening on the dorsal surface of the muzzle. – *Distribution:* Widely distributed over Eurasia, Africa, much of the Australian region and the Americas, but sparingly distributed on oceanic islands. – Two genera, three additional subgenera, and 85 species.

Genus ***Myotis*** KAUP 1829 (Fig. 146)

Dental formula usually i2/3, c1/1, p3/3, m3/3x2 = 38 but middle upper and lower premolars absent in a few species. Hypocone on anterior and middle upper molars absent or poorly developed. Tragus of ear long and slender. Uropatagium nearly naked dorsally. – *Distribution:* Same as for tribe. – Four subgenera and 84 species are recognized here, but placement of many species is uncertain, as are the number of species themselves.

Subgenus ***Myotis*** KAUP 1829

Anterior upper premolar usually in toothrow and not greatly reduced. Size medium to relatively large. Maxillary toothrow usually relatively long. Coronoid process of mandible relatively high. Upper molars with reduced accessory cusps. Wings and legs relatively long, but hind feet small. Ears relatively long. These bats forage mainly by gleaning off solid surfaces. – *Distribution:* Ranging across Eurasia and southeast through the Indo-Malayan region to the Philippines and Celebes; also northwestern, eastern and southern Africa, Madagascar and the Comoros; also North America south to southern Mexico. – Twenty species, 24 additional subspecies.

1. *M. bechsteini* (KUHL 1817) [*bechsteini* group]. – Size medium (forearm length, 39–45 mm). Ear relatively large. Braincase fairly low and rostrum relatively slender. Nasal emargination long and narrow. Middle upper premolar in toothrow. – *Distribution:* Ranging through western Eurasia from Sweden, England and Portugal to Iran. – No currently recognized subspecies.

2. *M. myotis* (BORKHAUSEN 1797) [*myotis* group]. – Size relatively large (forearm length, 58–71 mm; condylobasal length, 22–25 mm). Ear fairly large. Braincase relatively low and rostrum relatively broad. Nasal emargination medium. Middle upper premolar usually at least partly displaced me-

dially from the toothrow. Anterior and middle lower incisors three-cusped. – *Distribution:* Ranging through western Eurasia from England and Portugal to Ukrainia and Israel; also many Mediterranean islands and recorded from the Azores. – Two subspecies are recognized:

M. m. macrocephalicus (Israel and Lebanon), *M. m. myotis* (remainder of range).

3. <u>*M. blythii*</u> (TOMES 1857) [*myotis* group]. – Size fairly large (forearm length, 53–64 mm; condylobasal length, 18–23 mm). Rostrum somewhat shortened and narrowed. – *Distribution:* Ranging across the southern Palearctic from Portugal and Morocco to northern China and northeastern India including several Mediterranean islands. – Six subspecies are currently recognized:

M. b. punicus (northwestern Africa), *M. b. oxygnathus* (southern Europe), *M. b. lesviacus* (Lesvos island off Anatolia), *M. b. omari* (southwestern Asia), *M. b. blythi* (northwestern India and central Asia), *M. b. ancilla* (northern China).

4. <u>*M. chinensis*</u> (TOMES 1857) [*myotis* group]. – Size relatively large (forearm length, 64–66 mm; condylobasal length, 19–23 mm). Rostrum somewhat shortened. Anterior and middle lower incisors four-cusped. – *Distribution:* Confined to southern China and northern Thailand. – Two poorly defined subspecies.

5. <u>*M. sicarius*</u> (THOMAS 1915) [*myotis* group]. – Size fairly large (forearm length, 48–55 mm; condylobasal length, 17–19 mm). Rostrum shortened and narrowed. Middle upper premolar reduced and displaced medially from the toothrow. Anterior and middle lower incisors four-cusped. – *Distribution:* Confined to northeastern India. – No subspecies.

6. <u>*M. welwitschii*</u> (GRAY 1866) [*formosus* group]. – Size fairly large (forearm length, 54–59 mm; condylobasal length, 16–20 mm). Ear moderate in size. Braincase medium in height and rostrum somewhat shortened. Nasal emargination medium. Middle upper premolar displaced medially from the toothrow. Anterior and middle lower incisors four-cusped. Wing membranes conspicuously parti-colored in black and red. – *Distribution:* Ranging from Ethiopia to Angola and Transvaal. – No subspecies are currently recognized.

7. <u>*M. formosus*</u> (HODGSON 1835) [*formosus* group]. – Size medium (forearm length, 45–53 mm) Middle upper premolar may not be displaced medially from the toothrow. – *Distribution:* Known from Afghanistan, northern India, southern China, Korea, Sumatra, Java, Philippines, and Celebes. – Eight subspecies are here recognized:

M. f. formosus (Afghanistan and northern India), *M. f. rufoniger* (southern China), *M. f. tsuensis* (Korea and Tsushima islands), *M. f. watasei* (Taiwan), *M. f. hermani* (Sumatra), *M. f. bartelsi* (Java, Bali), *M. f. rufopictus* (Philippines), *M. f. weberi* (Celebes).

8. <u>*M. nattereri*</u> (KUHL 1817) [*nattereri* group]. – Size medium (forearm length, 38–42 mm; condylobasal length, 13–16 mm). Ear medium in length. Braincase relatively high but rostrum fairly long and narrow. Nasal emargination medium. Middle upper premolar in toothrow. Margin of uropatagium with a dense fringe of short hairs. – *Distribution:* Ranging from Ireland, Portugal, and Morocco to the Urals, Israel and Turkmenia. – Two subspecies are here recognized:

M. n. tschuliensis (Transcaucasia, Iraq, Turkmenia), *M. n. nattereri* (remainder or range).

9. <u>*M. schaubi*</u> KORMOS 1934 [*nattereri* group]. – Size medium (forearm length, 42–44 mm; condylobasal length, 15–17 mm). Braincase fairly high but rostrum fairly long and narrow. – *Distribution:* Confined to Transcaucasia and western Iran. – Originally described on the basis of fossil material from Europe, the sole living subspecies is *M. s. araxenus*.

10. <u>*M. bombinus*</u> THOMAS 1905 [*nattereri* group]. – Size medium (forearm length, 38–42 mm; condylobasal length, 13–15 mm). Margin of uropatagium with a dense fringe of long hairs. Tail and tibia relatively long. – *Distribution:* Confined to southeastern Siberia, extreme northeastern China, Korea, and Japan. – No subspecies are currently recognized.

11. <u>*M. pequinius*</u> THOMAS 1908 [*nattereri* group]. – Size medium (forearm length, 47–51 mm; condylobasal length, 17–18 mm). Rostrum relatively short and somewhat upturned. Middle upper premolar greatly reduced. – *Distribution:* Confined to northeastern China. – No subspecies.

12. <u>*M. thysanodes*</u> MILLER 1897 [*nattereri* group]. – Size medium (forearm length, 39–47 mm; condylobasal length, 15–18 mm). Braincase of medium height. – *Distribution:* Ranging through western North America from southwestern Canada to southern Mexico. – Three subspecies are currently recognized:

M. t. pahasapensis (a small area east of the Rocky mountains in Wyoming, South Dakota, and Nebraska), *M. t. az-*

tecus (southern Mexico), *M. t. thysanodes* (remainder of range).

13. <u>*M. emarginatus*</u> (GEOFFROY 1806) [*emarginatus* group]. – Size medium (forearm length, 37–43 mm; condylobasal length, 14–16 mm). Ear medium in length. Braincase relatively high but rostrum fairly long and narrow. Nasal emargination medium. Middle upper premolar in toothrow. No fringe of hair on margin of uropatagium. Posterior border of ear pinna with a conspicuous angular emargination. – *Distribution:* Ranging from the Netherlands, Portugal, and Morocco across southern Eurasia to Uzbekistan and Afghanistan. – Four subspecies are here recognized:

M. e. emarginatus (Europe, northwestern Africa, and southwestern Asia), *M. e. desertorum* (Oman to Afghanistan), *M. e. turcomanicus* (Turkmenia to Afghanistan), *M. e. saturatus* (Uzbekistan).

14. <u>*M. tricolor*</u> (TEMMINCK 1832) (= *loveni* GRANVIK 1924) [*emarginatus* group]. – Size fairly large (forearm length, 47–52 mm; condylobasal length, 15–18 mm). Rostrum relatively broad. Middle upper premolar reduced and displaced medially from toothrow. – *Distribution:* Ranging mostly in eastern Africa from Ethiopia to the Cape province, but west to western Zaire. – No subspecies.

15. <u>*M. morrisi*</u> HILL 1971 [*emarginatus* group]. – Size fairly large (forearm length, 45–46 mm; condylobasal length, 16–17 mm). Rostrum relatively narrow. Middle upper premolar somewhat displaced medially from the toothrow. – *Distribution:* Known only from northern Ethiopia. – No subspecies.

16. <u>*M. goudoti*</u> (A. SMITH 1834) [*emarginatus* group]. – Size medium (forearm length, 36–45 mm; condylobasal length, 13–14 mm). Posterior border of ear without a conspicuous emargination. – *Distribution:* Confined to Madagascar and the Comoro islands. – Two subspecies are recognized:

M. g. anjouanensis (Comoros), *M. g. goudoti* (Madagascar).

17. <u>*M. keenii*</u> (MERRIAM 1895) [*evotis* group]. – Size medium (forearm length, 34–39 mm; condylobasal length, 13–15 mm). Ear fairly long but not black. Braincase fairly low and rostrum fairly short but relatively narrow. Nasal emargination medium. Middle upper premolar in toothrow. No fringe of hair on margin of uropatagium. Sagittal crest absent or poorly developed. – *Distribution:* Ranging across central North America from southeastern Alaska to Newfoundland and south to Florida. – Two subspecies (which may be separate species):

M. k. septentrionalis (Mackensie and Newfoundland to Florida), *M. k. keenii* (Alaska to Washington).

18. <u>*M. auriculus*</u> BAKER & STAINS 1955 [*evotis* group]. – Size medium (forearm length, 35–41 mm; condylobasal length, 14–15 mm). Ears relatively long. Rostrum relatively long. Sagittal crest always present, though weak. – *Distribution:* Ranging from southwestern United States to central Mexico, also Guatemala. – Two subspecies are recognized:

M. a. apache (Arizona and New Mexico to Jalisco), *M. a. auriculus* (Nuevo Leon to Veracruz). The Guatemalan population has not been allocated subspecifically.

19. <u>*M. evotis*</u> (H. ALLEN 1864) [*evotis* group]. – Size medium (forearm length, 35–41 mm; condylobasal length, 14–16 mm). Ears relatively long and black. Rostrum relatively long. Margin of uropatagium with a sparse fringe of hair. Sagittal crest absent or poorly developed. – *Distribution:* Ranging through western North America from southwestern Canada to northwestern Mexico. – Two subspecies are currently recognized:

M. e. pacificus (British Colombia to northern California), *M. e. evotis* (British Colombia and Saskatchewan to Baja California and New Mexico).

20. <u>*M. milleri*</u> ELLIOT 1903 [*evotis* group]. – Size medium (forearm length, 34–37 mm; condylobasal length, 13–15 mm). Ears relatively long and black. Braincase relatively low and rostrum relatively long. – *Distribution:* Confined to the Sierra San Pedro Martir in northern Baja California, Mexico. – No subspecies.

Subgenus **Selysius** BONAPARTE 1841

Anterior upper premolar usually in toothrow and not greatly reduced. Size medium to relatively small. Maxillary toothrow relatively short. Coronoid process of mandible relatively low. Sagittal and lambdoidal crests poorly developed. Plagiopatagium broad, its margin tending to be attached well down on the small foot. Calcar usually with a well-developed keel. Hair on ventral side of uropatagium tending to converge toward the center forming an insect trap. These bats forage mainly by catching small insects in mid-air. – *Distribution:* Ranging across Eurasia and southeast through the Indo-Malayan region to the Moluccas and northwestern Australia, also dubious records from eastern Australia and Samoa; in Africa, known only from Morocco and Ethiopia; also North and South America south to northern Chile and northern Argentina, and north through the Lesser Antilles. – Twenty nine species, 33 additional subspecies.

21. M. mystacinus (KUHL 1817) [*mystacinus* group]. – Size small (forearm length, 31–35 mm; condylobasal length, 12–14 mm). Keel on calcar well developed. Tail and tibia relatively short. Braincase fairly low. Rostrum fairly long and slender. Middle upper premolar usually in toothrow. Penis thin and baculum with a convex margin. – *Distribution:* Widely distributed in the Palearctic from Ireland and Morocco to northeastern China and the Himalayas. – Six subspecies are recognized:

M. m. mystacinus (Morocco, Europe east at least to Transcaucasia), *M. m. transcaspicus* (western Soviet Central Asia), *M. m. sogdianus* (most of eastern Soviet Central Asia), *M. m. nipalensis* (Tadzhikistan and Tibet to Nepal), *M. m. davidi* (northeastern China), *M. m. przewalskii* (Mongolia and northwestern China).

22. M. brandti (EVERSMANN 1845) [*mystacinus* group]. – Size small (forearm length, 32–38 mm; condylobasal length, 13–14 mm). Penis club-shaped and baculum with a basal notch. – *Distribution:* Ranging across Eurasia from Britain and Spain to eastern Siberia and Korea, also Sakhalin, Japan, and the Kuriles. – Three subspecies are here recognized:

M. b. brandti (Europe and western Siberia), *M. b. gracilis* (eastern Siberia to Korea, Sakhalin, Kuriles, and Hokkaido), *M. b. fujiensis* (Honshu).

23. M. insularum (DOBSON 1878) [*mystacinus* group]. – Size small (forearm length, 34 mm). Middle upper premolar reduced and displaced medially from the toothrow. Outer margin of ear pinna relatively deeply emarginate and tragus relatively long. – *Distribution:* Supposedly from Samoa, but true provenance and taxonomic status uncertain. – No subspecies.

24. M. frater G. M. ALLEN 1923 [*frater* group]. – Size small to medium (forearm length, 36–41 mm; condylobasal length, 12–15 mm). Keel on calcar well developed. Tail and tibia relatively long. Braincase high and markedly inflated. Rostrum relatively short and broad; interorbital foramen near orbit. Middle upper premolar more or less displaced medially from the toothrow. – *Distribution:* Ranging from Tadzhikistan east to Japan and southern China. – Four subspecies are currently recognized:

M. f. bucharensis (Uzbekistan, Tadzhikistan, Afghanistan), *M. f. longicaudatus* (Altai mountains to southeastern Siberia and Korea), *M. f. kaguyae* (Japan), *M. f. frater* (southern China).

25. M. ozensis IMAIZUMI 1954 [*muricola* group]. – Size small (forearm length, 33–34 mm). No keel on calcar. Tail and tibia relatively short. Braincase relatively low and rostrum relatively short. Anterior and middle upper premolars apparently fused to form a single bicuspid tooth. Anterior lower premolar displaced medially. Tragus turned outwards. – *Distribution:* A poorly known species confined to Honshu, Japan. – No subspecies.

26. M. hosonoi IMAIZUMI 1954 [*muricola* group]. – Size small (forearm length, 34–36 mm). Keel on calcar well developed. Tail relatively short but extending slightly beyond edge of uropatagium. Middle upper premolar somewhat displaced medially from the toothrow. Posterior border or ear pinna without a distinct emargination, tragus relatively straight. – *Distribution:* A poorly known species confined to Honshu, Japan. – No subspecies.

27. M. yesoensis YOSHIYUKI 1984 [*muricola* group]. – Size small (forearm length, 32–36 mm; condylobasal length, 12–14 mm). Keel on calcar absent or poorly developed. Tail relatively long. Middle upper premolar in toothrow. Posterior border of ear with a distinct emargination. – *Distribution:* Confined to Hokkaido, Japan. – No subspecies.

28. M. ikonnikovi OGNEV 1912 [*muricola* group]. – Size small (forearm length, 32–35 mm). Keel on calcar poorly developed. Tail relatively long but extending to edge of uropatagium. Braincase fairly high. Middle upper premolar in toothrow. Posterior border of ear pinna with a distinct emargination, tragus relatively straight. – *Distribution:* Ranging from southeastern Siberia to Mongolia and Korea, also Sakhalin and Hokkaido in Japan. – No subspecies. May be a subspecies of *M. muricola*.

29. M. muricola (GRAY 1846) [*muricola* group]. – Size small to medium (forearm length, 30–40 mm; condylobasal length, 11–15 mm). Keel on calcar fairly well developed. Braincase usually fairly high. Rostrum relatively short but varying from relatively slender to relatively broad. Middle upper premolar varies from being in the toothrow to being greatly reduced and displaced medially. Possibly more than one species is represented here. – *Distribution:* Ranging from Afghanistan to Taiwan and southeastward to the Philippines,

Moluccas, and Lesser Sundas. – Nine subspecies are here recognized:

M. m. moupinensis (southern mainland China including Hainan), *M. m. latirostris* (Taiwan), *M. m. caliginosus* (Afghanistan to northeastern India), *M. m. muricola* (northeastern India and Vietnam to the Lesser Sundas and most of Borneo), *M. m. niasensis* (islands off the west coast of Sumatra), *M. m. nugax* (northeastern Borneo and the Palawan group of the Philippines), *M. m. herrei* (Luzon in the Philippines), *M. m. browni* (Mindanao in the Philippines), *M. m. ater* (Celebes and the Moluccas).

30. *M. australis* (DOBSON 1878) [*muricola* group]. – Size medium (forearm length, 35–39 mm). Keel on calcar well developed. Middle upper premolar in toothrow. Tragus turned outwards. – *Distribution:* A poorly known species definitely known only by the type which supposedly came from eastern Australia. A specimen from northwestern Australia may be referable to the species. – No currently recognized subspecies.

31. *M. annectans* (DOBSON 1871) (= *primula* THOMAS 1920) [*muricola* group]. – Size medium (forearm length, 43–47 mm; condylobasal length, 15–17 mm). Braincase moderately high and rostrum of medium width. Middle upper premolar always greatly reduced and displaced medially from the toothrow and may be absent, as is the middle lower premolar. Canines relatively short. No thumb or foot pads. – *Distribution*: Ranging from northeastern India to Thailand; the Sumatran record is apparently misidentified *M. ridleyi*. – No subspecies.

32. *M. ridleyi* (THOMAS 1898) [*muricola* group]. – Size small (forearm length, 28–32 mm; condylobasal length, 11–12 mm). Braincase relatively high; rostrum relatively short and of medium width. Middle upper and lower premolars absent. Canines relatively short. No thumb or foot pads. – *Distribution:* Known only from Malaya, Borneo and probably Sumatra. – No subspecies.

33. *M. rosseti* (OEY 1951) [*muricola* group]. – Size small (forearm length, 27–31 mm; condylobasal length, 11–12 mm). Braincase fairly high; rostrum relatively short and broad. Middle upper and lower premolars absent. Thumb and foot pads present. – *Distribution:* Known only from Thailand and Cambodia. – No subspecies.

34. *M. siligorensis* (HORSFIELD 1855) [*siligorensis* group]. – Size small (forearm length, 31–35 mm; condylobasal length, 11–12 mm). Braincase relatively high but rostrum relatively slender. Middle upper premolar in toothrow. Canines relatively short. Ears relatively long and tail extending slightly beyond edge of uropatagium. – *Distribution:* Ranging from northern India to southern China and south to Malaya and Borneo. – Four subspecies are recognized:

M. s. sowerbyi (southeastern China), *M. s. siligorensis* (northern India), *M. s. alticraniatus* (Burma, southwestern China, Vietnam), *M. s. thaianus* (Thailand). Malayan and Bornean populations have not been allocated subspecifically.

35. *M. californicus* (AUDUBON & BACHMAN 1842) [*leibii* group]. – Size fairly small (forearm length, 29–37 mm; condylobasal length, 11–14 mm). Braincase of moderate height and rostrum relatively slender. Foot very small with a keel on the calcar. Middle upper premolar in toothrow. Fur not glossy. – *Distribution:* Ranging from southwestern Canada to extreme southern Mexico. – Four poorly marked subspecies are recognized.

36. *M. leibii* (AUDUBON & BACHMAN 1842) [*leibii* group]. – Size fairly small (forearm length, 29–36 mm; condylobasal length, 12–14 mm). Braincase fairly low but rostrum of medium width. Middle upper premolar more or less displaced medially from toothrow. Fur glossy. – *Distribution:* Ranging from southern Canada to Central Mexico. – Three subspecies are recognized:

M. l. leibii (southeastern Canada to Oklahoma in the southcentral United States), *M. l. ciliolabrum* (southcentral Canada to central United States), *M. l. melanorhinus* (southwestern Canada to central Mexico). It is possible that *ciliolabrum* and *melanorhinus* are specifically distinct from *leibii*.

37. *M. planiceps* BAKER 1955 [*leibii* group]. – Size small (forearm length, 25–28 mm; condylobasal length, 13–15 mm). Braincase greatly flattened and rostrum relatively slender. Middle upper premolar in toothrow. – *Distribution*: Confined to a small area in northeastern Mexico. – No subspecies.

38. *M. altarium* THOMAS 1911 [*oreius* group]. – Size medium (forearm length, 45 mm; condylobasal length, 14–15 mm). Ears relatively long and narrow. Braincase relatively high and rostrum short and broad. Middle upper premolar in toothrow. – *Distribution:* Known only from southern China and northern Thailand. – No subspecies.

39. *M. oreias* (TEMMINCK 1840) [*oreias* group]. – Size fairly small (forearm length, 38 mm). – *Distribution:* A poorly known species recorded only from Malaya. – No subspecies.

40. *M. scotti* THOMAS 1927 [*scotti* group]. – Size medium (forearm length, 37–40 mm; con-

dylobasal length, 13–14 mm). Braincase relatively high. Canines relatively short. Middle upper premolar in toothrow. Tail relatively long. – *Distribution:* Confined to the highlands of Ethiopia. – No subspecies.

41. *M. sodalis* MILLER & G. M. ALLEN 1928 [*nigricans* group]. – Size medium (forearm length, 36–41 mm; condylobasal length, 13–15 mm). Braincase relatively low but occiput raised above it and rostrum relatively broad. Keel on calcar well developed. Middle upper premolar in toothrow. Sagittal crest absent or poorly developed. – *Distribution*: Confined to eastern United States. – No subspecies.

42. *M. elegans* HALL 1962 [*nigricans* group]. – Size fairly small (forearm length, 31–35 mm; condylobasal length, 12–13 mm). Rostrum of medium breadth. Sagittal crest absent. – *Distribution:* Ranging from northeastern Mexico to Costa Rica. – No subspecies.

43. *M. nigricans* (SCHINZ 1821) [*nigricans* group]. – Size medium to fairly small (forearm length, 31–40 mm; condylobasal length, 11–14 mm). Braincase of medium height, occiput not raised above it, and rostrum varying from medium to fairly broad. Keel on calcar absent or poorly developed. – *Distribution:* Ranging from tropical Mexico through Middle America and South America to Argentina, but west of the Andes not south of northern Peru; also Trinidad, Tobago, and Grenada. – Five subspecies are here recognized:

M. n. extremus (eastern Mexico), *M. n. nigricans* (Central and most of tropical South America north to Grenada), *M. n. caucensis* (eastern slopes of the Andes from northern Colombia to Peru), *M. n. punensis* (west of the Andes from Colombia to extreme northern Peru), *M. n. carteri* (west-central Mexico). More than one species may be represented within this complex.

44. *M. findleyi* BOGAN 1978 [*nigricans* group]. – Size relatively small (forearm length, 29–34 mm; condylobasal length, 11–13 mm). Rostrum relatively long and slender. Keel on calcar poorly developed. Sagittal crest fairly well developed. – *Distribution:* Confined to the Tres Marias islands off western Mexico. – No subspecies.

45. *M. dominicensis* MILLER 1902 [*nigricans* group]. – Size fairly small (forearm length, 32–36 mm; condylobasal length, 11–13 m). Occiput not raised above braincase and rostrum of medium width. Sagittal crest absent. – *Distribution:* Definitely known only from Dominica in the central Lesser Antilles, but *Myotis* from Montserrat and St. Martin in the northern Lesser Antilles are probably referable here. – No subspecies.

46. *M. atacamensis* (LATASTE 1891) [*nigricans* group]. – Size relatively small (forearm length, 30–34 mm; condylobasal length, 11–12 mm). Occiput not raised above braincase and rostrum relatively slender. Sagittal crest absent. – *Distribution:* Confined to the Pacific coastal region of southern Peru and northern Chile. – No subspecies.

47. *M. nesopolus* MILLER 1900 [*nigricans* group]. – Size fairly small (forearm length, 29–34 mm; condylobasal length, 12–13 mm). Rostrum of medium width. Sagittal crest absent. – *Distribution:* Confined to northwestern Venezuela and its offshore islands. – Two subspecies are recognized:

M. n. larensis (northwestern Venezuela), *M. n. nesopolus* (Curacao and Bonaire).

48. *M. martiniquensis* LA VAL 1973 [*nigricans* group]. – Size medium (forearm length, 34–39 mm; condylobasal length, 13–14 mm). Braincase fairly high and occiput not raised above it. – *Distribution:* Confined to the southern Lesser Antilles. – Two subspecies are recognized:

M. m. nyctor (Barbados), *M. m. martiniquensis* (Martinique).

49. *M. keaysi* J. A. ALLEN 1914 [*nigricans* group]. – Size medium (forearm length, 31–42 mm; condylobasal length, 12–14 mm). Sagittal crest well developed. – *Distribution:* Ranging from northeastern Mexico to Costa Rica, in Venezuela and Trinidad, and in the Andean region from Colombia to Argentina. – Two subspecies are recognized:

M. k. pilosotibialis (Middle America, Venezuela, and Trinidad, *M. k. keaysi* (Colombia to Bolivia).

Subgenus ***Leuconoe*** BOIE 1830

Anterior upper premolar usually in toothrow and not greatly reduced. Size fairly small to relatively large. Maxillary tooth row relatively long. Margin of plagiopatagium tending to be attached well up on the ankle or lower leg, leaving the large foot free. Wings and legs usually relatively short and hairy. Keel on calcar usually poorly developed. Usually with strong development of accessory molar cusps. These bats usually forage by gleaning insects (or in a few cases, fish) from on or near the surface of water. – *Distribution:* Virtually coextensive with that of the genus, but absent from the West Indies. – There are 33 species and 43 additional subspecies.

50. *M. horsfieldii* (TEMMINCK 1840) [*adversus* group]. – Size medium (forearm length, 35–40 mm; condylobasal length, 13–15 mm). Margin of plagiopatagium attached to side of foot, which is fairly large. Rostrum of medium width. Middle upper premolar in toothrow or somewhat displaced medially. Uropatagium without a fringe of hair. Postpalatal extension of skull well developed. – *Distribution:* Ranging from India and southeastern China to the Philippines and Celebes. – Five subspecies are currently recognized:

M. h. peshwa (India), *M. h. deignani* (southeastern China, northern Thailand), *M. h. dryas* (Andaman islands), *M. h. horsfieldii* (Malaya to Bali and Celebes), *M. h. jeannei* (Philippines).

51. *M. hasseltii* (TEMMINCK 1840) [*adversus* group]. – Size medium (forearm length, 35–42 mm; condylobasal length, 14–16 mm). Margin of uropatagium attached to ankle. Rostrum fairly broad. Middle upper premolar greatly reduced and displaced medially from the toothrow. Postpalatal extension of skull shortened. – *Distribution:* Ranging from Sri Lanka and Burma to Java and Borneo. – Four subspecies are currently recognized:

M. h. continentis (Burma to Cambodia), *M. h. hasseltii* (Malaya to Java), *M. h. macellus* (Borneo), *M. h. abbotti* (Mentawei islands west of Sumatra). The Sri Lankan population has not been allocated subpecifically.

52. *M. adversus* (HORSFIELD 1824) [*adversus* group]. – Size medium (forearm length, 36–44 mm; condylobasal length, 14–16 mm). Margin of uropatagium attached to ankle. – *Distribution:* Ranging from Malaya and Taiwan to the New Hebrides and Australia. – Six subspecies are here recognized:

M. a. taiwanensis (Taiwan, with a dubious record from Tibet), *M. a. adversus* (Java and apparently Malaya), *M. a. carimatae* (Sumatra, Borneo, and nearby islands), *M. a. moluccarum* (Celebes to the Solomons), *M. a. macropus* (northern and eastern Australia), *M. a. orientis* (New Hebrides).

53. *M. bocagei* (PETERS 1870) [*adversus* group]. – Size medium (forearm length, 36–40 mm; condylobasal length, 13–15 mm). Foot relatively small. – *Distribution:* Ranging from Senegal and Yemen to the Transvaal. – Three subspecies are currently recognized:

M. b. dogalensis (Yemen), *M. b. bocagei* (Ethiopia to Angola and Transvaal), *M. b. cupreolus* (Liberia to Zaire).

54. *M. riparius* HANDLEY 1960 [*ruber* group]. – Size fairly small (forearm length, 31–39 mm; condylobasal length, 13–14 mm). Margin of plagiopatagium attached to side of foot. Calcar with a keel. Uropatagium without a fringe of hair. Braincase usually with a well developed sagittal crest. Rostrum of medium width. Middle upper premolar usually at least somewhat displaced medially from the toothrow. – *Distribution:* Ranging from Honduras to Uruguay, including Trinidad, but west of the Andes not south of Ecuador. – No subspecies.

55. *M. simus* THOMAS 1901 [*ruber* group]. – Size fairly small (forearm length, 35–40 mm; condylobasal length, 12–14 mm). Calcar usually with a keel. Rostrum relatively short and broad. Middle upper premolar greatly reduced and displaced medially from the toothrow (anterior upper premolar also somewhat displaced). – *Distribution:* Restricted to tropical South America east of the Andes (chiefly Amazonian) from southern Colombia to northeastern Argentina and northeastern Brazil. – No subspecies.

56. *M. ruber* (E. GEOFFROY 1806) [*ruber* group]. – Size medium (forearm length, 39–41 mm; condylobasal length, 14–15 mm). Keel on calcar poorly developed. Foot relatively large. Braincase with a well developed sagittal crest. Middle upper premolar in toothrow. – *Distribution:* Confined to Paraguay, northeastern Argentina, and southeastern Brazil. – *No subspecies.*

57. *M. montivagus* (DOBSON 1874) [*montivagus* group]. – Size medium (forearm length, 39–47 mm; condylobasal length, 14–17 mm). Foot relatively small; calcar with a small keel. Braincase with a weak sagittal crest; rostrum relatively short and broad. Middle upper premolar displaced medially from the toothrow. – *Distribution:* Known from southern India, Burma, southern and eastern China, Malaya, and Borneo. – Four subspecies are currently recognized:

M. m. peytoni (southern India), *M. m. montivagus* (Burma, China), *M. m. federatus* (Malaya), *M. m. borneoensis* (Borneo).

58. *M. fortidens* MILLER & ALLEN 1928 [*montivagus* group]. – Size medium (forearm length, 35–39 mm; condylobasal length, 13–14 mm). Uropatagium without a fringe of hair. Middle upper and lower premolars absent. – *Distribution:* Confined to tropical Mexico and Guatemala. – Two subspecies are recognized:

M. f. sonoriensis (northwestern Mexico), *M. f. fortidens* (remainder of range).

59. M. grisescens HOWELL 1909 [*levis* group]. – Size medium (forearm length, 40–46 mm; condylobasal length, 14–16 mm). Margin of plagiopatagium attached to ankle of fairly large foot. No keel on calcar. Margin of uropatagium without a fringe of hair. Sagittal crest of braincase well developed. Rostrum of medium width. Middle upper premolar in toothrow. – *Distribution:* Confined to the south-central United States. – No subspecies.

60. M. velifer (J. A. ALLEN 1890) [*levis* group]. – Size fairly large (forearm length, 36–47 mm; condylobasal length, 14–17 mm). Margin of plagiopatagium attached to side of foot. Rostrum relatively broad. Middle upper premolar in toothrow or slightly displaced medially. – *Distribution:* Ranging from the southwestern United States to Honduras (but not in Baja California). – Three subspecies are currently recognized:

M. v. magnamolaris (Kansas to northern Texas), *M. v. incautus* (New Mexico to northeastern Mexico), *M. v. velifer* (Nevada to Honduras).

61. M. peninsularis MILLER 1898 [*levis* group]. – Size medium (forearm length, 37–41 mm; condylobasal length, 13–15 mm). Margin of plagiopatagium attached to side of foot. Sagittal crest of braincase fairly well developed. Rostrum fairly broad. – *Distribution:* Confined to southern Baja California. – No subspecies.

62. M. cobanensis GOODWIN 1955 [*levis* group]. – Size medium (forearm length, 41–42 mm; condylobasal length, 13–14 mm). Margin of plagiopatagium attached to side of foot. Sagittal crest of braincase poorly developed. Rostrum fairly broad and relatively short. Middle upper premolar reduced and displaced medially from toothrow. – *Distribution:* Known only from the highlands of Guatemala. – No subspecies.

63. M. oxyotus (PETERS 1867) [*levis* group]. – Size fairly small (forearm length, 36–44 mm; condylobasal length, 13–15 mm). Margin of plagiopatagium attached to side of foot. Sagittal crest of braincase absent or poorly developed. Rostrum of medium width or fairly slender. – *Distribution:* Known from Costa Rica and western Panama and in South America from Venezuela to Bolivia, mostly in mountain and Pacific coastal areas. – Two subspecies are recognized:

M. o. gardneri (Costa Rica, Panama), *M. o. oxyotus* (remainder of range except for coastal Peru, where subspecific allocation is uncertain).

64. M. levis (I. GEOFFROY 1824) [*levis* group]. – Size fairly small (forearm length, 35–41 mm; condylobasal length, 13–16 mm). Margin of plagiopatagium attached to side of foot. Uropatagium with a fringe of hair. Braincase relatively low, its sagittal crest absent or poorly developed. Rostrum relatively slender. – *Distribution:* Ranging from Bolivia and southeastern Brazil to southeastern Argentina. – Two subspecies are recognized:

M. l. levis (southern Brazil to northeastern Argentina), *M. l. dinelli* (Bolivia to southern Argentina).

65. M. aelleni BAUD 1979 [*levis* group]. – Size fairly small (forearm length, 37–42 mm; condylobasal length, 13–15 mm). Margin of plagiopatagium attached to side of foot. Braincase relatively broad, its sagittal crest poorly developed. Rostrum relatively slender. – *Distribution:* Known only from a small area in southwestern Argentina. – No subspecies.

66. M. chiloensis (WATERHOUSE 1840) [*levis* group]. – Size medium (forearm length, 37–39 mm; condylobasal length, 13–14 mm). Margin of plagiopatagium attached to side of foot. Sagittal crest virtually absent. Rostrum relatively slender. – *Distribution:* Confined to central and southern Chile. – No subspecies.

67. M. daubentoni (KUHL 1819) [*daubentoni* group]. – Size medium to fairly small (forearm length, 35–39 mm; condylobasal length, 12–15 mm). Margin of plagiopatagium attached to side of foot near the ankle. Margin of uropatagium without a fringe of hair. Braincase relatively low without a sagittal crest. Rostrum relatively slender. Middle upper premolar in toothrow. – *Distribution:* Ranging from western Europe to eastern Siberia and south to southern China and northeastern India, including Sakhalin, Kuriles, and Hokkaido in Japan. – Six subspecies are here recognized:

M. d. nathalinae (southwestern Europe), *M. d. daubentoni* (northwestern Europe), *M. d. volgensis* (eastern Europe to central Siberia), *M. d. ussuriensis* (eastern Siberia and northeastern China to Japan and the Kuriles), *M. d. petax* (Altai region of Siberia and Mongolia), *M. d. laniger* (southern China and northeastern India).

68. M. capaccinii (BONAPARTE 1837) [*daubentoni* group]. – Size medium (forearm length, 37–43 mm; condylobasal length, 13–15 mm). Margin of plagiopatagium attached to ankle. Braincase of medium height. – *Distribution:* Ranging from northwestern Africa through

southern Europe to southwestern Asia. – Two subspecies are here recognized:

M. c. capaccinii (northwestern Africa to Yugoslavia), *M. c. bureschi* (Bulgaria to Turkmenia).

69. *M. macrodactylus* (TEMMINCK 1840) [*daubentoni* group]. – Size medium (forearm length, 36–41 mm; condylobasal length, 13–14 mm). Margin of plagiopatagium attached to ankle. Braincase of medium height. – *Distribution:* Occurring in southern China, southeastern Siberia, Japan, and the Kurile islands. – Two subspecies are here recognized:

M. m. fimbriatus (southern China), *M. m. macrodactylus* (remainder of range).

70. *M. abei* YOSHIKURA 1944 [*daubentoni* group]. – Size fairly small (forearm length, 34 mm). Margin of plagiopatagium attached to ankle. – *Distribution:* A poorly known species recorded only from Sakhalin. – No subspecies.

71. *M. longipes* (DOBSON 1872) [*daubentoni* group]. – Size fairly small (forearm length, 33–39 mm; condylobasal length, 12–14 mm). Braincase of medium height. Middle upper premolar slightly displaced medially from the toothrow. – *Distribution:* Known only from Afghanistan and Kashmir. – No subspecies.

72. *M. pruinosus* YOSHIYUKI 1971 [*daubentoni* group]. – Size fairly small (forearm length, 30–33 mm; condylobasal length, 12–13 mm). Margin of uropatagium attached to side of foot, well removed from the ankle. – *Distribution:* Known only from Honshu in Japan. – No subspecies.

73. *M. lucifugus* (LE CONTE 1831) [*albescens* group]. – Size fairly small (forearm length, 33–41 mm; condylobasal length, 13–16 mm). Margin of plagiopatagium attached to side of foot. Margin of uropatagium without a fringe of hair. No keel on the calcar. Braincase relatively low, without a sagittal crest. Rostrum varying from medium width to relatively slender. Middle upper premolar varying from being within toothrow to absent. Fur usually glossy. – *Distribution:* Ranging from Alaska to eastern Canada and south to central Mexico. – Six subspecies are currently recognized:

M. l. lucifugus (central Alaska to eastern Canada and most of the eastern United States), *M. l. alascensis* (southeastern Alaska to California, mostly in coastal regions), *M. l. pernox* (a small area in western Alberta), *M. l. relictus* (a small area in eastern California), *M. l. carissima* (from southwestern Canada through much of western United States),

M. l. occultus (southwestern United States to central Mexico).

74. *M. austroriparius* (RHOADS 1897) [*albescens* group]. – Size fairly small (forearm length, 34–42 mm; condylobasal length, 13–14 mm). Braincase of medium height, a poorly developed sagittal crest usually present. Rostrum relatively slender. Middle upper premolar in toothrow. Fur not glossy. – *Distribution:* Confined to the southeastern United States. – No subspecies are currently recognized.

75. *M. yumanensis* (H. ALLEN 1864) [*albescens* group]. – Size fairly small (forearm length, 32–38 mm; condylobasal length, 12–14 mm). Braincase of medium height. Middle upper premolar in toothrow. Fur not glossy. – *Distribution:* Ranging from southwestern Canada to central Mexico. – Six subspecies are currently recognized:

M. y. saturatus (southwestern Canada to southern California, mostly in coastal regions), *M. y. oxalis* (a small area in central California), *M. y. sociabilis* (southwestern Canada to northern California, mostly in interior regions), *M. y. yumanensis* (southwestern United States and most of northwestern Mexico), *M. y. lambi* (a small area in central Baja California), *M. y. lutosus* (central Mexico).

76. *M. albescens* (E. GEOFFROY 1806) [*albescens* group]. – Size fairly small (forearm length 33–39 mm; condylobasal length, 12–14 mm). Margin of uropatagium usually with a poorly developed fringe of hair. Rostrum fairly slender. Middle upper premolar in toothrow. Fur white-tipped. – *Distribution:* Ranging from southern Mexico to central Argentina but west of the Andes not south of Ecuador. – No subspecies are currently recognized.

77. *M. volans* (H. ALLEN 1866) [*albescens* group]. – Size fairly small (forearm length, 33–42 mm; condylobasal length, 12–15 mm). Calcar with a keel. Rostrum of medium width. Middle upper premolar in toothrow. – *Distribution:* Ranging from southeastern Alaska to central Mexico. – Four subspecies are recognized:

M. v. amotus (central Mexico), *M. v. volans* (southern Baja California), *M. v. interior* (northwestern Mexico and most of the western United States), *M. v. longicrus* (southeastern Alaska to California).

78. *M. macrotarsus* (WATERHOUSE 1845) [*macrotarsus* group]. – Size fairly large (forearm length, 45–49 mm; condylobasal length, 16–17 mm). Foot unusually large. Margin of plagiopatagium attached to distal end of tibia. No fringe of hair on margin of uropatagium. Middle upper premolar more or less displaced medially from the tooth-

row. – *Distribution:* Known only from Borneo and the Philippines. – Two subspecies:

M. m. saba (Borneo), *M. m. macrotarsus* (Philippines).

79. *M. stalkeri* THOMAS 1910 [*macrotarsus* group]. – Size fairly large (forearm length, 48 mm). It is not clear how this species differs from *M. macrotarsus*. – *Distribution:* Confined to the Kei islands in the eastern Moluccas. – No subspecies.

80. *M. vivesi* MENEGAUX 1901 [*macrotarsus* group]. – Size relatively large (forearm length, 59–63 mm). Margin of plagiopatagium attached to side of unusually large foot, but greatly narrowed near the leg, thus largely freeing the foot for use in gaffing fish. Braincase fairly low. Rostrum relatively broad. Middle upper premolar in toothrow. – *Distribution:* Confined to coasts of northwestern Mexico. – No subspecies.

81. *M. pilosus* (PETERS 1869) (= *ricketti* THOMAS 1894) [*pilosus* group]. – Size relatively large (forearm length, 55–58 mm). Foot unusually large. Margin of plagiopatagium attached near middle of tibia. No fringe of hair on margin of uropatagium. Braincase of medium height. Rostrum relatively broad. Middle upper premolar displaced medially from the toothrow. – *Distribution:* Confined to southern and eastern China. – No subspecies.

82. *M. dasycneme* (BOIE 1825) [*dasycneme* group]. – Size fairly large (forearm length, 43–48 mm). Foot relatively large. Margin of plagiopatagium attached near ankle. No fringe of hair on margin of uropatagium. Braincase fairly low. Rostrum relatively broad. Middle upper premolar displaced medially from toothrow. – *Distribution:* Ranging form northwestern Europe east to central Siberia and northeastern China. – No currently recognized subspecies.

Subgenus ***Cistugo*** THOMAS 1912

Anterior upper premolar displaced medially from toothrow and greatly reduced. Size fairly small. Rostrum relatively short and broad. – *Distribution:* Confined to the arid southwestern corner of Africa. – Two species.

83. *M. seabrai* (THOMAS 1912). – Size relatively small (forearm length, 32–33 mm). – *Distribution:* Ranging from southwestern Angola to northwestern Cape Province. – No subspecies.

84. *M. lesueuri* (ROBERTS 1919). – Size fairly small (forearm length, 34–35 mm). – *Distribution:* Confined to central and southwestern Cape Province. – No subspecies.

Genus ***Lasionycteris*** PETERS 1866 (Fig. 147)

Dental formula i2/3, c1/1, p2/3, m3/3x2 = 36. Hypocone on anterior and middle upper molars fairly well developed. Tragus of ear short and bluntly rounded. Proximal half of uropatagium well-haired dorsally. – *Distribution:* Confined to the Nearctic, ranging from southeastern Alaska and southeastern Canada to northeastern Mexico. – A single species.

1. *L. noctivagans* (LE CONTE 1831). – Size medium (forearm length, 37–44 mm). – *Distribution:* Same as for genus. – No subspecies.

Tribe Plecotini GRAY 1866

Rostrum fairly elongate. Dental formula never less than i2/3, c1/1, p2/2, m3/3x2 = 34, but middle upper premolar always absent. Ears large and united by a band. Nostrils opening on dorsal surface of muzzle. – *Distribution:* Ranging widely across the Palearctic and Nearctic, south to the Cape Verde islands, Senegal, Ethiopia, northern India, and southern Mexico. – Three genera, two additional subgenera, and nine species.

Genus ***Barbastella*** GRAY 1821 (Fig. 148)

Dental formula i2/3, c1/1, p2/2, m3/3x2 = 34. Ear pinna and tympanic bulla relatively small. Lower canine relatively large. – *Distribution:* Ranging from northwestern Africa (and perhaps Senegal) through Europe and southwestern Asia and northeastern Africa to northern India and western China, also Japan. – Two species, one additional subspecies.

1. *B. barbastellus* (SCHREBER 1774). – Size relatively small (forearm length, 36–41 mm; condylobasal length, 13–14 mm). Outer margin of ear pinna with a prominent projecting lobe. – *Distribution:* Occurring in Europe south to the Canary islands, Morocco (and possibly Senegal) and the Caucasus. – No subspecies.

2. *B. leucomelas* (CRETZSCHMAR 1826). – Size relatively large (forearm length, 38–45 mm; condylobasal length, 14–15 mm). Outer margin of

ear pinna without a projecting lobe. – *Distribution:* Ethiopia to northern India and western China, also Japan and perhaps Senegal. – Two subspecies:

B. l. leucomelas (Africa to Iran), *B. l. darjelingensis* (Iran to western China and northern India, also Japan).

Genus *Euderma* H. ALLEN 1892 (Fig. 149)

Dental formula i2/3, c1/1, p2/2, m3/3 x 2 = 34. Ear pinna and tympanic bulla relatively large. Lower canine reduced in height. – *Distribution:* Ranging in North America from southwestern Canada to central Mexico. – A single species.

1. *E. maculatum* (J. A. ALLEN 1891). – Size relatively large (forearm length, 48–51 mm). Three large white spots on dorsum. – *Distribution:* Same as for genus. – No subspecies.

Genus *Plecotus* E. GEOFFROY 1818 (Fig. 150)

Dental formula i2/3, c1/1, p2/3, m3/3 x 2 = 36. Ear pinna and tympanic bulla relatively large. Lower canine relatively large. – *Distribution:* Virtually the same as for the tribe. – Three subgenera, six species.

Subgenus *Idionycteris* ANTHONY 1923

Calcar keeled. Zygomatic arch relatively thick and strong. Basal pits absent. Rostrum flattened with a median concavity. – *Distribution:* Ranging from the southwestern United States to Central Mexico. – A single species.

1. *P. phyllotis* (G. M. ALLEN 1916). – Size fairly large (forearm length, 44–45 mm). – *Distribution:* Same as for subgenus. – No subspecies.

Subgenus *Plecotus* E. GEOFFROY 1818

Calcar not keeled. Zygomatic arch relatively thick and strong. Basal pits absent. Rostrum arched, without a median concavity. – *Distribution:* Ranging widely across the Palearctic, south to the Cape Verde islands, Senegal, Ethiopia and northern India. – Two species, 10 additional subspecies.

2. *P. auritus* (LINNAEUS 1758). – Paired limbs of baculum about as long as wide. Size relatively small (forearm length, 35–40 mm; maxillary tooth row length usually less than 5.6 mm; greatest width of tragus usually about 5 mm). Proximal zone of dorsal pelage brown or brownish gray. – *Distribution:* Ranging from western Europe to Japan and south to the Himalayas. – Four subspecies are here recognized:

P. a. auritus (western Europe to central Siberia), *P. a. homochrous* (Himalayas), *P. a. uenoi* (Korea), *P. a. sacrimontis* (Japan). Subspecific allocation of most populations from eastern Asia (including northern China, eastern Siberia, and Sakhalin) is uncertain.

3. *P. austriacus* (FISCHER 1829). – Paired limbs of baculum considerably longer than wide. Size relatively large (forearm length, 37–41 mm; maxillary toothrow length usually greater than 5.6 mm; greatest width of tragus usually about 6 mm). Proximal zone of dorsal pelage dark gray or black. – *Distribution:* Ranging from western Europe and northwestern Africa (including Senegal) to Mongolia and the Himalayas, also the Canary and Cape Verde islands. – Eight subspecies are here recognized:

P. a. teneriffae (Canary islands), *P. a. austriacus* (western Europe and northwestern Africa including the Cape Verde islands, to southeastern Europe), *P. a. kolombatovici* (certain islands in the Adriatic), *P. a. christiei* (northeastern Africa to Ethiopia and the eastern end of the Mediterranean), *P. a. macrobullaris* (Caucasus region), *P. a. wardi* (Iran to Kashmir and Sinkiang), *P. a. ariel* (southwestern China). The subspecific, and even specific, allocation of many populations is uncertain.

Subgenus *Corynorhinus* H. ALLEN 1865

Calcar not keeled. Zygomatic arch relatively thin and fragile. Basal pits prominent. Rostrum flattened with a median concavity. – *Distribution:* Ranging through western and southeastern North America from southwestern Canada to Florida and southern Mexico. – Three species, five additional subspecies.

4. *P. mexicanus* (G. M. ALLEN 1916). – Brownish tips of hair on ventral side not contrasting with gray or brownish bases. Forearm length, 39–46 mm. Greatest length of skull, 14.7–15.9 mm. Tragus length usually less than 13 mm. – *Distribution:* Ranging from northwestern and northeastern to central Mexico, also Yucatan and Cozumel island off its eastern side. – No subspecies.

5. *P. townsendii* COOPER 1837. – Brownish tips of hair on ventral side not contrasting with gray or brownish bases. Forearm length, 39–48 mm. Greatest length of skull, 15.2–17.2 mm. Tragus length usually more than 13 mm. – *Distribution:*

Ranging from southwestern Canada to southern Mexico and east in the central United States to Virginia. – Five subspecies are currently recognized:

P. t. virginianus (Virginia to Kentucky), *P. t. ingens* (Missouri to Oklahoma), *P. t. pallescens* (southwestern Canada to northern Mexico), *P. t. townsendii* (coastal regions from extreme southwestern Canada to California), *P. t. australis* (central and southern Mexico).

6. *P. rafinesquii* LESSON 1827. – Whitish tips of hair on ventral side contrasting with blackish bases. Forearm length, 40–46 mm. Greatest length of skull, 15.3–16.7 mm. – *Distribution:* Confined to the southeastern and east central parts of the United States. – Two subspecies are recognized:

P. r. macrotis (southeastern United States), *P. r. rafinesquii* (east central United States).

Tribe Vespertilionini GRAY 1821

Rostrum fairly short. Dental formula never more than i2/3, c1/1, p2/3, m3/3 × 2 = 36 and rarely less than i2/3, c1/1, p1/2, m3/3 × 2 = 32. Middle lower premolar always greatly reduced and usually lost. Lateral upper incisor rarely lost. Ear pinna may be greatly lengthened but never greatly shortened. – *Distribution:* Widely distributed over Eurasia, Africa, the Australian region, and the Americas, but absent on some oceanic islands. – 14 genera, 10 additional subgenera, 123 species.

Genus *Eudiscopus* CONISBEE 1953 (Fig. 151)

Dental formula i2/3, c1/1, p2/3, m3/3x2 = 36, but middle lower premolar greatly reduced and displaced medially from the toothrow. Rostrum long and relatively narrow, shallow, and turned upward in relation to the flattened braincase. Hind feet with large plantar pads. – *Distribution:* Known only from Burma and Laos. – A single species.

1. *E. denticulus* (OSGOOD 1932). – Size fairly small (forearm length, 34–38 mm; condylobasal length, 12–14 mm). – *Distribution:* Same as for genus. – No subspecies.

Genus *Pipistrellus* KAUP 1829 (Fig. 152)

Dental formula almost always i2/3, c1/1, p1–2/2, m3/3x2 = 32–34, but lateral upper incisor may be absent in a few species. Ear pinna not greatly enlarged. Braincase neither markedly deepened or markedly flattened. Outer upper incisor cuspidate, without a flat crown, not pushed so far forward that its concavity points directly outwards, nor pushed inwards directly posterior to the anterior upper incisor. Fifth digit of wing not reduced, considerably longer than the third or fourth metacarpal. No thickened pads at the base of the thumb or on the plantar surface of the hind foot. Baculum not triangular. – *Distribution:* Virtually the same as for the tribe except that in the Americas it does not occur south of Honduras and is absent from the West Indies. – Seven subgenera, 60 species.

Subgenus *Pipistrellus* KAUP 1829 (= *Hypsugo* KOLENATI 1856; *Perimyotis* MENU 1984)

Outer upper incisors present, usually at least fairly well developed, and posterior to the inners. Inner upper incisors usually not deeply bicuspid. Anterior upper premolar normally present. Size small to fairly large (forearm length less than 45 mm). Penis rarely greatly enlarged (and if so, then braincase not flat on top). This subgenus is poorly defined, including all species not clearly belonging to other, more derived, subgenera. The groups within it are also very tentative. – *Distribution:* Same as for genus, but absent from all but the northern edge of Australia. – 39 species, 42 additional subspecies.

1. *P. pipistrellus* (SCHREBER 1774) [*pipistrellus* group]. – Size relatively small (forearm length, 27–35 mm). Inner upper incisor bicuspid. Outer upper incisor well developed. Anterior upper premolar in toothrow or slightly displaced medially. Rostrum fairly short, but fairly slender. Forehead slightly concave. Tragus without a sharp angle on the posterior border. Thumb relatively short. Lower canine relatively robust. – *Distribution:* Ranging from western Europe and northwestern Africa to Kashmir and northwestern China. – Two subspecies are currently recognized:

P. p. pipistrellus (northwestern Africa, Europe, and southwestern Asia to Iran), *P. p. aladdin* (Iran to Kashmir and northwestern China).

2. *P. sturdeei* THOMAS 1915 [*pipistrellus* group]. – Size relatively small (forearm length, 30 mm). Rostrum relatively slender. – *Distribution:* A poorly known species confined to the Bonin islands in northern Micronesia. – No subspecies.

3. *P. nathusii* (KEYSERLING & BLASIUS 1839) [*pipistrellus* group]. Size fairly small (forearm length, 30–37 mm). Outer upper incisor relatively long. Anterior upper premolar in toothrow. Rostrum relatively long. – *Distribution:* Confined to Europe, Asia Minor, and Trans-Caucasia. – No subspecies.

4. *P. permixtus* AELLEN 1957 [*pipistrellus* group]. – Size fairly small (forearm length, 33–34 mm). Anterior upper premolar on toothrow. Forehead almost flat. Thumb relatively long. Lower canine relatively robust. – *Distribution:* A poorly known species apparently confined to northeastern Tanzania. – No subspecies.

5. *P. subflavus* (F. CUVIER 1832) [*pipistrellus* group]. – Size relatively small (forearm length, 29–36 mm). Anterior upper premolar in toothrow. Lower canine fairly slender. – *Distribution:* Ranging through eastern North America from southeastern Canada to Honduras – Four subspecies are currently recognized:

P. s. subflavus (southeastern Canada to northeastern Mexico), *P. s. floridanus* (Florida and Georgia), *P. s. clarus* (a small area in Texas and Coahuila), *P. s. veraecrucis* (southeastern Mexico to Honduras).

6. *P. arabicus* HARRISON 1979 [*pipistrellus* group]. – Size relatively small (forearm length, 29–32 mm). Upper canine and anterior premolar relatively short. Rostrum fairly long and slender. Thumb relatively long. – *Distribution:* Confined to Oman. – No subspecies.

7. *P. javanicus* (GRAY 1838) [*pipistrellus* group]. – Size fairly small (forearm length, 27–36 mm). Anterior upper premolar displaced medially. Rostrum relatively short and broad. Forehead fairly flat. Upper canine with a poorly developed secondary cusp. – *Distribution:* Ranging from Japan and southeastern Siberia south through eastern Asia and east in the Malay archipelago to the Philippines, Celebes, and Java, with a single Australian record. – Four subspecies are here recognized:

P. j. abramus (Japan and Siberia to Vietnam), *P. j. meyeni* (Philippines), *P. j. javanicus* (Vietnam to Java and Celebes), *P. j. camortae* (Nicobars). More than one species may be represented in this complex.

8. *P. endoi* IMAIZUMI 1959 [*pipistrellus* group]. – Size fairly small (forearm length, 31–33 mm). Anterior upper premolar somewhat displaced medially. Rostrum relatively short and fairly broad. Forehead almost flat. Upper canine with the secondary cusp usually well developed. – *Distribution:* Confined to Japan. – No subspecies.

9. *P. peguensis* SINHA 1969 [*pipistrellus* group]. – Size fairly small (forearm length, 30–33 mm). Anterior upper incisor apparently unicuspid. Anterior upper premolar displaced medially. Rostrum relatively short and broad. Forehead almost flat. Lower canine slender but relatively short. – *Distribution:* A poorly known species apparently confined to Burma. – No subspecies.

10. *P. coromandra* (GRAY 1838) [*pipistrellus* group]. – Size relatively small (forearm length, 29–32 mm). Inner upper incisor bicuspid or unicuspid. Anterior upper premolar more or less displaced medially. Rostrum fairly broad. Forehead slightly concave to almost flat. – *Distribution:* Ranging from Afghanistan to southern China and south to Ceylon and Thailand. – Three subspecies are recognized:

P. c. coromandra (Afghanistan to Ceylon), *P. c. tramatus* (southern mainland China to Thailand), *P. c. portensis* (Hainan island).

11. *P. tenuis* (TEMMINCK 1840) [*pipistellus* group]. – Size fairly small (forearm length, 25–39 mm). Inner upper incisor usually bicuspid but may be unicuspid. Anterior upper premolar at least partly displaced medially. Rostrum fairly long and slender. Forehead more or less concave. – *Distribution:* Ranging from the Malay peninsula and the Philippines to Christmas island in the Indian Ocean, the northern edge of Australia, and the New Hebrides. – Ten subspecies are here recognized:

P. t. tenuis (Malay peninsula and Sumatra), *P. t. subulidens* (South Natuna islands), *P. t. nitidus* (Java, Bali, Borneo, and probably the Philippines), *P. t. sewelanus* (Lesser Sundas and Celebes), *P. t. murrayi* (Christmas island), *P. t. westralis* (extreme northern portions of Western Australia, Northern Territory, and perhaps northwestern Queensland), *P. t. adamsi* (Cape York peninsula), *P. t. papuanus* (Moluccas, lowlands of New Guinea and surrounding islands), *P. t. wattsi* (southeastern New Guinea), *P. t. collinus* (New Guinea highlands), *P. t. angulatus* (Bismarcks and perhaps parts of New Guinea), *P. t. ponceleti* (Solomons, New Hebrides). It is possible that more than one species is represented in this complex.

12. *P. africanus* (RÜPPELL 1842) [= *nanus* PETERS 1852] (*pipistrellus* group). – Size relatively small (forearm length, 25–33 mm). Inner upper incisor usually bicuspid but may be unicuspid. Anterior upper premolar more or less displaced medially. Rostrum fairly short but relatively slender. Forehead relatively concave. Tragus with a sharp angle on the posterior border. – *Distribution:* Ranging widely in sub-Saharan Africa from Senegal, Niger, and Ethiopia south to the Cape Province; also Pemba, Zanzibar, and Madagascar. – Six subspecies are recognized here:

P. a. africanus (Ethiopia to Zaire, mostly in the highlands), *P. a. nanus* (Zaire and Tanzania to the Cape Province), *P. a. helios* (southern Sudan and southern Somalia to northern Tanzania), *P. a. fouriei* (southern Angola, western Zambia, northern Namibia), *P. a. culex* (Nigeria and Ghana), *P. a. stampflii* (Ivory Coast to Sierra Leone).

13. *P. paterculus* THOMAS 1915 [*pipistrellus* group]. – Size relatively small (forearm length, 29–34 mm). Inner upper incisor bicuspid. Outer upper incisor well developed. Anterior upper premolar more or less displaced medially. Rostrum relatively short and broad. Forehead almost flat. Penis greatly enlarged. A species of uncertain relationships. – *Distribution:* Definitely known only from northern Burma, but has been recorded from Thailand and northern India. – A subspecies (*P. p. yunnanensis*), has been described from southwestern China.

14. *P. imbricatus* (HORSFIELD 1924) [*pipistrellus* group]. – Size fairly small (forearm length, 31–36 mm). Inner upper incisor bicuspid. Outer upper incisor biscuspid. Outer upper incisor well developed. Anterior upper premolar greatly reduced and displaced medially. Rostrum relatively short and broad. A poorly known species of uncertain relationships. – *Distribution:* Definitely known only from Java, surrounding islands, and Borneo, but has been recorded from the Philippines. – No subspecies.

15. *P. mimus* WROUGHTON 1899 [*pipistrellus* group]. – Size relatively small (forearm length, 26–30 mm). Inner upper incisor bicuspid or unicuspid. Outer upper incisor well developed. Anterior upper premolar more or less displaced medially. Rostrum relatively short but fairly slender. Forehead fairly concave. – *Distribution:* Ranging form Afghanistan to Vietnam and south to Ceylon. – Two subspecies are currently recognized:

P. m. principulus (Assam), *P. m. mimus* (remainder of range).

16. *P. babu* THOMAS 1915 [*pipistrellus* group]. – Size fairly small (forearm length, 33–35 mm). Inner upper incisor bicuspid. Outer upper incisor well developed. Anterior upper premolar displaced medially. Rostrum relatively short and broad. Forehead almost flat. – *Distribution:* Ranging from Afghanistan to southwestern China. – No subspecies.

17. *P. musciculus* THOMAS 1913 [*hesperus* group]. – Size relatively small (forearm length, 21–25 mm). Inner upper incisor bicuspid. Outer upper incisor well developed. Anterior upper premolar greatly reduced and displaced medially. Rostrum relatively short and fairly broad. Forehead slightly concave. Tragus without a sharp angle on the posterior border. – *Distribution:* Known only from forested regions in Cameroon, Gabon, and western Zaire. – No subspecies.

18. *P. hesperus* (H. ALLEN 1864) [*hesperus* group]. – Size relatively small (forearm length, 27–34 mm). Inner upper incisor faintly bicuspid. Rostrum fairly short and relatively broad. Forehead almost flat. – *Distribution:* Ranging from northwestern United States to central Mexico. – Two subspecies are currently recognized:

P. h. hesperus (chiefly west of the continental divide), *P. h. maximus* (chiefly east of the continental divide).

19. *P. pulveratus* (PETERS 1871) [*pulveratus* group]. – Size fairly small (forearm length, 33–35 mm). Inner upper incisor bicuspid. Outer upper incisor well developed. Anterior upper premolar displaced medially. Rostrum relatively long and fairly slender. Forehead relatively concave. No basicranial pits. – *Distribution:* Confined to southern China and northern Thailand. – No subspecies.

20. *P. lophurus* THOMAS 1915 [*pulveratus* group]. – Size fairly small (forearm length, 35 mm). Rostrum of medium width. Forehead somewhat concave. Basicranial pits well developed. – *Distribution:* Known only from Tenasserim (extreme southern Burma). – No subspecies.

21. *P. kitchneri* THOMAS 1915 [*pulveratus* group]. – Size fairly small (forearm length, 34–38 mm). Anterior upper premolar greatly reduced and displaced medially. Rostrum fairly broad. Forehead almost flat. Basicranial pits well developed. – *Distribution:* Confined to Borneo. – No subspecies.

22. *P. affinis* (DOBSON 1871) [*affinis* group]. – Size medium (forearm length, 38–41 mm). Inner upper incisor bicuspid. Outer upper incisor well developed. Anterior upper premolar displaced medially. Rostrum of medium width. Forehead almost flat. – *Distribution:* Ranging from northern India to southwestern China. – No subspecies.

23. *P. petersi* (MEYER 1899) [*affinis* group]. – Size medium (forearm length, 36–42 mm). Rostrum fairly broad. Forehead almost flat. Basicranial pits poorly developed. – *Distribution:* Ranging from Borneo and the Philippines through Celebes to the Moluccas. – No subspecies.

24. *P. mordax* (PETERS 1866) [*?affinis* group]. – Size fairly large (forearm length, 40–42 mm). Anterior upper premolar probably small and displaced medially. A poorly known species of uncertain relationships. – *Distribution:* Known only from Java (the records from India and Ceylon evidently being erroneous, at least in part misidentified *P. affinis*). – No subspecies.

25. *P. ceylonicus* (KELAART 1852) [*ceylonicus* group]. – Size medium (forearm length, 35–48 mm). Inner upper incisor more or less unicuspid. Outer upper incisor well developed. Anterior upper premolar more or less displaced medially. Rostrum fairly short and relatively broad. Forehead almost flat. Upper canine with a well-developed secondary cusp. – *Distribution:* Ranging from Pakistan to Ceylon and east to Hainan, also Borneo. – Seven subspecies are here recognized:

P. c. subcanus (Pakistan and northwestern India), *P. c. indicus* (remainder of India), *P. c. ceylonicus* (Ceylon), *P. c. shanorum* (Burma), *P. c. raptor* (Vietnam and extreme southeastern mainland China), *P. c. tonfangensis* (Hainan), *P. c. borneoensis* (Borneo).

26. *P. kuhlii* (KUHL 1817) [*kuhlii* group]. – Size fairly small (forearm length, 29–38 mm). Inner upper incisor unicuspid. Outer upper incisor reduced. Anterior upper premolar reduced and displaced medially. Rostrum fairly long and of medium width. Forehead slightly concave. – *Distribution:* Ranging from southwestern Europe and the Canary islands to Pakistan and south to the Cape Province. – Five subspecies are here recognized:

P. k. lepidus (Turkmenia and Iran to Pakistan), *P. k. kuhlii* (southern Europe and Turkey), *P. k. marginatus* (northern Africa and southwestern Asia), *P. k. fuscatus* (eastern Africa), *P. k. subtilis* (southeastern Africa). Allocation of many populations to subspecies is uncertain.

27. *P. aegyptius* (FISCHER 1829) (= *deserti* THOMAS 1902) [*kuhlii* group]. – Size relatively small (forearm length, 29–33 mm). – *Distribution:* Known only from Algeria, Libya, Egypt, northern Sudan, and Upper Volta. – No subspecies.

28. *P. aero* HELLER 1912 [*kuhlii* group]. – Size relatively small (forearm length, 31–33 mm). – *Distribution:* Probably a subspecies of *P. aegyptius*, but known only from the highlands of Kenya. – No subspecies.

29. *P. bicolor* (BOCAGE 1889) (= *anchietai* SEABRA 1900) [*kuhlii* group]. – Size relatively small (forearm length, 30–35 mm). Inner upper incisor more or less bicuspid. Anterior upper premolar reduced, displaced medially, and may be absent. – *Distribution:* Known only from Angola, Zambia, southern Zaire, and western Transvaal. – No subspecies.

30. *P. rusticus* (TOMES 1861) [*kuhlii* group]. – Size relatively small (forearm length, 26–30 mm). Rostrum relatively short but fairly slender. Forehead almost flat. – *Distribution:* Ranging from Liberia to Ethiopia and Kenya; also from Zambia to Namibia and Transvaal. – Two subspecies are recognized:

P. r. rusticus (southern segment of range), *P. r. marrensis* (northern segment of range).

31. *P. inexspectatus* AELLEN 1957 [*kuhlii* group]. – Size relatively small (forearm length, 31–32 mm). Inner upper incisor bicuspid. Rostrum relatively short and broad. Braincase relatively high. – *Distribution:* A poorly known species apparently confined to Benin, Cameroon, Zaire, Uganda, and Kenya. – No subspecies.

32. *P. eisentrauti* HILL 1968 [*kuhlii* group]. – Size fairly small (forearm length, 30–36 mm). Rostrum relatively broad. Braincase greatly inflated. – *Distribution:* Known from the Ivory Coast, Nigeria, Cameroon, and western Kenya. – Two subspecies are recognized:

P. e. bellieri (Ivory Coast), *P. e. eisentrauti* (Nigeria, Cameroon, and perhaps Kenya).

33. *P. maderensis* (DOBSON 1878) [*kuhlii* group]. – Size relatively small (forearm length, 31–32 mm). Outer margin of ear pinna deeply emarginated. Tail relatively long. – *Distribution:* Confined to Madeira and the Canary islands. – No subspecies.

34. *P. savii* (BONAPARTE 1837) [*savii* group]. – Size fairly small (forearm length, 31–39 mm). Inner upper incisor bicuspid. Outer upper incisor well developed. Anterior upper premolar greatly reduced (displaced medially) or absent. Rostrum of medium width and relatively broad. Forehead slightly concave or nearly flat. Basicranial pits poorly developed. No postorbital process on zygomatic arch. – *Distribution:* Ranging from southwestern Europe and northwestern Africa to Japan and Burma, also Canary and Cape Verde islands. – Five subspecies are here recognized:

P. s. savii (Cape Verdes, Canaries, northwestern Africa, and southern Europe), *P. s. caucasicus* (Crimea and southwestern Asia to northwestern China and Pakistan), *P. s. alaschanicus* (Mongolia to Hokkaido), *P. s. austenianus* (Burma and northeastern India).

35. *P. cadornae* THOMAS 1916 [*savii* group]. – Size fairly small (forearm length, 33–37 mm). Anter-

ior upper premolar greatly reduced and displaced medially. Forehead nearly flat. Basicranial pits well developed. A small postorbital process present on zygomatic arch. – *Distribution:* Ranging from northeastern India to northern Thailand. – No subspecies.

36. *P. macrotis* (TEMMINCK 1835) [*savii* group]. – Size relatively small (forearm length, 29–35 mm). Inner upper incisor more or less unicuspid. Anterior upper premolar greatly reduced and displaced medially. Rostrum relatively short but fairly slender. Forehead fairly concave. Braincase relatively inflated. Basicranial pits well developed. A small postorbital process present on zygomatic arch. – *Distribution:* Ranging from Malaya to Bali and Borneo. – Three subspecies are recognized:

P. m. macrotis (Malaya to Bali), *P. m. vordermanni* (Borneo and Billiton), *P. m. curtatus* (West Sumatran islands).

37. *P. bodenheimeri* HARRISON 1960 [*savii* group]. – Size relatively small (forearm length, 28–32 mm). Rostrum fairly slender. Uropatagium relatively extensive, completely enclosing tail. – *Distribution:* Ranging from Israel and Sinai to Socotra (off Somalia). – No subspecies.

38. *P. ariel* THOMAS 1904 [*savii* group]. – Size relatively small (forearm length, 28–31 mm). Inner upper incisor unicuspid. Anterior upper premolar greatly reduced and displaced medially. Rostrum relatively slender. Forehead distinctly concave. – *Distribution:* A poorly known species apparently ranging from extreme southern Egypt to central Sudan. – No subspecies.

39. *P. minahassae* MEYER 1899 [*minahassae* group]. – Size medium (forearm length, 35–37 mm). Inner upper incisor bicuspid. Outer upper incisor well developed. Anterior upper premolar displaced medially. Rostrum relatively short and broad with prominent supraorbital tubercles. Forehead slightly concave. Braincase inflated with a slight sagittal crest. – *Distribution:* A poorly known species confined to Celebes. – No subspecies.

Subgenus *Arielulus* HILL & HARRISON 1987.

Outer upper incisor present but greatly reduced and displaced to inners. Anterior upper premolar greatly reduced or absent. – *Distribution:* Ranging from India to Borneo and Java. – Three species, one additional subspecies.

40. *P. circumdatus* (TEMMINCK 1835). Size fairly large (forearm length, 37–44 mm). Inner upper incisor unicuspid or faintly bicuspid. Anterior upper premolar greatly reduced but present and displaced medially. Rostrum relatively short and broad, with prominent supraorbital ridges and median sulcus. Forehead fairly concave. Braincase inflated with well developed basicranial pits. – *Distribution:* Known from India, Burma, southwestern China, Malaya, and Java. – Two subspecies:

P. c. drungicus (southwestern China), *P. c. circumdatus* (remainder of range).

41. *P. societatis* HILL 1972. – Size medium (forearm length, 37–40 mm). Inner upper incisor faintly bicuspid. Anterior upper premolar may be absent. Supraorbital ridges and median sulcus of rostrum poorly developed. Braincase greatly inflated, but with the basicranial pits poorly developed. Last upper molar reduced. – *Distribution:* Known only from Malaya. – No subspecies.

42. *P. cuprosus* HILL & FRANCIS 1984. – Size fairly small (forearm length, 34–37 mm). Inner upper incisor faintly bicuspid. Anterior upper premolar may be absent. Supraorbital ridges and median sulcus of rostrum poorly developed. Braincase greatly inflated, but with the basicranial pits poorly developed. Last upper molar slightly reduced. Anterior palatal emargination slightly reduced. – *Distribution:* Known only from Borneo. – No subspecies.

Subgenus *Falsistrellus* TROUGHTON 1943

Outer upper incisor present but greatly reduced, and almost directly lateral to the inners. Inner upper incisor unicuspid. Anterior upper premolar present. Size relatively large (forearm length more than 45 mm). Penis not greatly enlarged. – *Distribution:* Confined to southwestern and southeastern Australia, includig Tasmania. – A single species is here recognized.

43. *P. tasmaniensis* (GOULD 1858). – Size relatively large (forearm length, 46–54 mm). *Distribution:* Same as for the subgenus. – Two subspecies are here recognized (recently considered separate species):

P. t. tasmaniensis (southeastern Australia, including Tasmania), *P. t. mackenziei* (southwestern Australia).

Subgenus *Vansonia* ROBERTS 1946

Outer upper incisor present, not greatly reduced, and more or less posterior to the inner. Inner upper incisor deeply bicuspid. Anterior upper premolar normally present. Size small to medium (forearm length less than 45 mm). Penis greatly enlarged. Braincase always at least somewhat inflated, but more or less flat on top. – *Distribution:* Ranging widely in tropical Africa, also Algeria, Egypt, and Iraq. – Four species, six additional subspecies.

44. *P. nanulus* THOMAS 1904. – Size relatively small (forearm length, 25–31 mm). Outer upper incisor well developed. Anterior upper premolar displaced medially. Rostrum relatively short but fairly slender. Forehead almost flat. Braincase inflation moderate. Fur on ventral side brownish. – *Distribution:* Ranging through the forested regions of tropical Africa from Sierra Leone to western Kenya, including Fernando Poo. – No subspecies.

45. *P. crassulus* THOMAS 1904. – Size relatively small (forearm length, 28–29 mm). Rostrum fairly broad. Braincase considerably inflated but flat on top. – *Distribution:* Ranging in the forested regions of tropical Africa from southern Cameroon and southwestern Sudan to northern Angola. – No subspecies.

46. *P. rueppelli* (FISCHER 1829). – Size fairly small (forearm length, 28–35 mm). Outer upper incisor more or less reduced. Anterior upper premolar in toothrow. Rostrum of medium length. Forehead slightly concave. Braincase considerably inflated but flat on top. Fur on ventral side whitish. – *Distribution:* Ranging from Algeria and Egypt south to Transvaal, including Zanzibar, but largely avoiding forested areas; also Iraq. – Six subspecies are recognized:

P. r. senegalensis (Algeria to Senegal), *P. r. rueppelli* (Egypt to Nigeria and central Sudan), *P. r. coxi* (Iraq), *P. r. fuscipes* (Ethiopia to northern Angola), *P. r. pulcher* (northeastern Tanzania), *P. r. leucomelas* (Malawi to Transvaal and southern Angola).

Subgenus *Scotozous* DOBSON 1875

Outer upper incisor greatly reduced or absent. Inner upper incisor unicuspid. Anterior upper premolars present. Size medium (forearm less than 45 mm). Penis not greatly enlarged. Braincase not inflated. – *Distribution:* Confined to India and Pakistan. A Taiwan record is almost certainly erroneous. – A single species.

47. *P. dormeri* (DOBSON 1875). – Size medium (forearm length, 33–36 mm). – *Distribution:* Same as for subgenus. – No currently recognized subspecies.

Subgenus *Vespadelus* TROUGHTON 1943

Braincase not flattened. Inner upper incisor bicuspid but outer upper incisor reduced. Anterior upper premolar absent. – *Distribution:* Confined to Australia, including Tasmania and Lord Howe island. – Five species, one additional subspecies.

48. *P. pumilus* (GRAY 1841) (= *finlaysoni* KITCHENER & al. 1987; *troughton* KITCHENER & al. 1987). – Size relatively to fairly small (forearm length, 28–35 mm). Forehead relatively concave. Rostrum relatively broad. Head, foot, and forearm brownish. – *Distribution:* Over most of Australia except the southern coastal areas. – Two subspecies:

P. p. pumilus (eastern coastal areas), *P. p. caurinus* (most of the remaining range). More than one species may be represented in this complex.

49. *P. douglasorum* KITCHENER 1976. – Size fairly small (forearm length, 34–38 mm), but larger than *pumilus* where sympatric with it. Forehead relatively concave. Rostrum relatively broad. Head, foot, and forearm yellowish. – *Distribution:* Confined to extreme northern parts of Western Australia and the Northern Territory. – No subspecies.

50. *P. vulturnus* THOMAS 1914 (= *baverstocki* KITCHENER & al. 1987). – Size relatively small (forearm length, 26–30 mm). Forehead relatively flat. Rostrum relatively slender. – *Distribution:* Confined to southeastern and southcentral portions of Australia including Tasmania. – No subspecies. More than one species may be represented in this complex.

51. *P. regulus* (THOMAS 1906). – Size fairly small (forearm length, 29–33 mm). Forehead relatively flat. Rostrum relatively slender. – *Distribution:* Confined to southwestern and southeastern Australia, including Tasmania. – No subspecies.

52. *E. darlingtoni* G. M. ALLEN 1933 (= *sagittula* MCKEAN, RICHARDS & PRICE 1978). – Size fairly small (forearm length, 32–36 mm). Forehead relatively flat. Rostrum fairly slender. – *Distribu-*

tion: Confined to southeastern Australia, including Tasmania, and Lord Howe island. – No subspecies.

Subgenus *Neoromicia* ROBERTS 1926

Braincase variable in height. Inner upper incisor bicuspid or unicuspid and outer upper incisor variable in length. Anterior upper premolar normally absent. – *Distribution:* Ranging widely in sub-Saharan Africa, including Zanzibar, also Madagascar. – Eight species, 12 additional subspecies.

53. *P. guineensis* (BOCAGE 1889) [*capensis* group]. – Wing membranes heavily pigmented. Rostrum relatively slender. Braincase relatively low. Size relatively small (forearm length, 26–30 mm). – *Distribution:* Ranging across the northern savanna regions of tropical Africa from Senegal to Ethiopia and perhaps south to Tanzania. – Two subspecies are recognized:

P. g. guineensis (Senegal to Central African Republic), *P. g. rectitragus* (Ethiopia, southern Sudan, and northeastern Zaire). Tanzanian records may pertain to *P. somalicus.*

54. *P. somalicus* (THOMAS 1901) [*capensis* group]. – Braincase relatively low. Size fairly small (forearm length, 26–32 mm). – *Distribution:* Ranging through tropical Africa, mostly in savannas, from Guinea (Bissau) and Ethiopia to Namibia and Natal; also probably Madagascar. – Four subspecies are here recognized:

P. s. ugandae (Sudan, Uganda, and northeastern Zaire), *P. s. somalicus* (Ethiopia and Kenya), *P. s. zuluensis* (Zambia to Namibia and Natal), *P. s. humbloti* (Madagascar). Subspecific allocations of West African and Tanzanian populations are uncertain.

55. *P. capensis* (A. SMITH 1829) [*capensis* group]. – Braincase relatively low. Size medium (forearm length, 28–36 mm). – *Distribution:* Ranging through most of sub-Saharan Africa from Guinea and Ethiopia to the Cape Province, including Zanzibar and Madagascar. – Seven subspecies are here recognized:

P. c. garambae (southern Sudan and northeastern Zaire), *P. c. grandidieri* (Ethiopia to southeastern Zaire, including Zanzibar), *P. c. damarensis* (Angola and Zambia to Namibia and Botswana), *P. c. gracilior* (Mozambique to Natal), *P. c. capensis* (eastern Cape Province), *P. c. notius* (western Cape Province), *P. c. matroka* (Madagascar). Most West African populations have not been allocated to subspecies.

56. *P. melckorum* (auctorum, probably not of ROBERTS 1919) [*capensis* group]. – Braincase relatively low. Size fairly large (forearm length, 34–38 mm). Molars unusually heavy. – *Distribution:* Ranging from Kenya to Zambia and Transvaal. – No subspecies. *P. melckorum* Roberts 1919 was described from western Cape Province, but appears to belong in *capensis*. The present (more northern) species is clearly distinct from *P. capensis* and should be renamed.

57. *P. brunneus* (THOMAS 1880) [*capensis* group]. – Braincase relatively high. Size medium (forearm length, 34–37 mm). – *Distribution:* Known only from a few localities in the West African forest belt (Ivory Coast to northwestern Zaire). – No subspecies.

58. *P. tenuipinnis* (PETERS 1872) [*tenuipinnis* group]. – Wing membrances lightly pigmented. Rostrum relatively broad. Inner upper incisor weakly bicuspid. Size fairly small (forearm length, 27–32 mm). – *Distribution:* Ranging through tropical forested regions from Senegal to western Kenya and northern Angola. – Two subspecies are here recognized:

P. t. tenuipinnis (Senegal to Congo), *P. t. ater* (Kenya and northeastern Zaire to Angola).

59. *P. rendalli* (THOMAS 1889) [*tenuipinnis* group]. – Inner upper incisor unicuspid. Size medium (forearm length, 31–38 mm). – *Distribution:* Ranging across tropical Africa (chiefly in savannas) from Gambia to southern Somalia and south to northern Botswana and northern Mozambique. – Two subspecies are here recognized:

P. r. rendalli (Gambia to Chad), *P. r. phasma* (Sudan and Somalia to Botswana and Mozambique).

60. *P. flavescens* (SEABRA 1900) [*tenuipinnis* group]. – Inner upper incisor strongly bicuspid. Size fairly large (forearm length, 34–37 mm). – *Distribution:* Known only from Angola. – No subspecies.

Genus *Nyctalus* BOWDITCH 1825 (Fig. 153)

Dental formula i 2/3, c 1/1, p 2/2, m 3/3 × 2 = 34. Braincase neither markedly deepened or markedly flattened. Outer upper incisor cuspidate, without a flat crown, not pushed so far forward that its concavity points directly outwards, nor pushed inwards directly posterior to the anterior upper incisor. Fifth digit of wing reduced, only a little longer than the third or fourth metacarpal. No thickened pads at the base of the thumb or on the plantar surface of the hind foot. – *Distribution:* Ranging from the Azores and extreme northern Africa across the Palearctic to Japan and in much of the Indo-Malayan region as far as Borneo and the Philippines. – Eight species, 11 additional subspecies. Of the two groups recognized here, the *stenopterus* group is often included in *Pipistrellus*.

1. *N. stenopterus* (DOBSON 1875) [*stenopterus* group]. – Size medium (forearm length, 38–42 mm). Rostrum fairly short and broad. Supraorbital tubercle fairly well developed. Accessory cusp on upper canine poorly developed. – *Distribution:* Ranging from Malaya through Sumatra and Borneo to the Philippines. – No subspecies.

2. *N. joffreyi* THOMAS 1915 [*stenopterus* group]. – Size fairly small (forearm length, 39 mm). Rostrum relatively short and broad. Supraorbital tubercle greatly developed. Accessory cusp on upper canine well developed. – *Distribution:* Known only from Burma. – No subspecies.

3. *N. anthonyi* (TATE 1942) [*stenopterus* group]. – Size fairly small (forearm length, 38 mm). Rostrum relatively short and broad. Supraorbital tubercle well developed. Accessory cusp on upper canine well developed. Anterior upper premolar greatly reduced. – *Distribution:* A poorly known species recorded only from Burma. – No subspecies.

4. *N. leisleri* (KUHL 1817) [*noctula* group]. – Size fairly small (forearm length, 35–46 mm). Rostrum of medium length and width. Supraorbital tubercle poorly developed. Accessory cusp on upper canine poorly developed. Anterior upper premolar displaced medially but not greatly reduced. – *Distribution:* Ranging from western Europe and extreme northern Africa to Pakistan, also Madeira and the Azores. – Three subspecies are recognized:

N. l. azoreum (Azores), *N. l. verrucosus* (Madeira), *N. l. leisleri* (remainder of range).

5. *N. montanus* (BARRETT-HAMILTON 1906) [*noctula* group]. – Size fairly small (forearm length, 42–44 mm). Anterior upper premolar displaced medially and greatly reduced. – *Distribution:* Ranging from Afghanistan to northern India. – No subspecies.

6. *N. noctula* (SCHREBER 1774) [*noctula* group]. – Size medium (forearm length, 45–57 mm). Anterior upper premolar displaced medially and greatly reduced. – *Distribution:* Ranging from western Europe and northwestern Africa to Japan and Malaya; also recorded from the Azores. A record from Mozambique is either accidental or erroneous. – Seven subspecies are here recognized:

N. n. noctula (northwestern Africa to eastern Europe and northern Iran), *N. n. lebanoticus* (southwestern Asia), *N. n. mecklenburzevi* (Soviet Central Asia), *N. n. plancei* (northern China), *N. n. furvus* (Japan), *N. n. velutinus* (southern China, including Taiwan), *N. n. labiatus* (northern India to Malaya). It is possible that there is more than one species represented in this complex.

7. *N. aviator* THOMAS 1911 [*noctula* group]. – Size fairly large (forearm length, 59–62 mm). Margin of plagiopatagium attached to ankle. – *Distribution:* Confined to Japan, Korea, and northeastern China. – No subspecies.

8. *N. lasiopterus* (SCHREBER 1780) [*noctula* group]. – Size relatively large (forearm length, 63–69 mm). Margin of plagiopatagium attached to metatarsus. – *Distribution:* Ranging from extreme northern Africa and western Europe to Iran and Uzbekistan. – No subspecies.

Genus *Ia* THOMAS 1902 (Fig. 154)

Dental formula i 2/3, c 1/1, p 2/2, m 3/3 x 2 = 34. Braincase fairly deep. Outer upper incisor greatly reduced, with a flat crown, and pushed forward so that it is directly lateral to the inner upper incisor. Anterior upper premolar greatly reduced. No thickened pads at the base of the thumb or on the plantar surface of the hind foot. – *Distribution:* Ranging from northeastern India and southern China to Thailand and Vietnam. – A single species.

1. *I. io* THOMAS 1902 (= *beaulieui* BOURRET 1942). – Size relatively large (forearm length, 71–80 mm). – *Distribution:* Same as for genus. – No subspecies.

Genus ***Glischropus*** DOBSON 1875 (Fig. 155)

Dental formula i 2/3, c 1/1, p 2/2, m 3/3 x 2 = 34. Braincase neither markedly deepened nor markedly flattened. Outer upper incisor well developed, but pushed so far forward that its concavity points directly outwards. Anterior upper premolar displaced medially, but not reduced. Fifth digit of wing not reduced, considerably longer than the third or fourth metacarpal. Thickened pads present at the base of the thumb and on the plantar surface of the foot. – *Distribution:* Ranging from Burma and Thailand to Java, Borneo, and the southwestern Philippines, also the northern

Moluccas. – Two species, one additional subspecies.

1. *G. tylopus* (DOBSON 1875). – Size relatively small (forearm length, 28–30 mm). Braincase not flattened. – *Distribution:* Ranging from Burma and Thailand to Sumatra, Borneo, and Palawan, also the northern Moluccas. – Two subspecies:

G. t. batjanus (northern Moluccas), G. t. tylopus (remainder of range).

2. *G. javanus* (CHASEN 1939). – Size fairly small (forearm length, 32–33 mm). Braincase somewhat flattened. – *Distribution:* Confined to Java. – No subspecies.

Genus **Eptesicus** RAFINESQUE 1820 (Fig. 156)

Dental formula i 2/3, c 1/1, p 1/2, m 3/3 x 2 = 32. Ear pinna not greatly enlarged. Braincase neither markedly deepened nor markedly flattened. Upper canine without a supplemental cusp. Outer upper incisor usually small and lateral to inner upper incisor. Rostrum not greatly broadened. Nasal and anterior palatal emarginations not greatly enlarged. Baculum triangular. – *Distribution:* Ranging through much of Africa, most of Eurasia, and the Americas. – Two subgenera, 16 species.

Subgenus **Eptesicus** RAFINESQUE 1820

Surfaces of forearms, legs, and tail without warts (though this effect may be simulated by nematode infection). Rostrum variable in width, but usually not greatly shortened; usually not markedly flattened dorsally. Braincase variable in height. Zygomatic arches varying from slender to relatively broad, often with postorbital processes. Inner upper incisor bicuspid or unicuspid, but outer upper incisor usually reduced. – *Distribution:* Ranging widely across the Palearctic, but spottily distributed in Africa, and not south of Thailand in the Indo-Malayan region. Also widely distributed in North, Middle, and South America, including the West Indies. – 15 species and 46 additional subspecies.

1. *E. nilssoni* (KEYSERLING & BLASIUS 1839) [*nilssoni* group]. – Rostrum of medium length but fairly broad; not flattened dorsally. Inner upper incisor bicuspid and outer upper incisor well developed. Last upper molar well developed. Braincase fairly low. Basicranial pits well developed. Size medium (forearm length, 37–43 mm). – *Distribution:* Ranging across the northern Palearctic from western Europe to Japan, south to northern Iran, Nepal, and western China. – Six subspecies are currently recognized:

E. n. nilssoni (western Europe at least to central and perhaps to eastern Siberia), E. n. gobiensis (Iran and Mongolia, perhaps to southeastern Siberia), E. n. kashgaricus (Afghanistan to northwestern China), E. n. centrasiaticus (west-central China), E. n. parvus (Korea and perhaps Sakhalin and Hokkaido in Japan), E. n. japonensis (Honshu in Japan). Subspecies boundaries, however, are far from certain.

2. *E. bobrinskoi* KUZYAKIN 1935 [*nilssoni* group]. – Rostrum relatively slender, flattened dorsally but with the edges not sharply defined. Inner upper incisor bicuspid and outer upper incisor well developed. Size fairly small (forearm length, 33–39 mm). – *Distribution:* Ranging across the southern region of former U.S.S.R. from the North Caucasus to central Siberia, but the Iran record is apparently erroneous. – No subspecies.

3. *E. nasutus* (DOBSON 1877) (= *walli* THOMAS 1919) [*nasutus* group]. – Rostrum relatively short and broad, flattened dorsally, its edges sharply defined. Inner upper incisor more or less unicuspid; outer upper incisor reduced. Last upper molar well developed. Braincase fairly low. No basicranial pits. Size medium to fairly small (forearm length, 33–41 mm). – *Distribution:* Ranging from southwestern Arabia and Iraq to Pakistan. – Four subspecies are currently recognized:

E. n. matschiei (southwestern Arabia), E. n. batinensis (southeastern Arabia), E. n. pellucens (Iraq and western Iran), E. n. nasutus (Pakistan, Afghanistan, and eastern Iran).

4. *E. tatei* ELLERMAN & MORRISON-SCOTT 1951 [*serotinus* group]. – Rostrum fairly long but of medium width; not flattened dorsally. Inner upper incisor bicuspid; outer upper incisor reduced. Last upper molar well developed. Braincase fairly high. Shallow basicranial pits present. Size fairly large (forearm length, 43–47 mm). – *Distribution:* Definitely known only from northeastern India. – No subspecies.

5. *E. pachyotis* (DOBSON 1871 [*serotinus* group]. – Last upper molar reduced. Braincase of medium height. Basicranial pits poorly developed. Size medium (forearm length, 38–39 mm). – *Distribution:* Ranging from northeastern India to northern Thailand. – No subspecies.

6. *E. demissus* THOMAS 1916 [*serotinus* group]. – Rostrum relatively broad. Last upper molar well

developed. Braincase relatively high. Basicranial pits well developed. Size fairly large (forearm length, 42 mm). – *Distribution:* A poorly known species recorded only from southern Thailand. – No subspecies.

7. *E. bottae* (PETERS 1869) [*serotinus* group]. – Rostrum of medium length but fairly broad; flattened dorsally. Inner upper incisor more or less bicuspid, but outer upper incisor reduced. Last upper molar reduced. Braincase of medium height. Basicranial pits poorly developed. Size fairly large (forearm length, 40–50 mm). – *Distribution:* Ranging from Egypt and southwestern Arabia to Pakistan and northwestern China. – Six subspecies are currently recognized:

E. b. innesi (Egypt, Israel), *E. b. bottae* (southwestern Arabia), *E. b. omanensis* (southeastern Arabia), *E. b. hingstoni* (Iraq), *E. b. anatolicus* (Anatolia to southwestern Iran), *E. b. ognevi* (Transcaucasia to Pakistan and northwestern China).

8. *E. kobayashii* MORI 1928 [*serotinus* group]. – Skull width across zygomatic arches relatively great. Third finger of wing relatively long. Size fairly large (forearm length, 45–47 mm). – *Distribution:* A poorly known form recorded only from Korea. Perhaps an eastern representative of *E. bottae.* – No subspecies.

9. *E. hottentotus* (A. SMITH 1833) [*serotinus* group]. – Rostrum fairly long, but broad. Inner upper incisor unicuspid. Braincase fairly high. Size relatively large (forearm length, 46–53 mm). – *Distribution:* Ranging from southern Angola and southern Kenya to the Cape Province. – Three subspecies are currently recognized:

E. h. portavernus (southern Kenya), *E. h. bensoni* (Zambia to Natal), *E. h. hottentotus* (southern Angola to southern Cape Province).

10. *E. serotinus* (SCHREBER 1774) [*serotinus* group]. – Rostrum fairly long, varying from fairly to relatively broad, with or without dorsal flattening. Inner upper incisor more or less bicuspid. Last upper molar more or less reduced. Braincase medium to fairly high. Size medium to relatively large (forearm length, 39–58 mm; but always larger than *E. bottae* where they are sympatric). – *Distribution:* Ranging from western Africa and England to China; also from southern Canada (and possibly Alaska) to Colombia and northeastern Brazil; also the Bahamas, Greater Antilles and on Dominica and Barbados in the Lesser Antilles. This species shows considerable diversity over its extensive range and New World representatives have usually been separated as *E. fuscus,* but a clear cut separation between the two has not been demonstrated. – Twenty two subspecies are here recognized:

E. s. platyops (tropical western Africa), *E. s. isabellinus* (northwestern Africa), *E. s. serotinus* (Europe to Israel and northern Iran), *E. s. turcomanus* (Soviet Central Asia to Mongolia and northeastern Iran), *E. s. shiraziensis* (southwestern Iran), *E. s. pashtonus* (Afghanistan), *E. s. pachyomus* (Pakistan and northwestern India), *E. s. pallens* (Korea and northern China), *E. s. horikawae* (Taiwan), *E. s. andersoni* (southern mainland China), *E. s. bernardinus* (southwestern Canada to central California), *E. s. pallidus* (central Canada to northern Mexico), *E. s. fuscus* (southeastern Canada to northeastern Mexico), *E. s. osceola* (Florida), *E. s. peninsulae* (southern Baja California), *E. s. miradorensis* (central Mexico to northern South America), *E. s. petersoni* (Isle of Pines off western Cuba), *E. s. dutertreus* (Cuba, southern Bahamas, Caymans), *E. s. hispaniolae* (Hispaniola), *E. s. wetmorei* (Puerto Rico, also Dominica and probably Barbados in the Lesser Antilles), *E. s. bahamensis* (northern Bahamas), *E. s. lynni* (Jamaica).

11. *E. guadeloupensis* GENOWAYS & BAKER 1975 [*serotinus* group]. – Rostrum relatively long and fairly narrow. Inner upper incisor apparently unicuspid. Last upper molar not reduced. Braincase fairly high. Ear and tibia relatively long. Size relatively large (forearm length, 49–52 mm). – *Distribution:* Confined to Guadeloupe in the Lesser Antilles. – No subspecies.

12. *E. brasiliensis* (DESMAREST 1819) [*serotinus* group]. – Rostrum of medium length and width, rounded to slightly flattened dorsally. Last upper molar usually not reduced. Braincase fairly high. Size medium (forearm length, 39–47 mm). – *Distribution:* Ranging from southern Mexico to central Argentina but west of the Andes not south of Ecuador and probably including Trinidad and Tobago. – Five subspecies are here recognized:

E. b. andinus (Middle American range and Andean range in South America to Venezuela and Peru), *E. b. melanopterus* (lowlands of northern South America from Colombia to Amazonian Brazil, including Trinidad and Tobago), *E. b. thomasi* (Amazonian Ecuador and Peru), *E. b. brasiliensis* (eastern Brazil), *E. b. argentinus* (Paraguay, Uruguay, and northern Argentina).

13. *E. furinalis* (D'ORBIGNY 1847) [*serotinus* group]. – Rostrum of medium length and fairly slender, rounded dorsally. Last upper molar not reduced. Braincase fairly high. Size medium (forearm length, 34–47 mm). – *Distribution:* Ranging from tropical Mexico to northern Argentina, but west of the Andes not south of Ecuador. – Seven subspecies are here recognized:

E. f. gaumeri (tropical Mexico to the Guianas except for the highlands of Costa Rica and western Panama), *E. f. carteri* (highlands of Costa Rica and western Panama), *E. f. chiralensis* (Andean region from Venezuela to Peru), *E. f. montosus* (upland regions of central Bolivia and central Brazil), *E. f. chapmani* (Amazonian regions of Colombia, Brazil, and Bolivia), *E. f. furinalis* (eastern Brazil to southern Bolivia and northeastern Argentina), *E. f. findleyi* (northwestern Argentina).

14. *E. innoxius* (GERVAIS 1841 [*innoxius* group]. – Rostrum of medium length and relatively slender, rounded dorsally. Inner upper incisor bicuspid, outer upper incisor reduced. Last upper molar not reduced. Braincase relatively high. Basicranial pits poorly developed. Size fairly small (forearm length, 34–39 mm). Fur relatively pale in color. – *Distribution:* Confined to the Pacific slope of Ecuador and northern Peru. – No subspecies.

15. *E. diminutus* OSGOOD 1915 [*innoxius* group]. – Rostrum of medium width. Braincase fairly high. Size fairly small (forearm length, 31–38 mm). Fur relatively dark in color. – *Distribution:* Ranging from eastern Brazil to northern Argentina. – Two subspecies are currently recognized:

E. d. diminutus (east-central Brazil), *E. d. fidelis* (extreme southeastern Brazil to northern Argentina).

Subgenus **Rhinopterus** MILLER 1906

Surfaces of forearms, legs, and tail with prominent warts. Rostrum shortened, broadened, and flattened. Braincase fairly high. Zygomatic arches relatively broad, but without postorbital processes. Inner upper incisor bicuspid but outer upper incisor reduced. – *Distribution:* Ranging in dry savannas (immediately south of the Sahara) from Mali to central Sudan. – A single species.

16. *E. floweri* (DE WINTON 1901) (= *lowei* THOMAS 1915). – Size fairly small (forearm length, 34–38 mm). – *Distribution:* Same as for subgenus. – No subspecies.

Genus **Vespertilio** LINNAEUS 1758 (Fig. 157)

Dental formula i 2/3, c 1/1, p 1/2, m 3/3 x 2 = 32. Ears not greatly enlarged. Braincase neither markedly deepened nor markedly flattened. Upper canine without a supplemental cusp. Outer upper incisor small and lateral to inner upper incisor. Rostrum considerably broadened and flattened. Nasal and anterior palatal emarginations greatly enlarged. – *Distribution:* Ranging across the Palearctic from western Europe to Afghanistan and Japan, south in eastern Asia to southern China. – Three species, two additional subspecies.

1. *V. murinus* LINNAEUS 1758. – Size relatively small (forearm length, 40–47 mm; condylobasal length, 14–15 mm). Rostrum not greatly broadened. – *Distribution:* Ranging from western Europe to Afghanistan and southern Siberia. – Two subspecies are recognized:

V. m. ussuriensis (southeastern Siberia, Korea, northeastern China), *V. m. murinus* (remainder of range).

2. *V. orientalis* WALLIN 1969. – Size medium (forearm length, 45–50 mm; condylobasal length, 16–17 mm). Rostrum greatly broadened. Ear relatively long with slender tragus. Hair of underparts dark in color. – *Distribution:* Confined to central and southern China, including Taiwan; also Japan. – No subspecies.

3. *V. superans* THOMAS 1899. – Size relatively large (forearm length, 45–52 mm; condylobasal length, 17–18 mm). Rostrum greatly broadened. Ear relatively short with broad tragus. Hair of underparts light in color. – *Distribution:* Ranging from southeastern Siberia and Japan to northern China. – Two subspecies are recognized:

V. s. anderssoni (Mongolia and northwestern China), *V. s. superans* (remainder of range).

Genus **Laephotis** THOMAS 1901 (Fig. 158)

Dental formula i 2/3, c 1/1, p 1/2, m 3/3 x 2 = 32. Ears relatively enlarged, but not united. Each auditory bulla only as wide as the space between them. Braincase slightly flattened. Upper canine without a supplemental cusp. Outer upper incisor small and lateral to inner upper incisor. – *Distribution:* Restricted to dryer portions of the sub-Saharan African mainland from Kenya to South Africa. – Four species.

1. *L. wintoni* THOMAS 1901. – Size relatively large (forearm length, 36–41 mm). Ear fairly long (21–22 mm). – *Distribution:* Known only from Ethiopia and Kenya. A western Cape Province specimen apparently belongs here. – No subspecies.

2. *L. namibensis* SETZER 1971. – Size relatively large (forearm length, 38–39 mm). Ear relatively long (22–25 mm). – *Distribution:* Definitely known only from Namibia. – No subspecies.

3. *L. angolensis* MONARD 1935. – Size relatively small (forearm length, 32–36 mm). Ear relatively short (15–16 mm). – *Distribution:* Known only from Angola and southern Zaire. – No subspecies.

4. *L. botswanae* SETZER 1971. – Size fairly small (forearm length, 34–38 mm). Ear fairly short

(16–18 mm). – *Distribution:* Ranging from southern Zaire and Malawi to Botswana and Transvaal. – No subspecies.

Genus *__Histiotus__* GERVAIS 1855 (Fig. 159)

Dental formula i 2/3, c 1/1, p 1/2, m 3/3 x 2 = 32. Ears greatly enlarged and united by a band connecting their posterior surfaces. Each auditory bulla considerably wider than the space between them. Upper canine without a supplemental cusp. Outer upper incisor small and lateral to inner upper incisor. – *Distribution:* Confined to continental South America, where widespread except for the northeastern portion. – Three species and five additional subspecies are here recognized.

1. *H. __montanus__* (PHILIPPI & LANDBECK 1861). – Ears relatively short and rounded. Connecting band poorly developed. Skull relatively broad. Forearm length, 44–51 mm. – *Distribution:* Ranging in the Andes from Venezuela to northwestern Argentina but south of the tropics occurring also in the lowlands south to Tierra del Fuego. – Five subspecies are here recognized:

H. m. colombiae (western Venezuela to Ecuador), *H. m. inambarus* (southern Peru to Paraguay), *H. m. montanus* (Uruguay, northern and central Argentina, central Chile), *H. m. magellanicus* (southern Chile and southern Argentina), *H. m. alienus* (definitely known only from southeastern Brazil and often separated specifically).

2. *H. __macrotus__* (POEPPIG 1835). – Ears relatively long but rounded. Connecting band well developed. Skull relatively broad. Forearm length, 46–51 mm. – *Distribution:* Ranging in the Andes and on the Pacific slope from southern Peru to southern Patagonia. – Two subspecies are recognized:

H. m. laephotis (southern Peru to northwestern Argentina), *H. m. macrotus* (Chile).

3. *H. __velatus__* (I. GEOFFROY 1824). – Ears more or less triangular. Connecting band well developed. Skull relatively narrow. Forearm length, 45–49 mm. – *Distribution:* Confined to eastern Brazil and eastern Paraguay. – No subspecies currently recognized.

Genus *__Philetor__* THOMAS 1902 (Fig. 160)

Dental formula i 2/3, c 1/1, p 1/2, m 3/3 x 2 = 32. Ears not greatly enlarged. Braincase neither markedly deepened nor markedly flattened, but is considerably inflated. Upper canine with a prominent supplemental cusp. Outer upper incisor small and lateral to the inner upper incisor. Rostrum considerably shortened with prominent supraorbital tubercles. – *Distribution:* Spotty; extending from Nepal to the Philippines and Bismarcks. – A single species with two additional subspecies usually recognized.

1. *P. brachypterus* (TEMMINCK 1840) (= *rohui* THOMAS 1902; = *veraecundus* CHASEN 1940). Forearm length, 31–38 mm. – *Distribution:* Same as for genus. – Three subspecies are usually recognized, but the actual pattern of variation is too erratic to fit a three subspecies treatment.

Genus *__Tylonycteris__* PETERS 1872 (Fig. 161)

Dental formula i 2/3, c 1/1, p 1/2, m 3/3 x 2 = 32. Ears not greatly enlarged. Braincase greatly flattened. Upper canine with a prominent secondary cusp. Outer upper incisor fairly long and lateral to the inner upper incisor. Rostrum of medium width, flattened, but with subraorbital tubercles more or less developed. Large fleshy pads present on the foot and at the base of the thumb. Adapted to roosting inside hollow bamboo stems. – *Distribution:* Ranging from India and southern China to the Philippines, Celebes, and the Lesser Sundas. – Two species, four additional subspecies.

1. *T. __pachypus__* (TEMMINCK 1840). – Size relatively small (forearm length, 24–29 mm). Supraorbital tubercle relatively poorly developed. Lambdoidal crest relatively well developed. – *Distribution:* Essentially same as for genus except for absence from Celebes. – Four subspecies are currently recognized:

T. p. aurex (southern India), *T. p. fulvida* (northeastern India and southern China to Thailand and the Andaman islands), *T. p. pachypus* (Malaya to Borneo and Bali), *T. p. meyeri* (Philippines), *T. p. bhaktii* (known only from Lombok in the Lesser Sundas).

2. *T. __robustula__* THOMAS 1915. Size relativley large (forearm length, 26–31 mm). Supraorbital tubercle relatively well developed. – *Distribution:* Ranging from southwestern China to the Philippines, Celebes, and the Lesser Sundas. – Two subspecies are here recognized:

T. r. malayana (southwestern China and Vietnam to Malaya), *T. r. robustula* (remainder of range).

Genus *__Mimetillus__* THOMAS 1905 (Fig. 162)

Dental formula i 2/3, c 1/1, p 1/2, m 3/3 x 2 = 32. Ears not greatly enlarged. Braincase greatly flattened. Upper canine without a secondary cusp. Outer upper incisor fairly long and lateral to the inner upper incisor. Rostrum relatively broad with well developed supraorbital tubercles. Wing greatly reduced in size. – *Distribution:* Confined to tropical Africa, ranging from Sierra Leone to Ethiopia and south to Angola and Tanzania, including Fernando Poo. – A single species with two additional subspecies.

1. *M. moloneyi* (THOMAS 1891). – Forearm length, 26–30 mm. – *Distribution:* Same as for genus. – Three subspecies:

M. m. moloneyi (Sierra Leone to Ethiopia and Kenya), *M. m. thomasi* (Zambia and southern Zaire to southern Tanzania), *M. m. berneri* (Angola).

Genus ***Hesperoptenus*** PETERS 1868 (Fig. 163)

Dental formula i 2/3, c 1/1, p 1/2, m 3/3 x 2 = 32. Ears not greatly enlarged. Braincase neither markedly deepened nor markedly flattened. Upper canine without a secondary cusp. Outer upper incisor of variable size but always short and more or less posterior to the inner upper incisor. Rostrum relatively short. – *Distribution:* Ranging from India and Ceylon to Celebes. – Two subgenera, five species.

Subgenus ***Hesperoptenus*** PETERS 1868

Ear pinna and tragus thin, not fleshy. Second phalanx of third digit of wing not reduced. Braincase somewhat inflated. Rostrum relatively slender with poorly developed supraorbital processes. Outer upper incisor only slightly displaced medially from toothrow. – *Distribution:* Known only from Malaya and Borneo. – A single species.

1. *H. doriae* (PETERS 1869). – Size medium (forearm length, 38–42 mm). – *Distribution:* Same as for subgenus. – No subspecies.

Subgenus ***Milithronycteris*** HILL 1976

Ear pinna and tragus fleshy. Second phalanx of third digit of wing reduced. Braincase not inflated. Rostrum relatively broad with well developed supraorbital processes. Outer upper incisor displaced medially from toothrow. – *Distribution:* Same as for genus. – Four species.

2. *H. tickelli* (BLYTH 1851). – Size relatively large (forearm length, 49–61 mm). No pad at base of thumb. Outer upper incisor fairly large. Dentition not especially massive. Lower incisors not greatly imbricated. Fur relatively pale. – *Distribution:* Ranging from India and Ceylon to Thailand and the Andaman islands. – No subspecies.

3. *H. tomesi* THOMAS 1905. – Size relatively large (forearm length, 50–54 mm). No pad at base of thumb. Outer upper incisor relatively large. Dentition relatively massive. Lower incisors greatly imbricated. Fur relatively dark. – *Distribution:* Known only from Malaya and Borneo. – No subspecies.

4. *H. gaskelli* HILL 1983. – Size medium (forearm length, 38–41 mm). No pad at base of thumb. Outer upper incisor relatively large. Lower incisors considerably imbricated. Fur relatively dark. – *Distribution:* Known only from Celebes. – No subspecies.

5. *H. blanfordi* (DOBSON 1877). – Size relatively small (forearm length, 24–29 mm). A broad pad present at the base of the thumb. Outer upper incisor greatly reduced. Dentition not especially massive. Lower incisors not imbricated. Fur fairly dark. – *Distribution:* Known from southern Burma, Thailand, Malaya, and Borneo. – No subspecies.

Genus ***Chalinolobus*** PETERS 1866 (Fig. 164)

Dental formula i 2/3, c 1/1, p 1–2/2, m 3/3 x 2 = 32–34. Ears not greatly enlarged but their lateral borders extended ventrally around the corners of the mouth, reaching the lower jaw. Braincase markedly deepened. Upper canine without a well developed secondary cusp. Outer upper incisor more or less reduced, but variable in position relative to the inner upper incisor. Rostrum fairly short and broad. – *Distribution:* Ranging widely in sub-Saharan Africa from Senegal to Ethiopia and south to Namibia and Natal; also Australia, Tasmania, southeastern New Guinea, New Caledonia, Norfolk island, and New Zealand. – Two subgenera, 15 species.

Subgenus ***Chalinolobus*** PETERS 1866

Anterior upper premolar usually present, though very small. Outer upper incisor more or less lateral to the inner upper incisor, and with its single cusp near the anterior face of the tooth. Post-palatal spine relatively narrow and pointed. Fifth metacarpal not much shorter than third. – *Distribution:*

Same as for genus but excluding the African portion of the range. – Six species, three additional subspecies.

1. *C. dwyeri* RYAN 1966. – Size medium (forearm length, 38–42 mm). Supraorbital swellings of skull well developed. Post-orbital constriction poorly developed. No median crest on braincase. Inner upper incisor bicuspid. Outer upper incisor relatively long. Pronounced antero-internal cusp on posterior upper premolar. Fur black with a white margin around ventral surface. – *Distribution:* Confined to mountains of southeastern mainland Australia. – No subspecies.

2. *C. nigrogriseus* (GOULD 1856). – Size relatively small (forearm length, 31–39 mm). Supraorbital swellings well developed. Post-orbital constriction well developed. Low median crest on braincase. Inner upper incisor unicuspid. Outer upper incisor relatively short. Antero-internal cusp on posterior upper premolar generally poorly developed. Fur dark gray with varying amounts of whitish frosting. – *Distribution:* Confined to coastal areas of northern and eastern Australia; also southeastern New Guinea including Fergusson island in the D'Entrecasteaux group. – Two subspecies are currently recognized:

C. n. rogersi (northeastern Western Australia to northwestern Queensland), *C. n. nigrogriseus* (New Guinea to northeastern New South Wales).

3. *C. tuberculatus* (GRAY 1843). – Size medium (forearm length, 41 mm). Supraorbital swellings of skull well developed. Low median crest on braincase. Inner upper incisor unicuspid or weakly bicuspid. Outer upper incisor relatively short. Anterointernal cusp present on last upper premolar. Fur black or brown. – *Distribution:* Confined to New Zealand. – No subspecies.

4. *C. picatus* (GOULD 1852). – Size relatively small (forearm length, 31–36 mm). Supraorbital swellings of skull poorly developed. Post-orbital constriction well developed. No median crest on braincase. Inner upper incisor bicuspid. Outer upper incisor of medium length. Antero-internal cusp present on last upper premolar. Fur black with lateral and posterior areas of ventral side white. – *Distribution:* Confined to dryer interior areas of southeastern Australia. – No subspecies.

5. *C. gouldii* (GRAY 1841). – Size relatively large (forearm length, 40–48 mm). Supraorbital swellings of skull well developed. Low median crest on braincase. Inner upper incisor unicuspid. Outer upper incisor greatly reduced. Antero-internal cusp generally present on last upper premolar. Fur on head black, usually grading to brown on rump. – *Distribution:* Widespread in Australia (except for a few northeastern areas), Tasmania, New Caledonia, and Norfolk island. – Three subspecies are currently recognized:

C. g. neocaledonicus (New Caledonia and probably by Norfolk island), *C. g. venatoris* (northern half of Australia), *C. g. gouldii* (southern half of Australia, including Tasmania).

6. *C. morio* (GRAY 1841). – Size fairly small (forearm length, 35–39 mm). Supraorbital swellings of skull poorly developed. Post-orbital constriction well developed. No median crest on braincase. Inner upper incisor bicuspid. Outer upper incisor relatively long. Antero-internal cusp present, but variably developed on last upper premolar. Fur chocolate brown. – *Distribution:* Largely confined to the southern half of Australia, including Tasmania. – No subspecies.

Subgenus *Glauconycteris* DOBSON 1875

Anterior upper premolar absent. Outer upper incisor more or less posterior to inner upper incisor, and with its cusp near the middle of the tooth. Post-palatal spine relatively broad and blunt. Fifth metacarpal much shorter than third. – *Distribution:* Occupying the African portion of the range of the genus. – Nine species, five additional subspecies.

7. *C. beatrix* (THOMAS 1901). – Plagiopatagium and uropatagium dark in color, without a reticulated pattern. Fur dark brown, often with a white shoulder spot. Ear pinna short and rounded. Size relatively small (forearm length, 33–41 mm). Inner upper incisor strongly bicuspid with subequal cusps. – *Distribution:* Ranging in forested areas from Ivory Coast to northern Angola and east to western Kenya. – Two subspecies are here recognized:

C. b. beatrix (Ivory Coast to Angola), *C. b. humeralis* (Zaire to Kenya).

8. *C. poensis* (GRAY 1842). – Plagiopatagium and uropatagium usually dark in color, without a reticulated pattern. Fur light grayish brown, usually with pale flank stripes and/or shoulder spots. Ear pinna large and rounded. Size relatively small (forearm length, 35–40 mm). Inner upper incisor weakly bicuspid with unequal cusps. – *Distribution:* Ranging in forested regions from Senegal to Tanzania, including Fernando Poo. – No subspecies.

9. *C. argentatus* DOBSON 1875. – Plagiopatagium and uropatagium usually pale in color, without a reticulated pattern. Fur light brownish gray, usually with pale flank stripes. Ear pinna rounded. Size fairly small (forearm length, 38–43 mm). Inner upper incisor unicuspid or weakly bicuspid. – *Distribution:* Ranging in woodlands from Cameroon to Kenya and south to Malawi and northern Angola. – No subspecies.

10. *C. alboguttatus* (J. A. ALLEN 1917). – Plagiopatagium and uropatagium dark in color, without a reticulated pattern. Fur dark brown with pale flank stripes and/or shoulder spots. Ear pinna small and rounded. Size fairly small (forearm length, 37–41 mm). Inner upper incisor weakly bicuspid with unequal cusps. – *Distribution:* Known only from forested regions in Cameroon and northeastern Zaire. – No subspecies.

11. *C. egeria* (THOMAS 1913). – Plagiopatagium and uropatagium dark in color, without a reticulated pattern. Fur dark brown. Ear pinna large and subquadrangular. Size relatively small (forearm length, 38 mm). Inner upper incisor strongly bicuspid with subequal cusps. – *Distribution:* Known only from forested regions in Cameroon and Uganda. – No subspecies.

12. *C. kenyacola* (PETERSON 1982). – Plagiopatagium and uropatagium pale in color without a distinct reticulated pattern. Fur dark brown with whitish markings on nose, chin, and at the bases of the ears. Ear pinna fairly large but more or less rounded. Size fairly small (forearm length, 40–41 mm). Inner upper incisor unicuspid. – *Distribution:* Known only from eastern Kenya. – No subspecies.

13. *C. variegatus* (TOMES 1861). – Plagiopatagium and uropatagium pale but with a prominent dark reticulated pattern. Fur varying from light to dark brown without markings. Ear pinna rounded. Size fairly large (forearm length, 38–46 mm). Inner upper incisor varying from unicuspid to bicuspid (but with unequal cusps). – *Distribution:* Ranging widely in the savanna regions of Africa from Senegal to Ethiopia and south to Natal and northern Namibia. – Three subspecies are here recognized:

C. v. phalaena (central Sudan), *C. v. machadoi* (east-central Angola), *C. v. variegatus* (remainder of range).

14. *C. gleni* (PETERSON & SMITH 1973). – Plagiopatagium and uropatagium pale with a poorly developed dark reticulated pattern. Fur pale brown without markings. Ear pinna relatively long and bluntly pointed. Size medium (forearm length, 38–42 mm). Inner upper incisor unicuspid or weakly bicuspid. – *Distribution:* Known only from forested regions in Cameroon and Uganda. – No subspecies.

15. *C. superbus* (HAYMAN 1939). – Plagiopatagium and uropatagium dark in color, without reticulations. Fur black, strikingly marked with white spots and stripes. Ear pinna relatively large and rounded. Size relatively large (forearm length, 46–47 mm). Inner upper incisor unicuspid. – *Distribution:* Ranging in forested regions from Ivory Coast to Uganda. – Two subspecies are currently recognized:

C. s. sheila (Ivory Coast, Ghana), *C. s. superbus* (Uganda, northeastern Zaire).

Tribe Nycticeini GERVAIS 1855

Rostrum fairly short. Dental formula almost always i 1/3, c 1/1, p 1/2, m 3/3 x 2 = 30. Ear pinna may be greatly lengthened but never greatly shortened. – *Distribution:* Widely distributed over Africa, southern Eurasia, the Malay archipelago, Australia, Madagascar, and the Mascarene islands; also eastern North America, including Cuba in the West Indies. – Six genera, four additional subgenera, 24 species.

Genus *Nycticeius* RAFINESQUE 1819 (Fig. 165)

Ears not greatly enlarged. Width of each auditory bulla less than space between them. Anterior and middle upper molars with the mesostyle unreduced, the W-shaped pattern not distorted. Anterior and middle lower molars with the talonid larger than the trigonid. Lateral lower incisor not reduced. Rostrum in lacrimal region not greatly broadened. Anterior palatal emargination not greatly reduced. – *Distribution:* Widely distributed in sub-Saharan Africa, also in Egypt and southwestern Arabia; most of Australia (but not Tasmania) and in New Guinea largely confined to the southeast; southeastern North America to eastern Mexico and western Cuba. – Four subgenera, six species.

Subgenus *Nycticeius* RAFINESQUE 1819

Anterior palatal emargination does not extend behind canine. Rostrum medium in width and length. Maxillary toothrows nearly parallel. Last upper molar not reduced. – *Distribution:* Ranging from the east-central United States to eastern Mexico and western Cuba. – A single species, three additional subspecies.

1. N. humeralis (RAFINESQUE 1818). – Size relatively large (forearm length, 28–39 mm). – *Distribution:* Same as for subgenus. – Four subspecies are here recognized:

N. h. mexicanus (southern Texas and eastern Mexico), N. h. subtropicalis (southern Florida), N. h. cubanus (western Cuba), N. h. humeralis (remainder of range).

Subgenus **Nycticeinops** HILL & HARRISON 1987

Anterior palatal emargination does not extend behind canine. Rostrum medium in width but relatively short. Maxillary toothrows converging anteriorly. Last upper molar not reduced. – *Distribution:* Ranging from Senegal to Egypt and southwestern Arabia, south to Natal and northern Namibia, chiefly in savanna regions. – A single species.

2. N. schlieffeni PETERS 1859. Size relatively small (forearm length, 28–34 mm). – *Distribution:* Same as for subgenus. – Several poorly defined subspecies are frequently recognized.

Subgenus **Scotorepens** TROUGHTON 1943

Anterior palatal emargination extends behind canine. Rostrum medium in width to relatively broad. Last upper molar somewhat reduced. – *Distribution:* Confined to continental Australia and southern and eastern New Guinea. – Three species and two additional subspecies are here recognized.

3. N. greyii (GRAY 1843) (= caprenus TROUGHTON 1937; aquilo TROUGHTON 1937). – Size relatively small (forearm length, 27–33 mm). Rostrum usually fairly slender. – *Distribution:* Ranging across Australia from northern Western Australia to New South Wales, but absent from the Cape York peninsula. – No currently recognized subspecies.

4. N. sanborni (TROUGHTON 1937). – Size medium (forearm length, 31–35 mm). Rostrum usually fairly broad. – *Distribution:* Confined to northwestern Australia, northeastern Australia and southeastern New Guinea. – No subspecies.

5. N. balstoni (THOMAS 1906). – Size fairly large (forearm length, 33–38 mm). Rostrum of variable width. – *Distribution:* Ranging widely in central and southern Australia. – Three subspecies are here recognized:

N. b. balstoni (Western Australia to western Victoria and southern Queensland), N. b. influatus (central Queensland), N. b. orion (Pacific coast of New South Wales and Victoria).

Subgenus **Scoteanax** TROUGHTON 1943

Anterior palatal emargination extends behind canine. Rostrum medium in width. Last upper molar greatly reduced. – *Distribution:* Confined to eastern Australia. – A single species.

6. N. rueppellii PETERS 1866. – Size relatively large (forearm length, 50–56 mm). – *Distribution:* Same as for genus. – No subspecies.

Genus **Rhogeesa** H. ALLEN 1866 (Fig. 166)

Ears not greatly enlarged. Width of each auditory bulla less than space between them. Anterior and middle upper molars with the mesostyle unreduced, the W-shaped pattern not distorted. Anterior and middle lower molars with the talonid larger than the trigonid. Lateral lower incisor reduced. Rostrum in lacrimal region not greatly broadened. Anterior palatal emargination not greatly reduced. – *Distribution:* Widely distributed in tropical America from Mexico to Bolivia and eastern Brazil (including the Tres Marias islands, Margarita, Trinidad, and Tobago, but west of the Andes not south of Ecuador). – Two subgenera, seven species.

Subgenus **Rhogeesa** H. ALLEN 1866

Outer lower incisors not greatly reduced. First phalanx of third digit usually relatively short. Skull relatively small. – *Distribution:* Same as for genus. – Six species.

1. R. tumida H. ALLEN 1866. – Ears relatively short. Hairs of dorsum dark with two bands. Size varying from large to small (forearm length, 25–34 mm). Lingual cingulum of upper canine always with at least some indication of cusps. Uropatagium mostly naked. Outer lower incisor usually not reduced. Postorbital constriction poorly developed. Third metacarpal relatively short. – *Distribution:* Ranging from northeastern

Mexico to western Ecuador and northeastern Brazil. – No currently recognized subspecies.

2. *R. genowaysi* BAKER 1984. – Except for slightly shorter ears, indistinguishable from *R. tumida* in gross morphology (forearm length, 27–31 mm), though markedly different in karyotype. – *Distribution:* Known only from southwestern Mexico (Chiapas). – No subspecies.

3. *R. minutilla* MILLER 1897. – Ears relatively short. Hairs of dorsum light with two bands. Size fairly small (forearm length, 26–29 mm). Lingual cingulum of upper canine always with at least some indication of cusps. Uropatagium relatively bare. Outer lower incisor usually not reduced. Postorbital constriction well developed. Third metacarpal relatively long. – *Distribution:* Confined to arid northeastern Colombia and northwestern Venezuela, also Margarita island. – No subspecies.

4. *R. parvula* H. ALLEN 1866. – Ears relatively short. Hairs of dorsum with two bands. Size varying from large to small (forearm length, 25–33 mm). Lingual cingulum of upper canine with cusps. Uropatagium relatively well haired. Outer lower incisor usually reduced. Postorbital constriction well developed. – *Distribution:* Confined to western Mexico, including the Tres Marias islands. – No currently recognized subspecies.

5. *R. mira* LA VAL 1973. – Ears relatively short. Hairs of dorsum with two bands. Size relatively small (forearm length, 24–27 mm). Lingual cingulum of upper canine without cusps. Uropatagium fairly well haired. Outer lower incisor only slightly reduced. – *Distribution:* Known only from southwestern Mexico (Michoacan). – No subspecies.

6. *R. gracilis* MILLER 1897. – Ears relatively long. Hairs of dorsum with three bands. Size relatively large (forearm length, 30–34 mm). Lingual cingulum of upper canine without cusps. Uropatagium fairly well haired. Outer lower incisor only slightly reduced. – *Distribution:* Confined to southwestern Mexico. – No subspecies.

Subgenus **Baeodon** MILLER 1906

Outer lower incisor greatly reduced. First phalanx of third digit relatively long. Skull relatively large. – *Distribution:* Restricted to southwestern Mexico. – A single species.

7. *R. alleni* THOMAS 1892. – Size relatively large (forearm length, 30–35 mm). – *Distribution:* Same as for subgenus. – No subspecies.

Genus **Scotoecus** THOMAS 1901 (Fig. 167)

Ears not greatly enlarged. Width of each auditory bulla less than space between them. Anterior and middle upper molars with the mesostyle unreduced, the W-shaped pattern not distorted. Anterior and middle lower molars with the talonid larger than the trigonid. Lateral lower incisor not reduced. Rostrum in lacrimal region greatly broadened. Anterior palatal emargination not reduced. Anterior upper premolar may be present, sometimes giving a dental formula of i 1/3, c 1/1, p 2/2, m 3/3 × 2 = 32. – *Distribution:* Widely distributed in sub-Saharan Africa south to Angola and Mozambique; also in Pakistan and northern India. – Three species with four additional subspecies are here recognized.

1. *S. pallidus* (DOBSON 1876). – Tragus relatively long, narrow, and straight. Anterior lower premolar reduced but not compressed in toothrow. Chiropatagium darkly pigmented. Ventral hair grayish white. Braincase slightly flattened. Anterior upper premolar absent. Size relatively large (forearm length, 34–38 mm). – *Distribution:* Confined to Pakistan and northern India. – No subspecies.

2. *S. albofuscus* (THOMAS 1890). – Tragus relatively short and broad, but straight. Anterior lower premolar not reduced but compressed in toothrow. Chiropatagium lightly pigmented. Ventral hair brown. Braincase slightly flattened. Anterior upper premolar absent. Size relatively small (forearm length, 28–31 mm). – *Distribution:* Ranging from Gambia to Cameroon and from Kenya to Mozambique. – Two subspecies:

S. a. woodi (eastern segment of range), *S. a. albofuscus* (western segment of range).

3. *S. hirundo* (DE WINTON 1899). – Tragus relatively short and broad, bent forward distally. Anterior lower molar not reduced, but compressed in toothrow. Chiropatagium darkly pigmented. Ventral hair grayish. Braincase slightly elevated. Anterior upper premolar present or absent. Size varying from relatively small to relatively large (forearm length, 29–38 mm). Two species may be represented in this complex. – *Distribution:* Ranging from Senegal to Ethiopia and south to Malawi and Angola, but largely avoiding forested regions. – Four subspecies are here recognized:

S. h. hirundo (Senegal to Benin, and possibly to Ethiopia), *S. h. falabae* (Nigeria and Cameroon), *S. h. hindei* (southern Sudan and Somalia to southeastern Zaire), *S. h. albigula* (Kenya to Angola).

Genus *Scotomanes* DOBSON 1875 (= *Scoteinus* DOBSON 1875) (Fig. 168)

Ears not greatly enlarged. Width of each auditory bulla less than space between them. Anterior and middle upper molars with the mesostyle unreduced, the W-shaped pattern not distorted. Anterior and middle lower molars with the talonid larger than the trigonid. Lateral lower incisor not reduced. Rostrum in lacrimal region greatly broadened. Anterior palatal emargination considerably reduced. Last upper molar considerably reduced. – *Distribution:* Ranging from northeastern India to southern China and Vietnam. – Two species are here recognized, with two additional subspecies.

1. *S. ornatus* (BLYTH 1851). – Size medium to relatively large (forearm length, 50–64 mm). White spots on head, throat, and sides of the body. – *Distribution:* Same as for genus. – Three subspecies are recognized:

S. o. ornatus (northeastern India), *S. o. imbrensis* (northeastern India and northern Burma), *S. o. sinensis* (southern China, Vietnam, northern Thailand).

2. *S. emarginatus* (DOBSON 1871). – Size medium (forearm length, 55–56 mm). No white spots on head, throat, and sides of body. Doubtfully distinct from *S. ornatus*, poorly known. – *Distribution:* Without definite locality. – No subspecies.

Genus *Scotophilus* LEACH 1821 (Fig. 169)

Ears not greatly enlarged. Width of each auditory bulla less than space between them. Anterior and middle upper molars with the mesostyle reduced, the W-shaped pattern distorted or nearly absent. Anterior and middle lower molars with the trigonid larger than the talonid. Lateral lower incisor not reduced. Rostrum in lacrimal region not greatly broadened. Anterior palatal emargination not reduced. Last upper molar considerably reduced. – *Distribution:* Widely distributed in sub-Saharan Africa, Madagascar, Reunion, and southwestern Arabia; also Afghanistan and Ceylon to Taiwan, Philippines, Celebes, Timor, and perhaps the Aru islands. – Five species and 19 additional subspecies are here recognized.

1. *S. kuhli* LEACH 1822. – Size relatively small (forearm length, 45–55 mm). Cingula of upper incisors relatively narrow. Thumb relatively long. – *Distribution:* Ranging from Pakistan and Ceylon to Taiwan, Philippines, Celebes, Timor, and perhaps to the Aru islands. – Six subspecies are here recognized:

S. k. temmincki (Java and the Lesser Sundas), *S. k. panayensis* (Philippines), *S. k. castaneus* (Celebes to the Malay peninsula and Nicobars), *S. k. gairdneri* (Thailand, Vietnam), *S. k. consobrinus* (southern China, including Hainan and Taiwan), *S. k. kuhli* (Pakistan, India, Burma, Ceylon).

2. *S. borbonicus* (E. GEOFFROY 1803). – Size relatively small (forearm length, 41–53 mm). Cingula of upper incisors relatively broad. Thumb relatively short. – *Distribution:* Ranging from Gambia to Ethiopia and from Kenya to Namibia and Natal; also Reunion in the Mascarenes and probably Madagascar – Four subspecies are here recognized:

S. b. nigritellus (Gambia to Ethiopia), *S. b. viridis* (Kenya to Natal), *S. b. damarensis* (Zambia to Namibia), *S. b. borbonicus* (Réunion, where probably now extinct).

3. *S. leucogaster* (CRETZSCHMAR 1826). – Size fairly small to medium (forearm length, 42–60 mm). Cingula of upper incisors relatively broad. Thumb relatively short. – *Distribution:* Ranging through forest and savanna, regions from Mauretania and Ethiopia to the Cape Province, including Yemen, Zanzibar and Madagascar. – Eight subspecies are here recognized:

S. l. nucella (forested regions of Ivory Coast and Ghana, possibly east to Uganda), *S. l. leucogaster* (savanna areas from Mauretania to Ethiopia), *S. l. colias* (savanna areas from Ethiopia to Tanzania and west to Senegal), *S. l. nux* (forested regions from Ivory Coast to Kenya), *S. l. dingani* (Zaire to Natal), *S. l. herero* (Angola to Namibia and Botswana), *S. l. pondoensis* (eastern Cape province), *S. l. robustus* (Madagascar). Often broken into two or more species, since *nucella* and *leucogaster* are broadly sympatric with *nux* and *colias* in western Africa. However, intergradation occurs in Ethiopia.

4. *S. heathi* (HORSFIELD 1831). – Size medium (forearm length, 54–69 mm). Cingula of upper incisors relatively narrow. Thumb relatively long. – *Distribution:* Ranging from Afghanistan and Ceylon to southern China and Vietnam; also Celebes. – Four subspecies are currently recognized:

S. h. heathi (Afghanistan and Ceylon to southwestern China and northern Thailand), *S. h. insularis* (southeastern China, including Hainan), *S. h. watkinsi* (southern Thailand), *S. h. celebensis* (Celebes). Populations from Vietnam have not been allocated subspecifically.

5. *S. nigrita* (SCHREBER 1774) (= *gigas* DOBSON 1875). – Size relatively large (forearm length,

70–89 mm). Cingula of upper incisors relatively broad. Thumb relatively long. – *Distribution:* Ranging from Senegal to Sudan; also from southeastern Zaire and Kenya to Mozambique. – Two subspecies are recognized:

S. n. nigrita (northern segment of range), S. n. alvenslebeni (southern segment of range).

Genus *Otonycteris* PETERS 1859 (Fig. 170)

Ears greatly enlarged. Width of each auditory bulla wider than space between them. Anterior and middle upper molars with the mesostyle unreduced, the W-shaped pattern not distorted. Anterior and middle lower molars with the talonid larger than the trigonid. Lateral lower incisor not reduced. Rostrum in lacrimal region not greatly broadened. Anterior palatal emargination considerably reduced. Last upper molar greatly reduced. – *Distribution:* Ranging across arid areas of northern Africa, chiefly north of the Sahara, southwestern and southern central Asia. – A single species with four additional subspecies.

1. *O. hemprichi* PETERS 1859. – Size medium (forearm length, 55–67 mm). – *Distribution:* Same as for genus. – Five subspecies are currently recognized:

O. h. hemprichi (Morocco to Egypt and south to northern Niger), O. h. jin (Turkey to Oman), O. h. petersi (Iraq), O. h. leucophaeus (Turkmenia to Kashmir), O. h. cinerea (eastern Iran and Afghanistan).

Tribe Lasiurini TATE 1942

Rostrum relatively short and deep. Dental formula i 1/3, c 1/1, p 1–2/2, m 3/3 × 2 = 30–32. Ear pinna at least somewhat shortened. Braincase high. – *Distribution:* Ranging widely in North, Middle, and South America, including the Greater Antilles, Bermuda, Galapagos, and Hawaii. – One genus, an additional subgenus, six species.

Genus *Lasiurus* GRAY 1831 (Fig. 171)

Uropatagium furred on dorsal surface in varying degree. – *Distribution:* Same as for tribe. – Two subgenera, six species.

Subgenus *Lasiurus* GRAY 1831

Anterior upper premolar usually present. Uropatagium usually furred over its entire surface. Ear pinna greatly shortened. Lateral wings of pre-sternum poorly developed. Rostrum greatly shortened. – *Distribution:* Same as for genus. – Four species, 11 additional subspecies.

1. *L. borealis* (MÜLLER 1776). – Size relatively small (forearm length, 37–44 mm). Basicranial plane parallel with palate. Lacrimal tubercle variably developed. Anterior upper premolar usually present. – *Distribution:* Same as for subgenus but absent from Hawaii and most of the Rocky Mountain region. – Ten subspecies are here recognized:

L. b. borealis (eastern and central North America from central Canada to northern Florida and northwestern Mexico), L. b. teliotis (southeastern Canada to south-central Mexico), L. b. frantzii (southern Mexico to Amazonian Brazil, including Trinidad and Tobago), L. b. pfeifferi (Cuba), L. b. degelidus (Jamaica), L. b. minor (Bahamas, Hispaniola, and Puerto Rico), L. b. blossevillii (eastern Brazil to northern Argentina), L. b. varius (Chile and southern Argentina), L. b. brachyotis (Galapagos), L. b. seminolus (southeastern United States and apparently Bermuda). More than one species is usually recognized in this complex but a skull character (lacrimal tubercle) does not clearly distinguish seminolus from all of the rest, coloration in some areas shows great variation, and migration obscures possible overlap at mating time. However, it is nevertheless possible that several species are represented in this complex.

2. *L. castaneus* HANDLEY 1960. – Size medium (forearm length, 44–45 mm). Basicranial plane tilted upward from palate. Lacrimal tubercle absent. Small anterior upper premolar present. – *Distribution:* From Costa Rica to Colombia. – No subspecies.

3. *L. egregius* (PETERS 1871). – Size fairly large (forearm length, 48–50 mm). Basicranial plane parallel with palate. Lacrimal tubercle present. Small anterior upper premolar absent. – *Distribution:* Known only from Panama and southeastern Brazil. – No subspecies.

4. *L. cinereus* (PALISOT DE BEAUVOIS 1796). – Size relatively large (forearm length, 46–57 mm). Basicranial plane parallel with palate. Lacrimal tubercle poorly developed. Small anterior upper premolar usually present. – *Distribution:* Ranging from central Canada to Guatemala, from Colombia and Venezuela to Chile and southeastern Brazil (mostly avoiding lowland tropical areas), Galapagos, and Hawaii. There are also records from Cuba, Hispaniola, Bermuda, Iceland, and the Orkney islands (north of Britain). – Three subspecies are recognized:

L. c. cinereus (North American range), L. c. villosissimus (South American range), L. c. semotus (Hawaii). The Galapagos population has not been subspecifically allocated.

Subgenus *Dasypterus* PETERS 1871

Anterior upper premolar absent. Uropatagium only furred proximally. Ear pinna not greatly shortened. Lateral wings of presternum well developed. Rostrum not greatly shortened. – *Distribution:* Ranging from the southern United States south to northern Argentina, including Trinidad, but west of the Andes not south of Ecuador; also Cuba. – Two species, six additional subspecies.

5. *L. ega* (GERVAIS 1856). – Size fairly large (forearm length, 42–53 mm). – *Distribution:* Ranging from southwestern United States to northern Argentina, including Trinidad, but west of the Andes not south of Ecuador. – Five subspecies are here recognized:

L. e. xanthinus (southwestern United States to Costa Rica), *L. e. panamensis* (Costa Rica and Panama), *L. e. fuscatus* (northern South America west of the Andes), *L. e. ega* (northern South America east of the Andes, including Trinidad), *L. e. caudatus* (eastern Brazil to northern Argentina). It is possible that more than one species is represented in this complex.

6. *L. intermedius* H. ALLEN 1862. – Size relatively large (forearm length, 45–63 mm). – *Distribution:* Ranging from southeastern United States (with a few records farther north) and western Mexico to Honduras; also Cuba. – Three subspecies are recognized:

N. i. floridanus (southeastern United States), *N. i. intermedius* (southern Texas and western Mexico to Honduras), *N. i. insularis* (Cuba).

Tribe Antrozoini MILLER 1897

Rostrum relatively elongate. Dental formula i 1/2–3, c 1/1, p 1/2, m 3/3 x 2 = 28–30, but outer lower incisor, if present greatly reduced. Ears large but not united by a band. Muzzle truncate with the nostrils opening beneath a horseshoe-shaped ridge. – *Distribution:* Ranging through western North America and northern Middle America from southwestern Canada to Honduras; also Cuba. – Two genera, two species.

Genus *Bauerus* VAN GELDER 1959 (Fig. 172)

Outer lower incisor greatly reduced but usually present. Forehead flat. Sagittal crest well developed posteriorly. Rostrum relatively broad. Auditory bulla relatively small. – *Distribution:* Ranging from tropical Mexico, including the Tres Marias islands, to Costa Rica. – A single species is currently recognized.

1. *B. dubiaquercus* VAN GELDER 1959 (= *meyeri* PINE 1966). – Forearm length, 47–58 mm. – *Distribution:* Same as for genus. – No subspecies are currently recognized.

Genus *Antrozous* H. ALLEN 1862 (Fig. 174)

Outer lower incisor absent. Forehead slightly convex. Sagittal crest relatively poorly developed posteriorly. Rostrum relatively slender. Auditory bulla relatively large. – *Distribution:* Ranging from southwestern Canada to central Mexico; also Cuba. – A single species and five additional subspecies are currently recognized.

1. *A. pallidus* (LE CONTE 1856). Forearm length, 45–65 mm. – *Distribution:* Same as for genus. – Six subspecies are currently recognized:

A. p. pacificus (coast of Oregon and California), *A. p. pallidus* (mostly in interior areas from British Columbia to northwestern Baja California and northeastern Mexico), *A. p. minor* (southern Nevada to southern Baja California), *A. p. bunkeri* (Kansas, Oklahoma), *A. p. packardi* (mostly in coastal areas of northwestern Mexico), *A. p. koopmani* (Cuba).

Tribe Nyctophilini PETERS 1865

Rostrum of medium length to fairly short. Dental formula i 1/3, c 1/1, p 1/2, m 3/3 x 2 = 30 and outer lower incisor not reduced. Ears large and usually united by a band. Muzzle truncate with the nostrils opening beneath a noseleaf, developed to a greater or lesser degree. – *Distribution:* Ranging widely in Australia, including Tasmania, New Guinea, and possibly Timor. Known subfossil from Lord Howe island (off the eastern coast of Australia). – Two genera, eight species.

Genus *Nyctophilus* LEACH 1822 (= *Lamingtona* MCKEAN & CALABY 1968 (Fig. 174)

Anterior noseleaf poorly developed; posterior noseleaf of variable development. Ears of variable length. Rostrum of medium length. – *Distribution:* Same as for tribe. – Seven species with six additional subspecies are recognized here.

1. *N. walkeri* THOMAS 1892. – Size relatively small (forearm length, 31–35 mm; condylobasal length, 11–13 mm). Noseleaves poorly developed. Band connecting ears fairly high. Teeth relatively large.

– *Distribution:* Confined to extreme northern Australia from northeastern Western Australia to northwestern Queensland. – No subspecies.

2. N. *microtis* THOMAS 1888 (= *lophorhina* MCKEAN & CALABY 1968). – Size fairly small (forearm length, 38–40 mm; condylobasal length, 13–15 mm). Noseleaves poorly developed. Band connecting ears (which are relatively short) absent or poorly developed. Teeth relatively large. Auditory bullae relatively small. – *Distribution:* Known only from eastern New Guinea. – No subspecies here recognized.

3. N. *arnhemensis* JOHNSON 1959. – Size fairly small (forearm length, 34–41 mm; condylobasal length, 13–15 mm). Noseleaves poorly developed. Band connecting ears (which are relatively long) relatively low. Teeth relatively large. Auditory bullae relatively large. – *Distribution:* Confined to extreme northern Australia from northeastern Western Australia to northwestern Queensland. – No subspecies.

4. N. *gouldi* TOMES 1858. – Size fairly large (forearm length, 37–43 mm; condylobasal length, 14–17 mm). Noseleaves poorly developed. Band connecting ears fairly high. Teeth relatively large. – *Distribution:* Ranging widely in coastal regions of eastern, northern, and western Australia, but apparently not Tasmania, also known from two localities in central New Guinea. – Three subspecies are here recognized:

N. g. bifax (New Guinea, eastern Queensland, northeastern New South Wales), *N. g. daedalus* (northern Northern Territory, northern Western Australia), *N. g. gouldi* (eastern Queensland to Victoria). There may be more than one species in this complex.

5. N. *timoriensis* (GEOFFROY 1806). – Size relatively large (forearm length, 41–48 mm; condylobasal length, 16–19 mm). Noseleaves poorly developed. Band connecting ears relatively high. Teeth relatively large. – *Distribution:* Widely distributed in coastal regions of Australia except the northeast. Also known from western New Guinea and perhaps Timor. – Three subspecies are here recognized:

N. t. sherrini (southern Queensland to South Australia, including Tasmania), *N. t. major* (Western Australia), *N. t. timoriensis* (Northern Territory, New Guinea, and perhaps Timor).

6. N. *microdon* LAURIE & HILL 1954. – Size medium (forearm length, 37–41 mm; condylobasal length, 13–15 mm). Posterior noseleaf fairly well developed. Band connecting ears relatively high. Teeth relatively small. – *Distribution:* Known only from the highlands of eastern New Guinea. – No subspecies.

7. N. *geoffroyi* LEACH 1821. – Size fairly small (forearm length, 30–41 mm; condylobasal length, 13–16 mm). Posterior noseleaf relatively well developed. Band connecting ears relatively high. Teeth relatively small. – *Distribution:* Ranging widely across Australia (except Cape York) in both coastal and inland areas, including Tasmania. – Three poorly marked subspecies.

Genus ***Pharotis*** THOMAS 1914 (Fig. 175)

Anterior and posterior noseleaves well developed. Ears relatively large and connected by a high band. Rostrum relatively short. – *Distribution:* Known only from eastern New Guinea. – A single species.

1. *P. imogene* THOMAS 1914. – Size fairly small (forearm length, 37–38 mm). – *Distribution:* Same as for genus. – No subspecies.

Subfamily **Murininae** MILLER 1907 (Fig. 176)

Dental formula i 2/3, c 1/1, p 2/2, m 3/3 x 2 = 34. Anterior upper premolar relatively large, not conspicuously simpler than posterior upper premolar. Nostrils elongated as tubes. Ears somewhat funnel-shaped. Second phalanx of third digit of wing not greatly elongated. – *Distribution:* Ranging from Japan and south-central Siberia to Pakistan and Ceylon, east to the Philippines, New Guinea, and northeastern Australia. – Two genera, one additional subgenus, and 16 species.

Genus ***Murina*** GRAY 1842 (Fig. 176)

Anterior and middle upper molars relatively unmodified, retaining a W-shaped cusp pattern. Last upper molar not greatly reduced. – *Distribution:* Same as for subfamily. – Two subgenera, 14 species.

Subgenus ***Murina*** GRAY 1842

Margin of plagiopatagium attached to ungual phalanx of outer digit of foot. Metacones of anterior and middle upper molars not reduced. – *Distribution:* Same as for genus. – Thirteen species, eight additional subspecies.

1. *M. leucogaster* MILNE-EDWARDS 1872 [*suilla* group]. – Anterior rostrum relatively narrow, the upper tooth rows strongly convergent anteriorly. Anterior upper premolar relatively small. Rostrum relatively deep and massive. Size relatively large (forearm length, 40–44 mm). Dorsal surface of uropatagium densely haired. – *Distribution:* Ranging from south-central Siberia to Sakhalin and south through Japan and China to northeastern India. – Three subspecies are currently recognized:

M. l. hilgendorfi (Siberia, Japan, Korea, and northern China), *M. l. leucogaster* (southern China), *M. l. rubex* (northeastern India). The named taxon *fuscus* (currently in the synonymy of *M. l. hilgendorfi*) may be a separate species.

2. *M. tenebrosa* YOSHIYUKI 1970 [*suilla* group]. – Rostrum relatively deep and massive. Size fairly small (forearm length, 34–35 mm). Dorsal surface of uropatagium scantily haired. – *Distribution:* Known only from the Tsushima islands (between Japan and Korea). – No subspecies.

3. *M. silvatica* YOSHIYUKI 1983 [*suilla* group]. – Rostrum not particularly massive. Size relatively small (forearm length, 28–33 mm). Dorsal surface of uropatagium densely haired. Canine relatively short. Anterior upper premolar relatively high. Fissure present between cochlea and basioccipital. Ear relatively long. – *Distribution:* Confined to Japan. – No subspecies.

4. *M. ussuriensis* OGNEV 1913 [*suilla* group]. – Rostrum not particularly massive. Size relatively small (forearm length, 30–32 mm). Canine relatively short. Anterior upper premolar relatively low. No fissure between cochlea and basioccipital. Ear relatively short. – *Distribution:* Known from the Kuriles, Sakhalin, southeastern Siberia, and Korea. – No subspecies.

5. *M. aurata* MILNE-EDWARDS 1872 [*suilla* group]. – Rostrum not particularly massive. Size relatively small (forearm length, 28–32 mm). Canine relatively short. Anterior upper premolar relatively low. Fissure present between cochlea and basioccipital. Ear relatively short. – *Distribution:* Ranging from northeastern India to southwestern China and northern Thailand. – No subspecies.

6. *M. tubinaris* (SCULLY 1881) [*suilla* group]. – Rostrum not particularly massive. Size relatively small (forearm length, 29–35 mm). Canines relatively long. – *Distribution:* Ranging from Pakistan to Vietnam. – No subspecies. Probably a subspecies of *M. suilla*.

7. *M. suilla* (TEMMINCK 1840) (= *balstoni* Thomas 1908) [*suilla* group]. – Rostrum not particularly massive. Size relatively small (forearm length, 29–31 mm). Canines relatively long. – *Distribution:* Ranging from Malaya to Borneo and Java. – Two subspecies are here recognized:

M. s. canescens (Nias island, off the west coast of Sumatra), *M. s. suilla* (remainder of range).

8. *M. florium* THOMAS 1908 [*suilla* group]. – Rostrum not particularly massive. Size fairly small (forearm length, 32–37 mm). Canines relatively long. – *Distribution:* Ranging from Celebes and the Lesser Sundas to New Guinea and northeastern Australia. – Three poorly marked subspecies are usually recognized.

9. *M. huttoni* (PETERS 1872) [*cyclotis* group]. – Anterior rostrum elongate but relatively broad, the upper toothrows almost parallel. Anterior upper premolar relatively large. Size medium (forearm length, 29–38 mm). Ears usually bluntly pointed. Talonids of anterior and middle lower molars and their cusps well developed. Upper canine relatively long. Teeth not greatly enlarged. – *Distribution:* Ranging from northern India to southeastern China and south to Malaya. – Two subspecies:

M. h. huttoni (northern India, northern Burma and Tibet, presumeably its southeastern corner), *M. h. rubella* (southeastern China to Malaya).

10. *M. puta* KISHIDA 1924 [*cyclotis* group]. – Size medium (forearm length, 35–36 mm). Ears rounded. Talonids of anterior and middle lower molars and their cusps well developed. Upper canine relatively long. Teeth considerably enlarged. – *Distribution:* Confined to Taiwan. – No subspecies, but may be a subspecies of *M. huttoni*.

11. *M. cyclotis* DOBSON 1872 [*cyclotis* group]. – Size medium (forearm length, 30–38 mm). Ears rounded. Talonids of anterior and middle lower molars and their cusps reduced. Upper canine relatively long. Teeth not greatly enlarged. – *Distribution:* Ranging from northeastern India to Hainan in southeastern China, south to Ceylon and Malaya and east through Borneo to Mindanao in the southern Philippines. – Three subspecies:

M. c. eileenae (Ceylon), *M. c. cyclotis* (northeastern India to Hainan and Vietnam), *M. c. peninsularis* (Malaya to Mindanao).

12. *M. rozendaali* HILL & FRANCIS 1984 [*cyclotis* group]. – Size fairly small (forearm length, 32–34 mm). Ears rounded. Talonids of anterior

and middle lower molars and their cusps well developed. Upper canine relatively short. Teeth not greatly enlarged. – *Distribution:* Known only from Borneo. – No subspecies.

13. *M. aenea* HILL 1963 [*cyclotis* group]. – Size medium (forearm length, 33–38 mm). Ears more or less rounded. Talonids of anterior and middle lower molars and their cusps somewhat reduced. Upper canine relatively short. Teeth greatly enlarged. – *Distribution:* Known only from Malaya and Borneo. – No subspecies.

Subgenus **Harpiola** THOMAS 1915

Margin of plagiopatagium attached to base of outer digit of foot. Metacones of anterior and middle upper molars reduced. – *Distribution:* Known only from northwestern India. – A single species.

14. *M. grisea* PETERS 1872. – Size fairly small (forearm length, 33 mm). – *Distribution:* Same as for subgenus. – No subspecies.

Genus **Harpiocephalus** GRAY 1842 (Fig. 177)

Anterior and middle upper molars highly modified with loss of mesostyle, the W-shaped pattern obliterated. Last upper molar greatly reduced. – *Distribution:* Ranging from India to Taiwan and Vietnam; also Sumatra, Borneo, Java, and a dubious record from the Moluccas. – Two species and three additional subspecies are currently recognized.

1. *H. harpia* (TEMMINCK 1840). – Muzzle not greatly broadened nor zygomata greatly expanded. Incisors and canines not enlarged. Size fairly large (forearm length, 43–50 mm). – *Distribution:* Essentially same as for genus. – Four subspecies are currently recognized:

H. h. madrassius (southern India), *H. h. lasyurus* (northeastern India), *H. h. rufulus* (Thailand, Vietnam), *H. h. harpia* (Sumatra, Borneo, Java, Moluccas). The Taiwan record, if valid, has not been allocated subspecifically.

2. *H. mordax* THOMAS 1923. – Muzzle greatly broadened and zygomata greatly expanded. Incisors and canines much enlarged. Size relatively large (forearm length, 48–55 mm). – *Distribution:* A poorly known species recorded only from Burma and Borneo. – No subspecies.

Subfamily **Miniopterinae** DOBSON 1875

Dental formula i 2/3, c 1/1, p 2/3, m 3/3 x 2 = 36. Anterior upper premolar relatively small and conspicuously simpler than posterior upper premolar. Nostrils not elongated as tubes. Sternum slender, its length more than twice as great as the breadth of the presternum, which has a relatively large median lobe. Six or seven ribs connected with sternum. Coracoid process straight and directed inwards. Ears not funnel-shaped. Second phalanx of third digit of wing greatly elongated. – *Distribution:* Ranging widely in Africa, including Madagascar, and southern Eurasia, through the Indo-Australian archipelago to the New Hebrides and Australia. – A single genus and 13 species are recognized here.

Genus **Miniopterus** BONAPARTE 1837 (Fig. 178)

Characters and *distribution* same as for genus. Of the 13 species and 31 additional subspecies recognized here, most have not been clearly distinguished from one another and there is great uncertainty as to how many species should actually be recognized and what populations should be allocated to each.

1. *M. manavi* THOMAS 1906 [*australis* group]. – Size relatively small (forearm length, 32–40 mm; condylobasal length, 12–14 mm). Braincase not markedly inflated. Rostrum relatively slender. – *Distribution:* Confined to Madagascar and the Comoro islands. – Two subspecies are recognized:

M. m. manavi (Madagascar), *M. m. grivaudi* (Comoros).

2. *M. minor* PETERS 1867 [*australis* group]. – Size relatively small (forearm length, 37–40 mm; condylobasal length, 13–14 mm). – *Distribution:* Ranging across central Africa from Kenya and Tanzania to Congo-Brazzaville, São Tomé in the Gulf of Guinea, and perhaps Madagascar. – Two subspecies are recognized:

M. m. minor (mainland range), *M. m. newtoni* (São Tomé).

3. *M. paululus* HOLLISTER 1913 [*australis* group]. – Size relatively small (forearm length, 34–38 mm; condylobasal length, 12–14 mm). Rostrum relatively slender. – *Distribution:* Known from the Philippines, Java, Lesser Sundas, east Papuan islands, Rennell in the Solomons, New Hebrides, Loyalties, and New Caledonia. – Two subspecies are recognized:

M. p. shortridgei (Java), *M. p. paululus* (Philippines). Other populations have not been allocated subspecifically.

4. *M. australis* TOMES 1858 [*australis* group]. – Size relatively small (forearm length, 34–47 mm; condylobasal length, 12–15 mm). Rostrum relatively broad. – *Distribution:* Ranging from Borneo to New Guinea, the Solomons, eastern Australia, and the New Hebrides. – Four subspecies are recognized here:

M. a. witkampi (Borneo), *M. a. tibialis* (Celebes, Moluccas), *M. a. australis* (eastern New Guinea, eastern Australia to the New Hebrides), *M. a. solomonensis* (Solomons). Allocation of western New Guinea populations subspecifically is uncertain.

5. *M. pusillus* DOBSON 1876 [*australis* group]. – Size relatively small (forearm length, 39–45 mm; condylobasal length, 12–14 mm). Braincase greatly inflated. – *Distribution:* Ranging from India and southern China to the Philippines, Lesser Sundas, and New Hebrides, but probably absent from New Guinea, the Bismarcks, and Australia. – Two subspecies are here recognized:

M. p. pusillus (India to the Aru islands), *M. p. macrocneme* (Solomons to the New Hebrides and New Caledonia).

6. *M. fuscus* BONHOTE 1902 [*fuscus* group]. – Size fairly small (forearm length, 38–45 mm; condylobasal length, 13–15 mm). Rostrum relatively broad. – *Distribution:* Ranging from the Riukiu islands to Java and Celebes and perhaps to New Guinea. – Two subspecies are here recognized:

M. f. fuscus (Riukius, southeastern China, and the Philippines), *M. f. medius* (Vietnam to Java and Celebes).

7. *M. fraterculus* THOMAS & SCHWANN 1906 [*fuscus* group]. – Size fairly small (forearm length, 41–44 mm; condylobasal length, 13–15 mm). – *Distribution:* Ranging through central and southern Africa from Cameroon and Kenya to the Cape Province and Madagascar. – Two subspecies are here recognized:

M. f. vicinior (Cameroon to Kenya and south to Zambia), *M. f. fraterculus* (Malawi to Cape Province and Madagascar).

8. *M. natalensis* (A. SMITH 1834) [*schreibersi* group]. – Size medium (forearm length, 41–49 mm; condylobasal length, 13–16 mm). – *Distribution:* Ranging from Sudan and southwestern Arabia to the Cape Province. – Two subspecies are here recognized:

M. n. arenarius (Sudan and Arabia to Zimbabwe), *M. n. natalensis* (South Africa).

9. *M. schreibersi* (KUHL 1817) [*schreibersi* group]. – Size medium (forearm length, 42–50 mm; condylobasal length, 14–16 mm). Braincase relatively broad. – *Distribution:* Ranging widely in Madagascar, sub-Saharan Africa, northern Africa (chiefly northwestern), southern Eurasia from Europe and through the Indo-Australian archipelago to Australia and the Solomons. – Fifteen subspecies are here recognized:

M. s. majori (Madagascar, Comoros), *M. s. dasythrix* (Malawi to South Africa), *M. s. smitianus* (Ethiopia to Namibia), *M. s. villiersi* (Guinea to Zaire), *M. s. schreibersi* (Europe, northern Africa), *M. s. pallidus* (southwestern Asia to Afghanistan), *M. s. fuliginosus* (Afghanistan to Ceylon and Burma), *M. s. parvipes* (southern China, Vietnam), *M. s. chinensis* (northeastern China), *M. s. japoniae* (Japan), *M. s. harardai* (Thailand), *M. s. blepotis* (Malaya to the Moluccas), *M. s. eschscholtzii* (Philippines), *M. s. orianae* (northwestern Australia), *M. s. oceanensis* (eastern Australia). However, boundaries are often unclear and some populations have not been allocated subspecifically.

10. *M. inflatus* THOMAS 1903 [*inflatus* group]. – Size fairly large (forearm length, 45–51 mm; condylobasal length, 15–17 mm). Braincase relatively broad. – *Distribution:* Ranging from Liberia to Ethiopia and south to Namibia and Mozambique. – Three subspecies are currently recognized:

M. i. inflatus (Liberia to Gabon), *M. i. rufus* (Zaire to Kenya), *M. i. africanus* (Ethiopia to Namibia and Mozambique).

11. *M. magnater* SANBORN 1931 [*inflatus* group]. – Size fairly large (forearm length, 47–53 mm; condylobasal length, 15–18 mm). Braincase relatively broad. – *Distribution:* Burma and southeastern China to Borneo and Java and east to the Lesser Sundas and New Guinea; possibly also in Madagascar. – Two subspecies are here recognized:

M. m. magnater (New Guinea), *M. m. macrodens* (remainder of range).

12. *M. robustior* REVILLIOD 1913 [*tristis* group]. – Size fairly small (forearm length, 39–43 mm; condylobasal length, 13–16 mm). Braincase relatively narrow. – *Distribution:* Confined to the Loyalty islands (east of New Caledonia). – No subspecies.

13. *M. tristis* (WATERHOUSE 1845) [*tristis* group]. – Size relatively large (forearm length, 43–59 mm; condylobasal length, 15–20 mm). Braincase relatively narrow. – *Distribution:* Ranging from the Philippines and Celebes to the New Hebrides. – Five subspecies are currently recognized:

M. t. insularis (New Hebrides to the Bismarcks, including the East Papuan islands), *M. t. propritristis* (eastern mainland New Guinea), *M. t. grandis* (western mainland New Guinea), *M. t. tristis* (Philippines), *M. t. celebensis* (Celebes).

Subfamily **Tomopeatinae** MILLER 1907

Anterior and middle upper premolars absent. Nostrils not elongated as tubes. Sternum slender, its length more than twice as great as the breadth of the presternum, which has a relatively small median lobe. Six or seven ribs connected with sternum. Coracoid process of scapula curved outwards. Ears not funnel-shaped, with a rudimentary keel, but no anterior basal lobe. Second phalanx of third digit of wing not greatly elongated. Seventh cervical vertebra fused with first thoracic. – *Distribution:* Confined to the arid zone of western Peru. – A single genus and species.

Genus *Tomopeas* MILLER 1900 (Fig. 179)

Dental formula i 1/2, c 1/1, p 1/2, m 3/3 x 2 = 28. – *Distribution:* Same as for subfamily. – A single genus and species.

1. *T. ravus* MILLER 1900. – Forearm length, 31–35 mm. – *Distribution:* Same as for genus. – No subspecies.

Family **Mystacinidae** DOBSON 1875
(New Zealand Shorttailed Bats)

Structure: Second digit of wing reduced to the metacarpal and a single minute phalanx. Tail short, not reaching edge of extensive uropatagium. Trochiter of humerus large, making an extensive articulation with the scapula. Last cervical and first two thoracic vertebrae not fused with one another. Rostrum not shortened. The premaxillaries retain both nasal and palatal branches. The ears are not funnel shaped. The pollex is not modified.

Ecology: Confined to forested areas. Primarily insectivorous, but also eat fruit, nectar, and pollen; occasionally are even scavengers. Roost in crevices and tree hollows, which may be modified by gnawing. Adept at moving rapidly on solid surfaces.

Distribution (Fig. 19): Confined to New Zealand.

Systematics: A single genus, two species.

Genus *Mystacina* GRAY 1843 (Fig. 180)

Dental formula i 1/1, c 1/1, p 2/2, m 3/3 × 2 = 28. – *Distribution:* Same as for family. – Two species, three additional subspecies.

1. *M. tuberculata* GRAY 1843. – Size relatively small (forearm length, 40–46 mm; condylobasal length, 17–20 mm). Ears relatively long. – *Distribution:* Same as for genus. – Three subspecies are currently recognized, all occurring on North island, but only *M. t. tuberculata* occurring on South island.

2. *M. robusta* DWYER 1962. – Size relatively large (forearm length, 45–48 mm; condylobasal length, 21–23 mm). Ears relatively short. – *Distribution:* Confined in historic times to a few islands off the southern end of New Zealand and probably now extinct. Known fossil from much of New Zealand.

Family **Molossidae** GERVAIS 1855
(Free-tailed Bats)

Structure: Second digit of wing reduced to the metacarpal and a single vestigial phalanx. Tail long, extending a considerable distance beyond the somewhat shortened uropatagium. Trochiter of humerus large and making an extensive articulation with the scapula. Last cervical and first thoracic vertebrae fused with one another. Rostrum may be relatively long, or shortened in varying degrees. Premaxillae with nasal branches, but palatal branches may be greatly reduced. Ears usually greatly broadened or lengthened. Pollex unmodified.

Ecology: Largely confined to tropical and warm temperate areas. Usually roost in caves, rock crevices, or man-made structures, but occasionally in trees, almost always under conditions where they are closely pressed against a substrate or other individuals. As far is known, always insectivorous.

Distribution (Fig. 19): Widely distributed in Africa, including Madagascar and the Mascarenes, southern Eurasia, and through the Indo-Australian archipelago to Fiji and Australia; also southern North America, Middle America, West Indies and South America except the extreme south.

Systematics: No subfamilies are here recognized. Twelve genera, six additional subgenera, 77 species.

Genus *Mormopterus* PETERS 1865 (Fig. 181)

Anterior palatal emargination well developed. Last upper molar not reduced. Ears relatively short, erect, and well separated. Basisphenoid pits absent or poorly developed. Wrinkles on upper lip absent or poorly developed. Wing tip usually relatively broad. Dental formula i 1/2–3, c 1/1, p 1–2/2, m 3/3 x 2 = 28–30. – *Distribution:* Occurring in the Mascarene islands, Madagascar, much of southern and eastern Africa, Sumatra, Moluccas, New Guinea, Australia, Cuba, Peru, and northern Chile. – Three subgenera and 11 species are here recognized.

Subgenus *Mormopterus* PETERS 1865

Skull not extremely flattened. Rostrum relatively long, usually with only moderately developed lacrimal tubercles. Anterior palatal emargination not greatly narrowed. No wartlike granulations on forearm. Ears usually widely separated. – *Distribution:* Same as for genus except that the known African mainland distribution is limited to two localities on the eastern side. – Nine species and five additional subspecies are here recognized.

1. *M. norfolkensis* GRAY 1839) [*norfolkensis* group]. – Lateral lower incisor absent. Anterior upper premolar present. Size medium (forearm length, 36–37 mm). The distinction of *M. norfolkensis* from *M. planiceps* (of which it may be a senior synonym) is not clear, at least to me. The distribution of *M. norfolkensis* is equally uncertain. – *Distribution:* Originally described on the basis of a specimen supposedly from Norfolk island (east of Australia), specimens from coastal eastern Australia (particularly New South Wales and southeastern Queensland) have been referred to it by some, but denied by others. The provenience of the type has also been questioned. – No subspecies.

2. *M. planiceps* (PETERS 1866) [*norfolkensis* group]. – Anterior upper premolar present but considerably reduced. Size medium (forearm length, 31–39 mm). Skull with varying amounts of flattening. No gular sac. – *Distribution:* Ranging over most of Australia (but not Tasmania) and southeastern New Guinea. – Four subspecies are here recognized:

M. p. planiceps (southern half of Australia), *M. p. coburgiana* (northern Australia except northeast), *M. p. ridei* (northeastern Queensland), *M. p. loriae* (southeastern New Guinea).

3. *M. beccarii* PETERS 1881 [*norfolkensis* group]. – Anterior upper premolar vestigial or absent. Size fairly large (forearm length, 33–39 mm). Skull not particularly flattened. No gular sac. – *Distribution:* Ranging from the Moluccas through New Guinea to the East Papuan islands (Ferguson) and in much of the northern half of Australia. – Two subspecies are recognized:

M. b. astrolabiensis (New Guinea region), *M. b. beccarii* (known only from Amboina in the Moluccas). – Australian populations have not been allocated subspecifically.

4. *M. minutus* (MILLER 1899) [*kalinowskii* group]. – Lateral lower incisor absent. Anterior upper premolar absent. Size relatively small (forearm length, 28–32 mm). Rostrum fairly low. – *Distribution:* Restricted to Cuba. – No subspecies.

5. *M. kalinowskii* (THOMAS 1893) [*kalinowskii* group]. – Anterior upper premolar absent. Size medium (forearm length, 34–39 mm). Rostrum fairly low. – *Distribution:* Restricted to arid areas of western Peru and extreme northern Chile. – No subspecies.

6. *M. phrudus* HANDLEY 1956 [*kalinowskii* group]. – Anterior upper premolar vestigial or absent. Size medium (forearm length, 33–35 mm). Rostrum relatively deep. Wing tip relatively narrow. – *Distribution:* Known only from the highlands of southern Peru. – No subspecies.

7. *M. doriae* ANDERSEN 1907 [*acetabulosus* group]. – Lateral lower incisor present. Anterior upper premolar absent. Size fairly large (forearm length, 38 mm). Gular sac well developed in males. No emargination in anterior margin of ear pinna, which closely approaches the ear of the opposite side. Rostrum relatively broad with well developed lacrimal tubercles. – *Distribution:* Known only from Sumatra. – No subspecies.

8. *M. jugularis* (PETERS 1865 [*acetabulosus* group]. – Size fairly large (forearm length, 36–39 mm). No emargination in anterior margin of ear pinna, which is well separated from the ear of the opposite side. Rostrum medium in width with only moderately developed lacrimal tubercles. – *Distribution:* Confined to Madagascar. – No subspecies.

9. *M. acetabulosus* (HERMANN 1804) [*acetabulosus* group]. – Size medium (forearm length, 39 mm). Emargination present in anterior margin of ear pinna, which is well separated from the ear of the opposite side. Rostrum relatively slender with only moderately developed lacrimal tubercles. – *Distribution:* Known from Mauritius and Reunion in the Mascarenes, Madagascar, and from two widely separated localities (Natal and Ethiopia), in eastern Africa. – No subspecies.

Subgenus *Sauromys* ROBERTS 1917

Skull extremely flattened. Rostrum relatively long with only moderately developed lacrimal tubercles. Anterior palatal emargination not greatly narrowed. No wart-like granulations on forearm. Ears not widely separated. – *Distribution:* Confined to southern Africa. – A single species and four additional subspecies.

10. *M. petrophilus* (ROBERTS 1917). – Size relatively large (forearm length, 36–42 mm). Anterior upper premolar present. No gular sac. – *Distribution:* Same as for subgenus. – Five subspecies are recognized:

M. p. petrophilus (Mozambique to Transvaal and Botswana), *M. p. erongensis* (northern Namibia), *M. p. haagneri* (southern Namibia), *M. p. umbratus* (northwestern Cape Province), *M. p. fitzsimonsi* (southwestern Cape Province).

Subgenus *Platymops* THOMAS 1906

Skull extremely flattened. Rostrum relatively short with greatly developed lacrimal tubercles. Anterior palatal emargination greatly narrowed. Wart-like granulations present on forearm. Ears widely separated. – *Distribution:* Confined to Kenya, southeastern Sudan, and southwestern Ethiopia. – A single species with one additional subspecies.

11. *M. setiger* PETERS 1878. – Size medium (forearm length, 29–36 mm). Anterior upper premolar vestigial or absent. Gular sac well developed in males. – *Distribution:* Same as for subgenus. – Two subspecies:

M. s. setiger (southern Kenya), *M. s. macmillani* (remainder of range).

Genus *Molossops* PETERS 1865 (Fig. 182)

No anterior palatal emargination. Last upper molar variably developed. Ears relatively short, erect, and usually well separated. Basisphenoid pits usually absent or poorly developed. Wrinkles on upper lip absent or poorly developed. Wing tips of variable breadth. Dental formula i 1/1–2, c 1/1, p 1–2/2, m 3/3 x 2 = 26–30. – *Distribution:* Ranging from southwestern Mexico to Uruguay, including Trinidad, but west of the Andes, not south of Ecuador. – Four subgenera and seven species are here recognized.

Subgenus *Cynomops* THOMAS 1920

Last upper molar greatly reduced. Ears well separated. Basisphenoid pits absent. No wrinkles on upper lip. Wing tips relatively narrow. Lateral lower incisor usually present. Anterior upper premolar absent. Upper incisors fairly long and in contact only near their bases. – *Distribution:* Essentially same as for genus, but not extending south of extreme northern Argentina. – Three species, six additional subspecies.

1. *M. abrasus* (TEMMINCK 1827). – Size relatively large (forearm length, 41–46 mm). – *Distribution:* Ranging from southwestern Colombia and Suriname to extreme northern Argentina and eastern Brazil. – Four subspecies are recognized:

M. a. mastivus (Venezuela and the Guianas), *M. a. abrasus* (eastern Brazil), *M. a. brachymeles* (eastern Peru), *M. a. cerastes* (Paraguay and northern Argentina).

2. *M. greenhalli* (GOODWIN 1958). – Size fairly small (forearm length, 32–39 mm). Outer lower incisor sometimes absent. – *Distribution:* Ranging from southwestern Mexico to western Ecuador and northeastern Brazil, including Trinidad. – Two subspecies:

M. g. mexicanus (Mexico to Costa Rica), *m. g. greenhalli* (remainder of range).

3. *M. planirostris* (PETERS 1865). – Size relatively small (forearm length, 29–34 mm). – *Distribution:* Ranging from Panama to extreme northern Argentina and eastern Brazil but not known west of the Andes in South America. – Three subspecies are recognized here:

M. p. planirostris (Panama to French Guiana), *M. p. paranus* (Amazonian Brazil), *M. p. milleri* (eastern Peru). Populations from south of the Amazon basin have not been allocated subspecifically.

Subgenus *Molossops* PETERS 1865

Last upper molar somewhat reduced. Ears well separated. Basisphenoid pits absent. No wrinkles on upper lip. Wing tips relatively broad. Lateral lower incisor absent. Upper incisors fairly long and in contact only near their bases. – *Distribution:* Occurring in much of tropical South America east of the Andes, south to Uruguay but avoiding most of the Amazon basin. – Two species, two additional subspecies.

4. *M. neglectus* WILLIAMS & GENOWAYS 1980. – Size medium (forearm length, 35–36 mm). Last upper molar moderately reduced. Sagittal crest

fairly well developed. – *Distribution:* Known from Suriname, eastern Amazonian Brazil, and eastern Peru. – No subspecies.

5. M. temminckii (BURMEISTER 1854). – Size relatively small (forearm length, 25–32 mm). Last upper molar considerably reduced. Sagittal crest poorly developed. – *Distribution:* Ranging from Venezuela and Colombia south through eastern Peru and Bolivia to Uruguay and eastern Brazil. – Three subspecies:

M. t. griseiventer (northern Venezuela and Colombia), M. t. temmincki (central Venezuela to northwestern Argentina and eastern Brazil), M. t. sylvia (Uruguay and northeastern Argentina).

Subgenus **Cabreramops** IBANEZ 1980

Last upper molar greatly reduced. Ears barely separated. Basisphenoid pits rather well developed. Upper lip somewhat wrinkled. Lateral lower incisor present. Anterior upper premolar absent. Upper incisors relatively long and in contact near their tips. – *Distribution:* Known only from western Ecuador. – A single species.

6. M. aequatorianus CABRERA 1917. – Size medium (forearm length, 35–38 mm). – *Distribution:* Same as for subgenus. – No subspecies.

Subgenus **Neoplatymops** PETERSON 1965

Last upper molar not reduced. Ears well separated. Basisphenoid pits present but shallow. Upper lip wrinkles few. Lateral lower incisors present. Anterior upper premolar present. Upper incisors long and not in contact. Forearm with wartlike granulations. Skull considerably flattened. – *Distribution:* Ranging from Venezuela and Guyana to southwestern and northeastern Brazil. – A single species.

7. M. mattogrossensis VIERA 1942. – Size relatively small (forearm length, 26–31 mm). – *Distribution:* Same as for subgenus. – No subspecies.

Genus **Myopterus** GEOFFROY 1818 (Fig. 184)

No anterior palatal emargination. Last upper molar reduced. Ears relatively short, erect, and well separated. Basisphenoid pits well developed. Wrinkles on upper lip absent. Wing tip broad. Dental formula i 1/1, c 1/1, p 1/2, m 3/3 x 2 = 26. – *Distribution:* Confined to tropical Africa, from Senegal to Uganda. – Two species are recognized here.

1. M. daubentonii DESMAREST 1820. – Size relatively large (forearm length, 44–56 mm). Basisphenoid pits relatively deep. – *Distribution:* Ranging from Senegal to northeastern Zaire. – Two subspecies are here recognized:

M. d. daubentonii (definitely known only from Senegal), M. d. albatus (Central African Republic and northeastern Zaire). The subspecific allocation of specimens from Ivory Coast is uncertain.

2. M. whitleyi (SCHARFF 1900). – Size relatively small (forearm length, 33–37 mm). Basisphenoid pits fairly deep. – *Distribution:* Ranging from Ghana to Uganda. – No subspecies.

Genus **Cheiromeles** HORSFIELD 1824 (Fig. 183)

No anterior palatal emargination. Last upper molar reduced. Ears unusually small, narrow, erect, and widely separated. Basisphenoid pits absent. Wrinkles on upper lip absent. Wing tip broad. Dental formula i 1/1, c 1/1, p 1/2, m 3/3 x 2 = 26. Hair very sparse. – *Distribution:* Ranging from the Malay peninsula to Java, Celebes, and the Philippines. – A single species is recognized here with three additional subspecies.

1. C. torquatus HORSFIELD 1824. – Size relatively large (forearm length, 73–90 mm). – *Distribution:* Same as for genus. – Four subspecies are here recognized:

C. t. torquatus (Malay peninsula and Sumatra to Borneo and the southwestern Philippines), C. t. caudatus (Banka and Java), C. t. jacobsoni (West Sumatran islands), C. t. parvidens (Celebes and main Philippines).

Genus **Tadarida** RAFINESQUE 1814 (= *Nyctinomus* E. GEOFFROY 1818) (Fig. 185)

Anterior palatal emargination well developed. Last upper molar not greatly reduced. Ears fairly large, with some slouch, somewhat separated or almost joined. Basisphenoid pits variably developed. Wing tips variably developed. Dental formula i 1/2–3, c 1/1, p 2/2, m 3/3 x 2 = 30–32. – *Distribution:* Ranging widely over much of Africa, including Madagascar; across the southern Palearctic, including India and Ceylon; New Guinea and the southern half of Australia; also widely distributed in southern North America, Middle America, the West Indies, and much of

South America. – Seven species and 15 additional subspecies are recognized here.

1. *T. brasiliensis* (I. GEOFFROY 1824) [*aegyptiaca* group]. Ears definitely separate. Basisphenoid pits shallow. Wing tips relatively broad. Outer lower incisor usually present. Anterior upper premolar reduced. Upper lip wrinkles well developed but relatively few in number. Size relatively small (forearm length, 31–47 mm). – *Distribution:* Ranging from the southern half of the United States south through Middle America and western South America to eastern Brazil and the northern edge of Patagonia; also through the West Indies south to St. Lucia. – Nine subspecies are currently recognized:

T. b. cynocephala (southeastern United States), *T. b. mexicana* (central and southwestern United States and most of Mexico). *T. b. intermedia* (extreme southern Mexico and northern Central America), *T. b. brasiliensis* (southern Central America and the entire South American range), *T. b. bahamensis* (Bahamas), *T. b. muscula* (Cuba), *T. b. murina* (Jamaica), *T. b. constanzae* (Hispaniola), *T. b. antillularum* (Puerto Rico to central Lesser Antilles).

2. *T. aegyptiaca* (E. GEOFFROY 1818) [*aegyptiaca* group]. – Basisphenoid pits of medium depth. Outer lower incisor absent. Anterior upper premolar slightly reduced. Size medium to fairly small (forearm length, 44–56 mm). – *Distribution:* Ranging from Algeria and Nigeria east to India and south to the Cape Province and Ceylon. – Five subspecies are here recognized:

T. a tragata (northeastern India), *T. a. thomasi* (remainder of India and Ceylon), *T. a. sindica* (Pakistan to Iran), *T. a. aegyptiaca* (Arabia and most of the African range), *T. a. bocagei* (western Zambia to Angola and Namibia).

3. *T. teniotis* (RAFINESQUE 1814) [*teniotis* group]. – Ears barely joined. Basisphenoid pits of medium depth. Wing tips relatively broad. Outer lower incisor present. Anterior upper premolar unreduced. Upper lip wrinkles well developed but relatively few in number. Rostrum relatively narrow. Size relatively large (forearm length, 54–64 mm). – *Distribution:* Ranging from the Canaries and Madeira through southern Europe, northern Africa, and southern Palearctic Asia to Japan and southern China. – Two subspecies are currently recognized:

T. t. insignis (eastern Asia to northeastern India), *T. t. teniotis* (Central Asia west).

4. T. *lobata* (THOMAS 1891) [*teniotis* group]. – Wing tips relatively narrow. Outer lower incisor absent. No well defined lip wrinkles. Rostrum relatively narrow. Size relatively large (forearm length, 56–63 mm). – *Distribution:* Known only from Kenya and Zimbabwe. – No subspecies.

5. *T. fulminans* (THOMAS 1903) (= *mastersoni* ROBERTS 1946) [*teniotis* group]. Wing tips relatively narrow. Outer lower incisor absent. No well defined lip wrinkles. Rostrum of medium width. Size relatively large (forearm length, 56–62 mm). – *Distribution:* Known from eastern Zaire and southern Kenya to Transvaal, also in Madagascar. – No subspecies are currently recognized.

6. *T. ventralis* (HEUGLIN 1861) [*teniotis* group]. – Basisphenoid pits relatively deep. Wing tips relatively narrow. Outer lower incisor absent. Anterior upper premolar greatly reduced. No well defined lip wrinkles. Rostrum relatively broad. Size relatively large (forearm length, 62–66 mm). – *Distribution:* Ranging from Ethiopia to Transvaal. – Two subspecies are here recognized:

T. v. ventralis (Ethiopia to northern Tanzania and northeastern Zaire), *T. v. africana* (Zambia and Malawi to Transvaal).

7. *T. australis* (GRAY 1838) [*australis* group]. – Ears barely joined. Basisphenoid pits represented by long shallow grooves. Wing tips relatively broad. Outer lower incisor absent. Anterior upper premolar variable. Upper lip wrinkles well developed and numerous. Size relatively large (forearm length, 57–61 mm). – *Distribution:* Known from eastern New Guinea (where possibly confined to mountains) and the southern half to two thirds of mainland Australia. – Two subspecies are here recognized:

T. a. kuboriensis (New Guinea), *T. a. australis* (Australia).

Genus **Chaerephon** DOBSON 1874 (Fig. 186)

Anterior palatal emargination narrowed or largely closed. Last upper molar not greatly reduced. Ears fairly large, with some slouch, almost always fully joined. Basisphenoid pits always present, but variably developed. Dental formula i 1/2, c 1/1, p 2/2, m 3/3 x 2 = 30. – *Distribution:* Widely distributed in sub-Saharan Africa, including Madagascar and nearby islands, also southwestern Arabia, and ranging from India and Ceylon to Fiji and northern Australia. – Thirteen species and 22 additional subspecies are here recognized:

1. *C. bemmelini* (JENTINCK 1879) [*bivittata* group]. – Anterior palatal emargination narrow but definitely present. Basisphenoid pits relatively shallow. Anterior upper premolar relatively large.

Size relatively small (forearm length, 40–47 mm). Upper lip wrinkles relatively few. – *Distribution:* Ranging from Liberia to southern Sudan and south to northeastern Zaire and northern Tanzania. – Two subspecies are recognized:

C. b. bemmelini (known from Liberia and Cameroon), C. b. cisturus (eastern African range).

2. C. ansorgei (THOMAS 1913) [*bivittata* group]. – Basisphenoid pits of moderate depth. Anterior upper premolar somewhat reduced. Size medium (forearm length, 43–49 mm). Rostrum relatively slender. – *Distribution:* Ranging from Cameroon and Ethiopia south to Angola and Transvaal. – No subspecies.

3. C. bivittata (HEUGLIN 1862) [*bivittata* group]. – Basisphenoid pits of moderate depth. Anterior upper premolar somewhat reduced. Size medium (forearm length, 46–52 mm). Upper lip wrinkles relatively numerous. Rostrum relatively broad. – *Distribution:* Ranging from Ethiopia and southern Sudan to Zimbabwe and Mozambique. – No subspecies.

4. C. nigeriae THOMAS 1913 [*plicata* group]. – Anterior palatal emargination virtually obliterated. Basioccipital pits of moderate depth. Anterior upper premolar somewhat reduced. Size medium (forearm length, 44–50 mm). Upper lip wrinkles relatively numerous. Forehead relatively flat. Rostrum relatively slender. – *Distribution:* Ranging from Ghana to southwestern Arabia and south to Namibia and Zimbabwe. – Two subspecies:

C. n. nigeriae (Ghana to southwestern Arabia and south to northeastern Zaire), C. n. spillmani (Angola and Tanzania south).

5. C. major (TROUESSART 1897) [*plicata* group]. – Basisphenoid pits shallow. Size relatively small (forearm length, 39–44 mm). Upper lip wrinkles relatively few. Forehead relatively flat. Rostrum relatively broad. Ears connected by a separate lappet extending backwards. – *Distribution:* Ranging from Mali and Liberia to Sudan and Tanzania. – No subspecies.

6. C. pumila (CRETZSCHMAR 1830) [*plicata* group]. – Anterior upper premolar greatly reduced. Size relatively small (forearm length, 32–42 mm). Forehead relatively elevated. Rostrum relatively broad. – *Distribution:* Ranging from Gambia to southwestern Arabia and south to the Cape Province, including Fernando Poo, Pemba, Zanzibar, Madagascar, Comoros, Aldabra, and the Amirante islands (near the Seychelles). – Twelve subspecies are here recognized:

C. p. gambianus (Senegal to Guinea-Bissau), C. p. nigri (Mali), C. p. websteri (Nigeria, Chad), C. p. pumila (Sudan to southwestern Arabia), C. p. faini (northeastern Zaire to northwestern Tanzania), C. p. hindei (southern Somalia to northeastern Tanzania), C. p. frater (southern Congo-Brazzaville to western Zambia), C. p. limbata (southern Tanzania to Angola and central Mozambique), C. p. langi (Botswana), C. p. elphicki (southern Mozambique and South Africa), C. p. leucogaster (Madagascar), C. p. pusillus (Aldabras, Amirantes). Subspecific allocation of many populations remains uncertain, however.

7. C. chapini J. A. ALLEN 1917 [*plicata* group]. – Size relatively small (forearm length, 34–39 mm). Upper lip wrinkles relatively few. Forehead relatively elevated. Rostrum relatively slender. Band connecting ears with a long bicolored crest of hair in males. – *Distribution:* Ranging from Ethiopia and Zaire to Zimbabwe and Namibia. – Three subspecies:

C. c. chapini (Ethiopia to Zaire), C. c. lancasteri (northeastern Angola to Botswana and Zimbabwe), C. c. shortridgei (southwestern Angola, northwestern Namibia).

8. C. johorensis (DOBSON 1873) [*plicata* group]. – Size medium (forearm length, 47–48 mm). Ears connected by a separate lappet extending backwards. – *Distribution:* Known only from Malaya and Sumatra. – No subspecies..

9. C. plicata (BUCHANNAN 1800) [*plicata* group]. – Anterior upper premolar relatively large. Size medium (forearm length, 40–53 mm). – *Distribution:* Ranging from India and Ceylon to southern China, the Philippines, Bali, and Cocos-Keeling (in the Indian Ocean south of Sumatra). – Five subspecies often recognized.

10. C. jobensis (MILLER 1902) [*plicata* group]. – Size medium (forearm length, 40–53 mm). Upper lip wrinkles relatively few. Forehead relatively elevated. – *Distribution:* Ranging from New Guinea and the northern half of Australia to Fiji. – Four subspecies:

C. j. jobensis (New Guinea), C. j. colonicus (Australia), C. j. solomonis (Solomons), C. j. bregullae (New Hebrides, Fiji).

11. C. russata J. A. ALLEN 1917 [*plicata* group]. – Basisphenoid pits fairly deep. Size medium (forearm length, 42–47 mm). Forehead relatively elevated. – *Distribution:* Confined to forested areas from Ghana to Kenya. – No subspecies.

12. C. aloysiisabaudiae (FESTA 1907) [*plicata* group]. – Basisphenoid pits relatively deep. Size fairly large (forearm length, 49–53 mm). Fore-

head relatively elevated. – *Distribution:* Ranging from Ghana to Uganda. – No subspecies.

13. C. gallagheri (HARRISON 1975) [*plicata* group]. – Basisphenoid pits relatively deep. Size relatively small (forearm length, 37–38 mm). Upper lip wrinkles relatively few. Forehead relatively elevated. Ears connected by a separate lappet extending backwards. Premaxillary region greatly inflated. – *Distribution:* Known only from central Zaire. – No subspecies.

Genus *Mops* LESSON 1842 (Fig. 187)

Anterior palatal emargination either open or largely closed. Last upper molar considerably, even greatly, reduced. Ears fairly large, with some slouch, almost always fully joined. Basisphenoid pits always present, but variably developed. Dental formula i 1/1–2, c 1/1, p 1–2/2, m 3/3 x 2 = 28–30. – *Distribution:* Ranging widely in sub-Saharan Africa, as well as Madagascar and southwestern Arabia. Also Malaya to the Philippines, Celebes, and possibly Java. – Two subgenera and 12 species are recognized here.

Subgenus *Xiphonycteris* DOLLMAN 1911

Anterior palatal emargination present. Anterior upper premolar moderate to relatively large. – *Distribution:* Ranging across tropical Africa from Sierra Leone to Kenya and Mozambique, including Fernando Poo and Zanzibar. – Five species, one additional subspecies.

1. M. nanulus J. A. ALEN 1917. – Last upper molar greatly reduced. Basisphenoid pits shallow. Anterior upper premolar. moderate in size. Upper incisor procumbent. Cingulum of upper canine not enlarged. Size relatively small (forearm length, 27–31 mm). – *Distribution:* Ranging from Sierra Leone to Ethiopia and southeastern Zaire. – No subspecies.

2. M. spurrelli (DOLLMAN 1911). – Anterior upper premolar relatively large. Upper incisor not procumbent. Cingulum of upper canine enlarged. Lateral lower incisor usually absent. Size relatively small (forearm length, 27–28 mm). – *Distribution:* Ranging from Guinea to the Central African Republic and southwestern Zaire, including Fernando Poo. – No subspecies.

3. M. petersoni (EL-RAYAH 1981). – Last upper molar considerably reduced. Basisphenoid pits of moderate depth. Size medium (forearm length, 32–35 mm). – *Distribution:* Known only from Ghana and Cameroon. – No subspecies.

4. M. brachypterus (PETERS 1852). – Last upper molar considerably reduced. Basisphenoid pits relatively shallow. Size relatively large (forearm length, 35–38 mm). – *Distribution:* Ranging from Sierra Leone to northern Mozambique. – Two subspecies are here recognized:

M. b. leonis (Sierra Leone to eastern Zaire, including Ferrando Poo), *M. b. brachypterus* (Uganda to northern Mozambique, including Zanzibar).

5. M. thersites (THOMAS 1903). – Last upper molar considerably reduced. Basisphenoid pits of medium depth. Size relatively large (forearm length, 38–41 mm). – *Distribution:* Ranging from Sierra Leone to Rwanda. – No subspecies are currently recognized.

Subgenus *Mops* LESSON 1842

Anterior palatal emargination largely closed. Anterior upper premolar absent or greatly reduced. – *Distribution:* Essentially same as for genus. – Seven species and seven additional subspecies are here recognized.

6. M. condylurus (A. SMITH 1833). – Last upper molar considerably reduced. Anterior upper premolar vestigial but present. Basisphenoid pits relatively shallow. Upper lip wrinkles relatively numerous. Size medium (forearm length, 44–51 mm). – *Distribution:* Ranging from Mauretania to Ethiopia and south to Angola and Natal. – Five subspecies are recognized here:

M. c. wonderi (Mauretania and Guinea-Bissau at least to Mali), *M. c. orientis* (Tanzania), *M. c. osborni* (Congo-Brazzaville and southwestern Zaire), *M. c. condylurus* (Angola and southeastern Zaire to Natal), *M. c. leucostigma* (Madagascar). Many populations remain unallocated subspecifically.

7. M. demonstrator (THOMAS 1903). – Last upper molar greatly reduced. Basisphenoid pits relatively deep. Upper lip wrinkles relatively few. Size relatively small (forearm length, 39–41 mm). – *Distribution:* Ranging from Upper Volta to Sudan and south to Angola and Mozambique. – Two subspecies are here recognized:

M. d. demonstrator (Upper Volta to Sudan and Uganda), *M. d. niveiventer* (Rwanda to Angola and Mozambique).

8. M. trevori J. A. ALLEN 1917 (= *niangarae* J. A. ALLEN 1917). – Anterior upper premolar of mode-

rate size. Basisphenoid pits relatively deep. Size fairly large (forearm length, 51–54 mm). Ears probably always joined. – *Distribution:* Known only from Uganda and northeastern Zaire. – No subspecies.

9. M. congicus J. A. ALLEN 1917. – Last upper molar greatly reduced. Basisphenoid pits relatively deep. Size fairly large (forearm length, 52–59 mm). – *Distribution:* Ranging in forested regions from Ghana to Uganda. – No subspecies.

10. M. midas (SUNDEVALL 1843). – Last upper molar greatly reduced. Basisphenoid pits of moderate depth. Upper lip wrinkles relatively few. Size relatively large (forearm length, 57–66 mm). – *Distribution:* Ranging from Senegal to southwestern Arabia, south in eastern Africa to Botswana and Transvaal; also on Madagascar. – Two subspecies:

M. m. midas (southwestern Arabia and continental African range), *M. m. miarensis* (Madagascar).

11. M. mops (DE BLAINVILLE 1840). – Last upper molar greatly reduced. Anterior upper premolar absent. Basisphenoid pits of moderate depth. Upper lip wrinkles relatively few. Size medium (forearm length, 43–48 mm). – *Distribution:* Known from Malaya, Sumatra, Borneo, and possibly Java. – No subspecies.

12. M. sarasinorum (MEYER 1899). – Last upper molar greatly reduced. Anterior upper premolar absent. Basisphenoid pits of moderate depth. Size relatively small (forearm length, 39–45 mm). – *Distribution:* Confined to the Philippines and Celebes. – Two subspecies are recognized here:

M. s. lanei (Philippines), *M. s. sarasinorum* (Celebes).

Genus **Otomops** THOMAS 1913 (Fig. 188)

Anterior palatal emargination well developed but narrow. Last upper molar not reduced. Ears relatively large, with considerable slouch, and joined by a band. Basisphenoid pits unusually deep. Dental formula i 1/2, c 1/1, p 2/2, m 3/3 x 2 = 30. – *Distribution:* Central, eastern, and southern Africa, Madagascar, southern India, Java, and eastern New Guinea. – Five species with two additional subspecies are recognized here.

1. O. papuensis LAWRENCE 1948. – Rostrum relatively short. Size relatively small (forearm length, 49–51 mm). – *Distribution:* Known only from southeastern New Guinea. – No subspecies.

2. O. secundus HAYMAN 1952. – Size fairly small (forearm length, 57–58 mm). – *Distribution:* Known only from northeastern New Guinea. – No subspecies.

3. O. formosus CHASEN 1939. – Size medium (forearm length, 59–60 mm). – *Distribution:* Known only from Java. – No subspecies.

4. O. wroughtoni (THOMAS 1913). Size fairly large (forearm length, 62–68 mm). – *Distribution:* Known only from southern India. – No subspecies.

5. O. martiensseni (MATSCHIE 1897). – Rostrum relatively long. Size relatively large (forearm length, 62–73 mm). – *Distribution:* Ranging from the Central African Republic to Djibouti and south to Angola and Natal; also on Madagascar. – Three subspecies are currently recognized:

O. m. martiensseni (Central African Republic and Djibouti to Tanzania and Zaire), *O. m. icarus* (Angola and Malawi to Natal), *O. m. madagascariensis* (Madagascar).

Genus **Nyctinomops** MILLER 1902 (Fig. 189)

Anterior palatal emargination well developed but narrow. Last upper molar not reduced. Ears fairly large with considerable slouch, usually well joined. Basisphenoid pits always present but variable as to depth. Dental formula i 1/2, c 1/1, p 2/2, m 3/3 x 2 = 30. – *Distribution:* Ranging from the southwestern United States (with scattered records in southwestern Canada and central United States) through Middle America and northern South America to Peru and Uruguay, including Trinidad, Jamaica, Cuba, and Hispaniola. – Four species and five additional subspecies.

1. N. femorosaccus (MERRIAM 1889). – Ears barely joined. Basisphenoid pits relatively shallow. Anterior upper premolar relatively large. Upper lip wrinkles relatively few. Size fairly small (forearm length, 45–50 mm). – *Distribution:* Confined to the southwestern United States and northern Mexico. – No subspecies.

2. N. laticaudatus (GEOFFROY 1805) (= *espiritosantensis* RUSCHI 1951). – Ears well joined. Anterior upper premolar somewhat reduced. Upper lip wrinkles relatively numerous. Size relatively small (forearm length, 39–47 mm). – *Distribution:* Ranging from northeastern Mexico to northwestern Peru, northern Argentina, and eastern Brazil, including Trinidad and Cuba. – Five subspecies are here recognized:

N. l. ferruginea (northeastern and southwestern Mexico), *N. l. yucatanica* (Cuba and southeastern Mexico at least to Panama and perhaps to northwestern Peru), *N. l europs* (lowlands from eastern Colombia, Trinidad, and Suriname to northern Brazil and Bolivia), *N. l macarenensis* (Macarena mountains in central Colombia and perhaps Mount Roraima in extreme northern Brazil), *N. l laticaudatus* (southern Brazil to northern Argentina).

3. *N. aurispinosus* (PEALE 1848) (= *similis* SANBORN 1941). Ears well joined. Basisphenoid pits of medium depth. Size medium (forearm length, 47–52 mm). – *Distribution:* Ranging from tropical Mexico to southern Peru and eastern Brazil. – No subspecies.

4. *N. macrotis* (GRAY 1839). – Ears well joined. Basisphenoid pits relatively deep. Size relatively large (forearm length, 58–64 mm). – *Distribution:* Ranging from the southwestern United States (with outliers in central United States and southwestern Canada) to central Mexico; also Cuba, Jamaica, and Hispaniola. In South America, from Colombia and Suriname to northwestern Peru, northwestern Argentina, and Uruguay. – No subspecies.

Genus ***Eumops*** MILLER 1906 (Fig. 190)

Anterior palatal emargination absent. Last upper molar variable. Ears fairly large, with considerable slouch, usually barely joined. Basisphenoid pits well developed but of variable depth. Dental formula i 1/2, c 1/1, p 2/2, m 3/3 x 2 = 30, but anterior upper premolar usually greatly reduced and sometimes absent. – *Distribution:* Ranging from the southwestern United States and Florida through Middle America and South America to northern Argentina, including Trinidad, Jamaica, and Cuba, but west of the Andes, not south of Peru. – Eight species and nine additional subspecies are recognized here.

1. *E. hansae* SANBORN 1932 (= *amazonicus* HANDLEY 1955). – Last upper molar unreduced. Basisphenoid pits unusually deep. Size relatively small (forearm length, 36–42 mm). – *Distribution:* Ranging from Costa Rica to eastern Peru and southeastern Brazil, but not west of the Andes in South America. – No subspecies.

2. *E. bonariensis* (PETERS 1874). – Last upper molar somewhat reduced. Basisphenoid pits relatively deep. Size relatively small (forearm length, 36–50 mm). – *Distribution:* Ranging from southeastern Mexico to northern Argentina, but west of the Andes no further south than northern Peru. A Patagonian record may be erroneous. – Four subspecies are currently recognized:

E. b. nanus (Mexico to the Guianas and northwestern Peru), *E. b. delticus* (Amazon basin), *E. b. beckeri* (Bolivia to northern Argentina), *E. b. bonariensis* (southeastern Brazil to east-central Argentina).

3. *E. maurus* (THOMAS 1901) (= *geijskesi* HUSSON 1962). – Last upper molar considerably reduced. Basisphenoid pits of moderate depth. Size fairly small (forearm length, 51–53 mm). With a band of white hair along the medial side of the plagiopatagium anterior to the femur. Anterior upper premolar may be absent. – *Distribution:* Known only from Guyana and Suriname. – No subspecies.

4. *E. auripendulus* (SHAW 1800). – Last upper molar greatly reduced. Basisphenoid pits relatively deep. Size medium (forearm length, 55–68 mm). Ears relatively short. Tragus small and pointed. – *Distribution:* Ranging from southern Mexico to northeastern Argentina, including Trinidad and Jamaica, but west of the Andes not south of northern Peru. – Two subspecies:

E. a. auripendulus (Mexico and Jamaica to Trinidad and Bolivia), *E. a. major* (northeastern Brazil to Paraguay and Argentina).

5. *E. glaucinus* (WAGNER 1843). – Last upper molar considerably reduced. Basisphenoid pits relatively deep. Size medium (forearm length, 55–68 mm). Ears relatively short. Tragus broad and square. – *Distribution:* Occurring on Florida, Cuba, and Jamaica, and from central Mexico to northwestern Argentina and southeastern Brazil, but west of the Andes not south of northern Peru. – Two subspecies:

E. g. floridanus (Florida), *E. g. glaucinus* (remainder of range).

6. *E. underwoodi* GOODWIN 1940. – Last upper molar considerably reduced. Basisphenoid pits of moderate depth. Size fairly large (forearm length, 64–77 mm). Ears relatively short and heavily keeled. – *Distribution:* Ranging from the southwestern United States (Arizona) to Belize and Honduras. – Two subspecies:

E. u. sonoriensis (Arizona and northwestern Mexico), *E. u. underwoodi* (remainder of range).

7. *E. dabbenei* THOMAS 1914 (= *mederai* MASSOIA 1976). – Last upper molar considerably reduced. Basisphenoid pits of moderate depth. Size relatively large (forearm length, 74–79 mm). Ears relatively short and heavily keeled. – Distribution

Ranging from Colombia and Venezuela to northern Argentina. – No subspecies. A poorly known species closely related to *E. underwoodi*.

8. *E. perotis* (SCHINZ 1821). – Last upper molar considerably reduced. Basisphenoid pits relatively deep. Size relatively large (forearm length, 67–84 mm). Ears relatively long. Tragus broad and square. – *Distribution:* Ranging from the southwestern United States to central Mexico; also from northern Venezuela to southwestern Peru, northern Argentina, and eastern Brazil; probably also in Cuba. – Four subspecies are recognized here:

E. p. trumbulli (Amazon-Orinoco basin and Guianas), *E. p. perotis* (remaining South American range), *E. p. californicus* (North American range), *E. p. gigas* (Cuba, if the record is valid.)

Genus ***Promops*** GERVAIS 1855 (Fig. 191)

Anterior palatal emargination absent. Last upper molar greatly reduced. Ears fairly large, with considerable slouch, barely joined. Basisphenoid pits of moderate depth. Dental formula i 1/2, c 1/1, p 2/2, m 3/3 x 2 = 30, but anterior upper premolar greatly reduced. Upper incisors not greatly shortened and broadened. – *Distribution:* Ranging from southwestern Mexico to northern Argentina and eastern Brazil, including Trinidad, but west of the Andes not south of Peru. – Two species and six additional subspecies are recognized here.

1. *P. centralis* THOMAS 1915. – Size relatively large (forearm length, 48–56 mm). – *Distribution:* Ranging from southwestern Mexico to Suriname, southern Peru, and northeastern Argentina, including Trinidad. – Three subspecies are here recognized:

P. c. centralis (Mexico to Trinidad and Suriname), *P. c. occultus* (eastern Peru to northern Argentina), *P. c. davisoni* (Pacific slopes of Ecuador and Peru), but *davisoni* may be better referred to *P. nasutus*.

2. *P. nasutus* (SPIX 1823). – Size relatively small (forearm length, 43–50 mm). – *Distribution:* Ranging east of the Andes from Trinidad and Suriname to northern Argentina and eastern Brazil, but avoiding much of the Amazon basin. – Five subspecies are recognized:

P. n. downsi (known from Trinidad, Suriname, and southeastern Venezuela), *P. n. pamana* (eastern Ecuador to western Brazil), *P. n. fosteri* (Paraguay), *P. n. ancilla* (northern Argentina), *P. n. nasutus* (eastern Brazil), but subspecies boundaries are uncertain.

Genus ***Molossus*** E. GEOFFROY 1805 (Fig. 192)

Anterior palatal emargination absent. Last upper molar greatly reduced. Ears fairly large, with considerable slouch, barely joined. Basisphenoid pits of moderate depth. Dental formula i 1/1, c 1/1, p 1/2, m 3/3 x 2 = 26. Upper incisors greatly shortened and broadened. – *Distribution:* Ranging from tropical Mexico through Central America and through much of South America to Uruguay, but west of the Andes not south of northern Peru; also throughout the West Indies. – Five species and 15 additional subspecies are recognized here.

1. *M. ater* E. GEOFFROY 1805. – Size relatively large (forearm length, 47–53 mm). Dorsal hairs with dark bases. – *Distribution:* Ranging from tropical Mexico to northern Argentina and eastern Brazil, but not west of the Andes in South America. Also recorded from Trinidad, but specimens from this island agree better with *M. pretiosus*. – Three subspecies are currently recognized:

M. a. nigricans (Middle America), *M. a. ater* (northern South America to southeastern Brazil), *M. a. castaneus* (Paraguay and northern Argentina).

2. *M. pretiosus* MILLER 1902. – Size fairly large (forearm length, 41–50 mm). Dorsal hairs with dark bases. – *Distribution:* Ranging from southwestern Mexico to Colombia, Guyana and probably Trinidad. – Two subspecies are recognized:

M. p. macdougalli (southwestern Mexico), *M. p. pretiosus* (Nicaragua to South America).

3. *M. sinaloae* J. A. ALLEN 1906. – Size fairly large (forearm length, 46–51 mm). Dorsal hairs with pale bases. – *Distribution:* Ranging from southwestern Mexico to Colombia and Suriname, including Trinidad. – Two subspecies are here recognized:

M. s. sinaloae (Mexico to Costa Rica), *M. s. trinitatis* (Panama and South American range).

4. *M. bondae* J. A. ALLEN 1904. – Size fairly small (forearm length, 39–43 mm). Dorsal hairs with dark bases. – *Distribution:* Ranging from Nicaragua to western Ecuador and northern Venezuela. – No subspecies.

5. *M. molossus* (PALLAS 1766). – Size relatively small, at least within the range of *M. bondae* (forearm length 33–41 mm). Dorsal hairs with pale bases. – *Distribution:* Ranging from tropical Mexico to Uruguay and throughout the West In-

dies except the Bahamas. – Twelve subspecies are here recognized:

M. m. aztecus (most of tropical Mexico and Central America south to Costa Rica), *M. m. lambi* (Pacific coastal Chiapas and Guatemala), *M. m. coibensis* (Panama), *M. m. daulensis* (western Ecuador), *M. m. pygmaeus* (Curacao, Bonaire), *M. m. crassicaudatus* (southeastern Colombia and Guyana to Uruguay), *M. m. barnesi* (French Guiana), *M. m. molossus* (= *debilis*) (central Colombia to Trinidad and north through the Lesser Antilles to St. Croix in the Virgin Islands), *M. m. fortis* (Puerto Rico and the Virgin Islands except St. Croix), *M. m. verrillii* (Hispaniola), *M. m. milleri* (Jamaica), *M. m. tropidorhynchus* (Cuba).

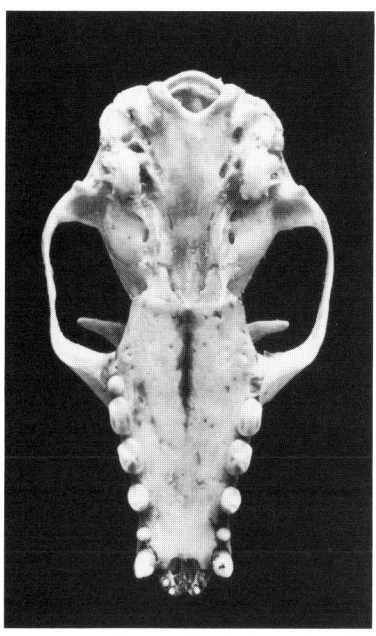 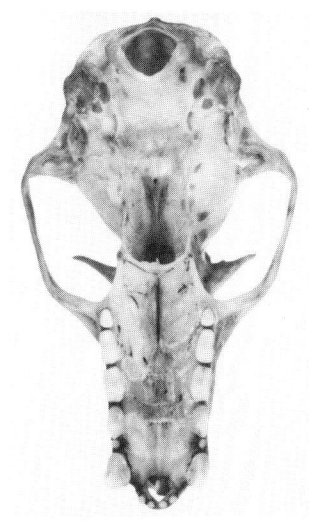

Fig. 20. *Eidolon helvum* – AMNH 86758 – ♂ (Zaire)

Fig. 21. *Rousettus celebensis* – AMNH 222857 – ♂ (Celebes)

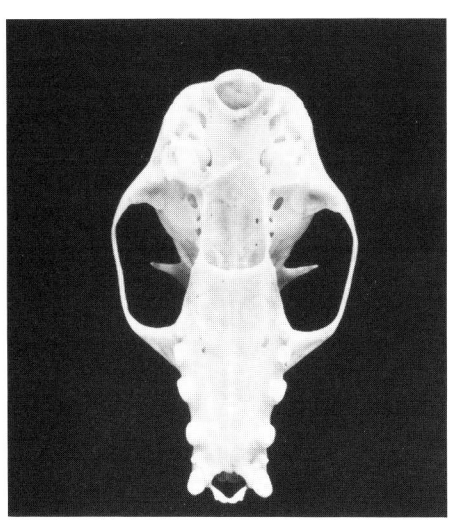 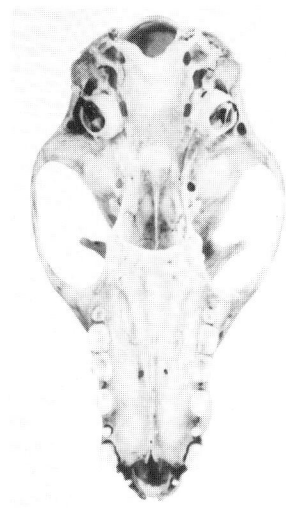

Fig. 22. *Myonycteris torquata* – AMNH 239351 – ♂ (Liberia)

Fig. 23. *Boneia bidens* – NAMRU – 2 (DJM 6707) – ♂ (Celebes)

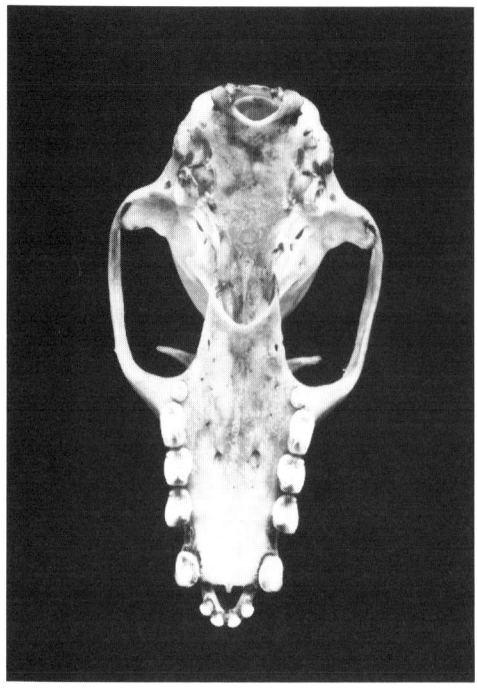

Fig. 24. *Pteropus mariannus* – AMNH 256895 – ♀ (Carolines: Palau)

Fig. 25. *Acerodon leucotis* – AMNH 207591 – ♀ (Philippines: Balabac)

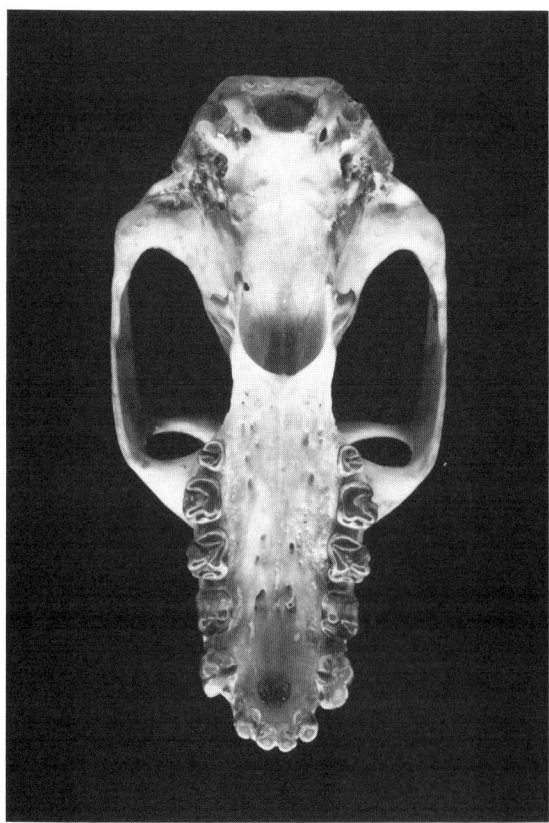

Fig. 26. *Pteralopex anceps* – USNM 276974 – ♂ (Solomons: Bougainville)

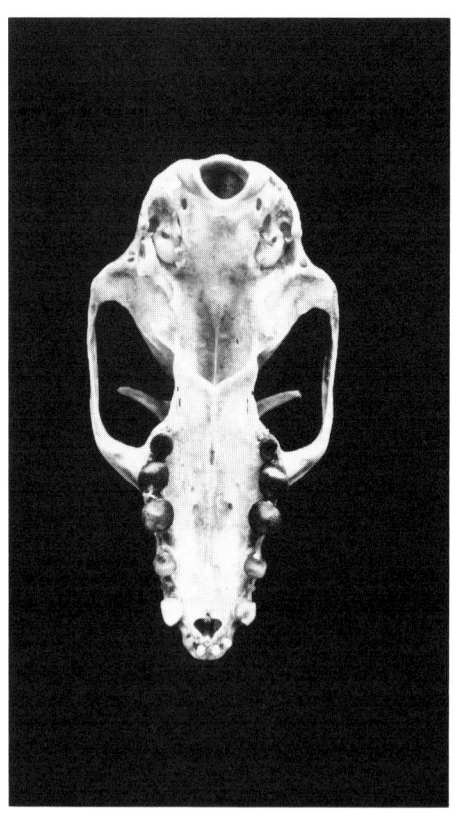

Fig. 27. *Styloctenium wallacei* – AMNH 222981 – ♀ (Celebes)

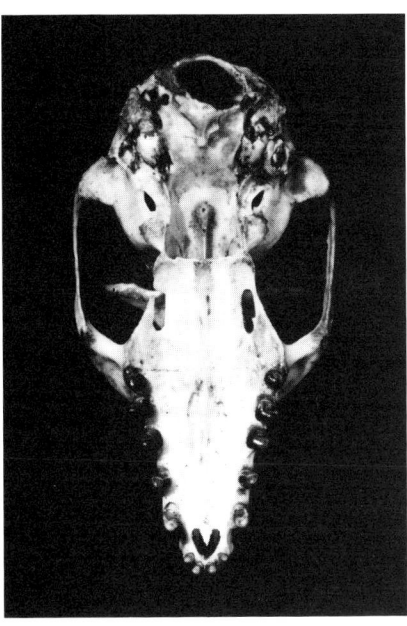

Fig. 28. *Neopteryx frosti* – BM 40.961 k – ♀ (Celebes – type)

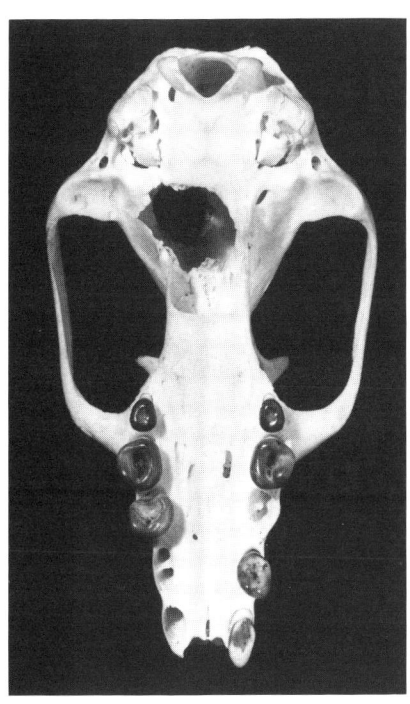

Fig. 29. *Aproteles bulmerae* – BM 78.1084 – unsexed (Papua New Guinea)

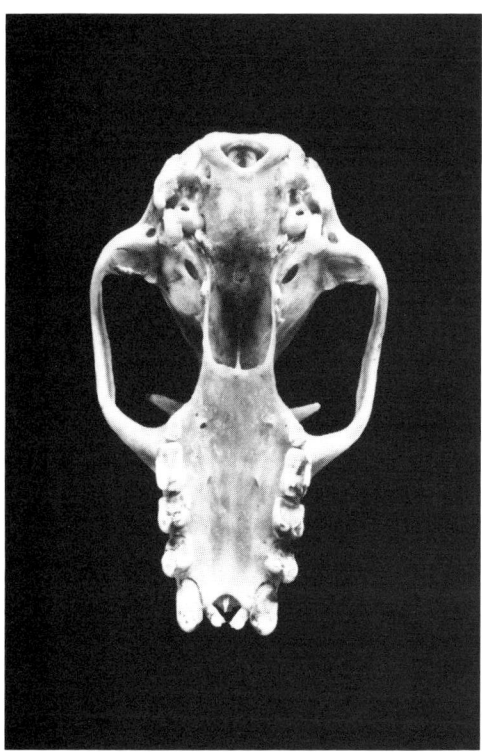

Fig. 30. *Dobsonia pannietensis* – AMNH 157368 – ♂ (New Guinea: D'Entrecasteaux)

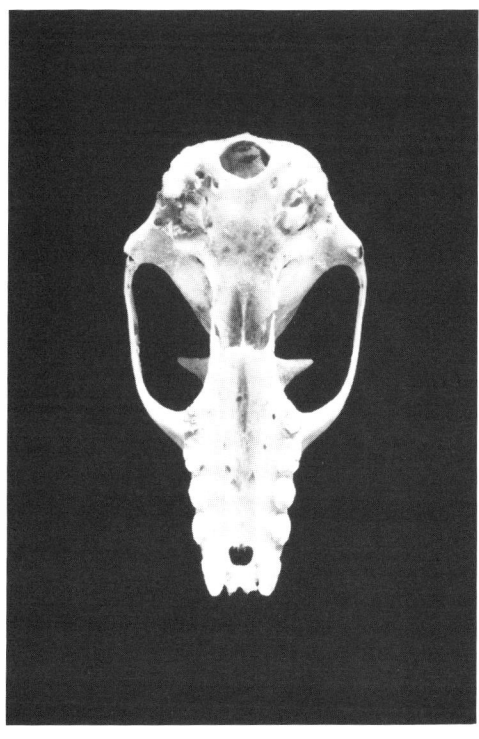

Fig. 31. *Harpyionycteris whiteheadi* – AMNH 153590 – ♂ (Celebes)

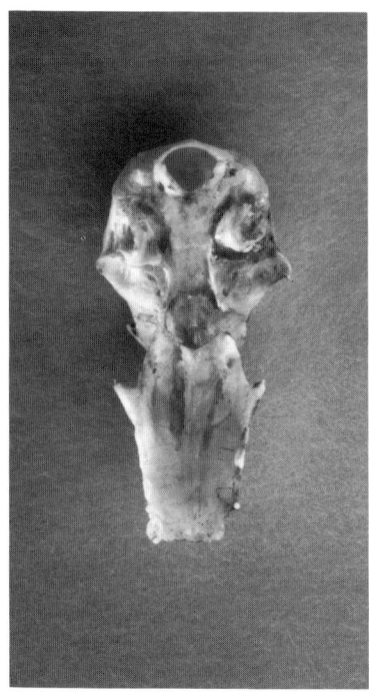

Fig. 32. *Plerotes anchietai* – CM 6971 – ♂ (imm.) (Angola)

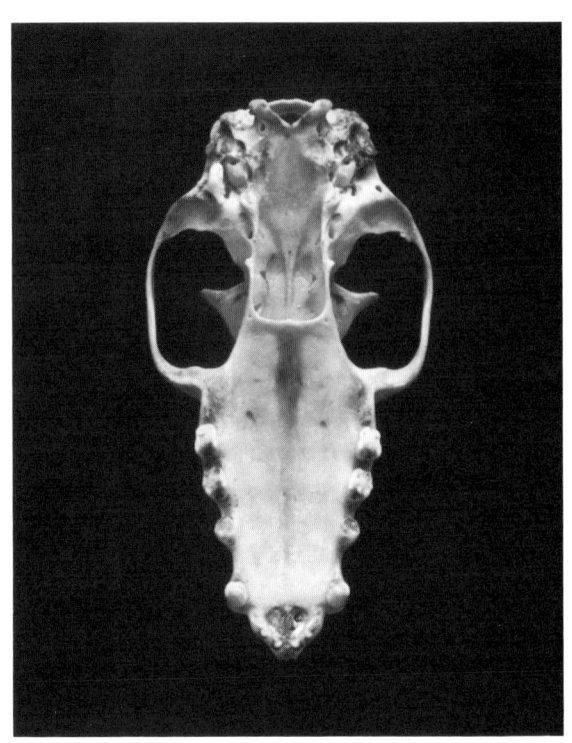

Fig. 33. *Hypsignathus monstrosus* – AMNH 48642 – ♂ (Zaire)

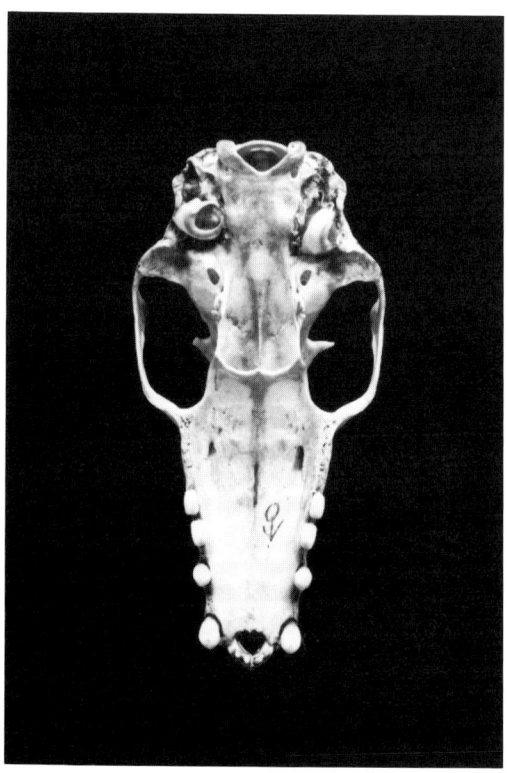

Fig. 34. *Epomops dobsoni* – AMNH 88068 – ♂ (Angola)

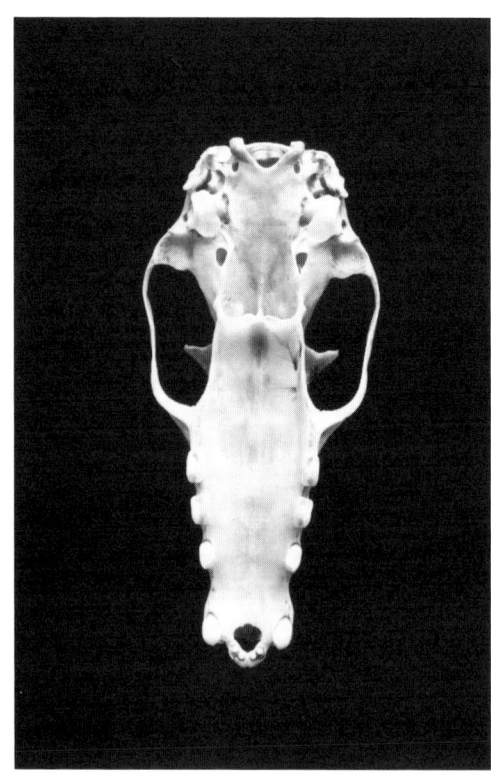

Fig. 35. *Epomophorus gambianus* – AMNH 241011 – ♂ (Ghana)

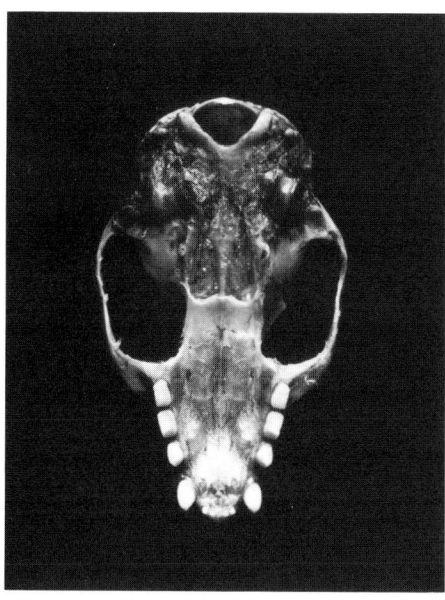

Fig. 36. *Micropteropus pusillus* – AMNH 206836 – ♀ (Zaire)

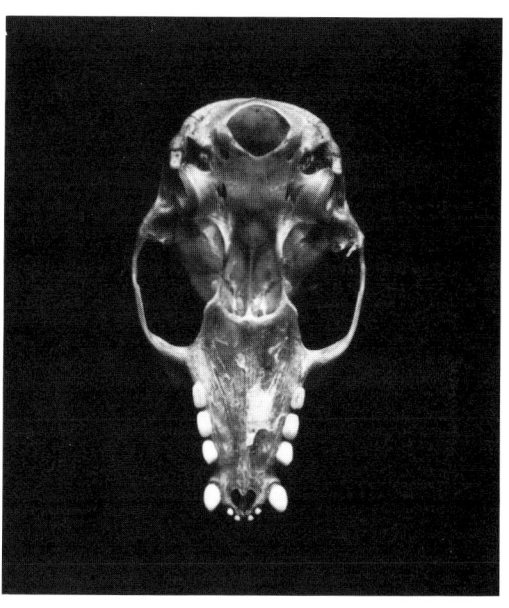

Fig. 37. *Nanonycteris veldkampi* –AMNH 241023 – ♀ (Cameroon)

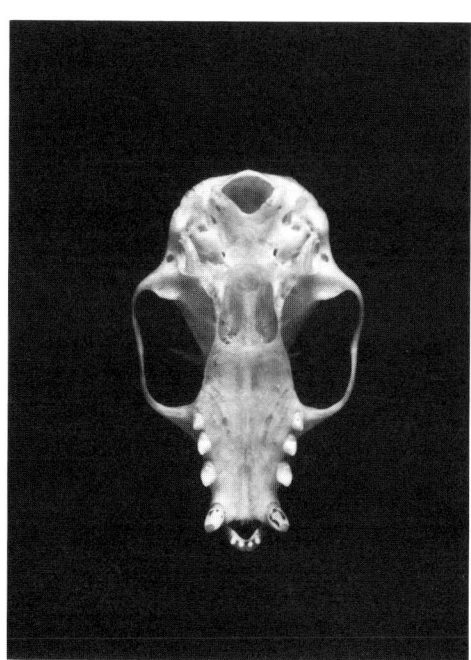

Fig. 38. *Scotonycteris zenkeri* – AMNH 239378 – ♂ (Ivory Coast)

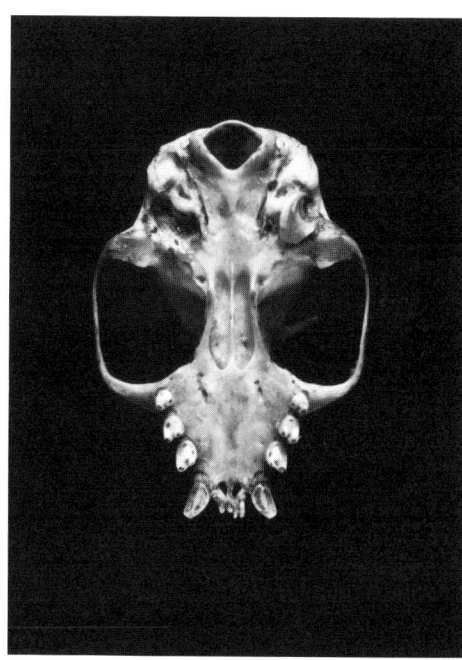

Fig. 39. *Casinycteris argynnis* – AMNH 48751 – ♀ (Zaire)

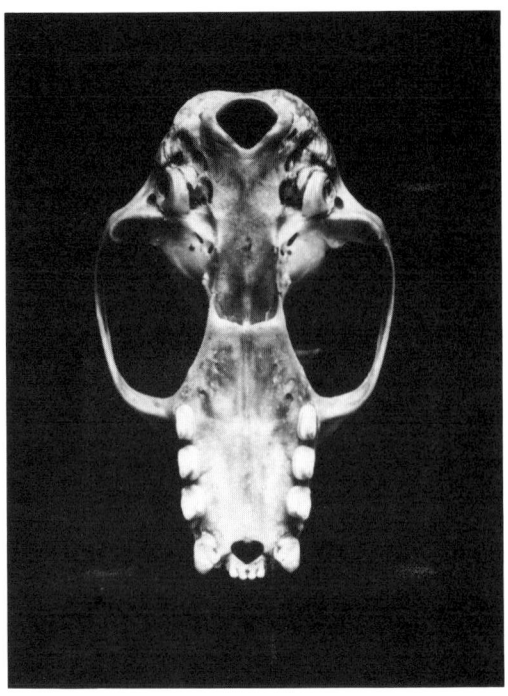

Fig. 40. *Cynopterus titthaecheilus* – AMNH 250056 – ♀ (Java)

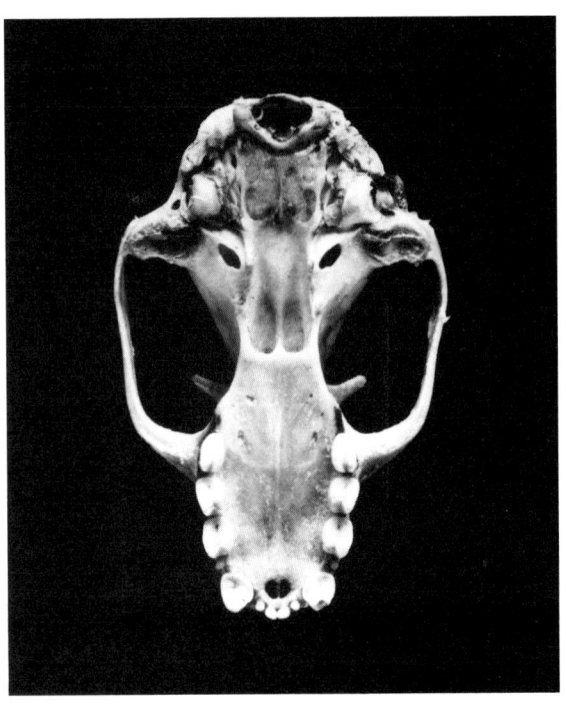

Fig. 41. *Ptenochirus jagorii* – AMNH 207857 – ♀ (Philippines: Cebu)

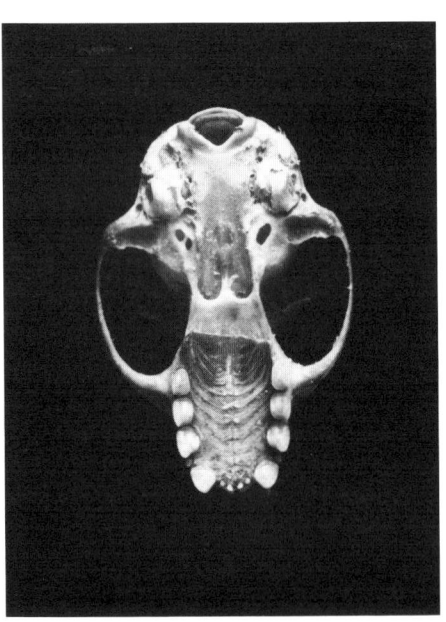

Fig. 42. *Megaerops ecaudatus* – AMNH 232507 – ♂ (Malaya)

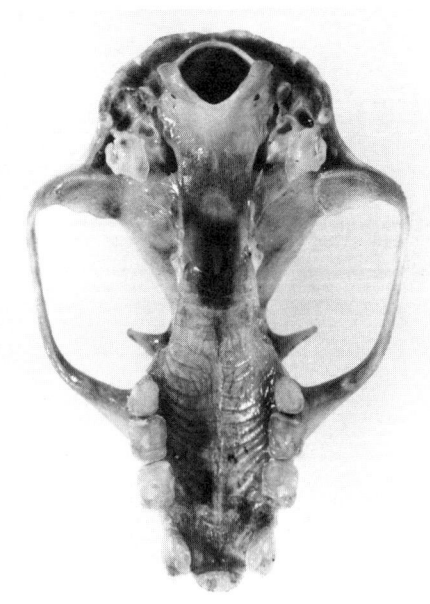

Fig. 43. *Dyacopterus spadiceus* – ROM 48163 – ♂ (Borneo)

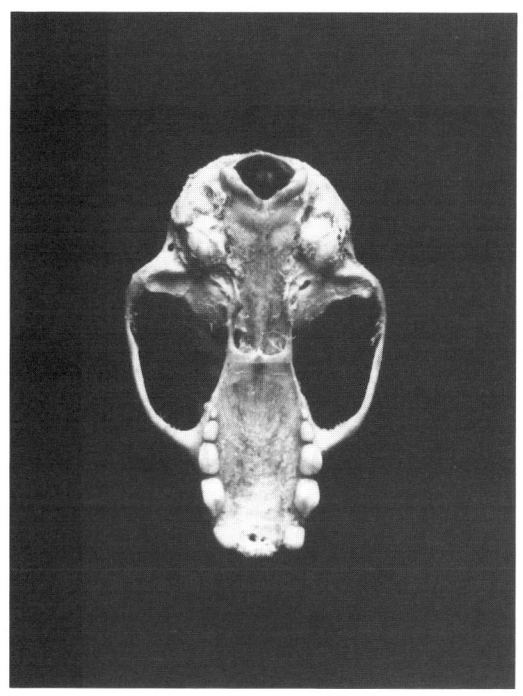

Fig. 44. *Balionycteris maculata* – AMNH 216755 – ♀ (Malaya)

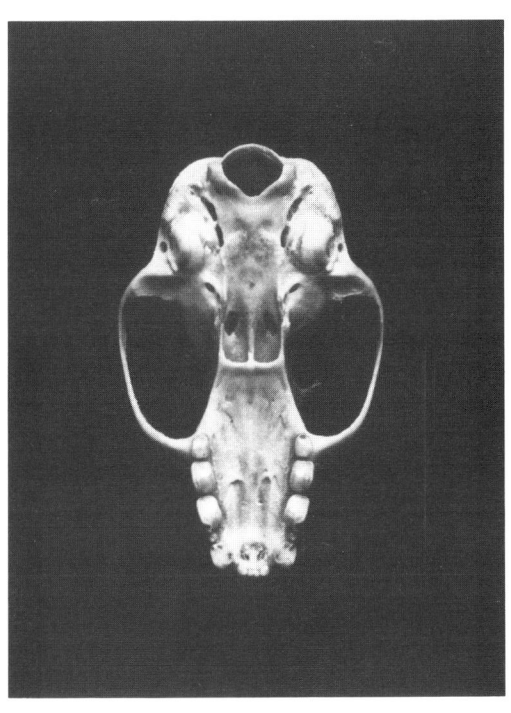

Fig. 45. *Chironax melanocephalus* – AMNH 216739 – ♀ (Malaya)

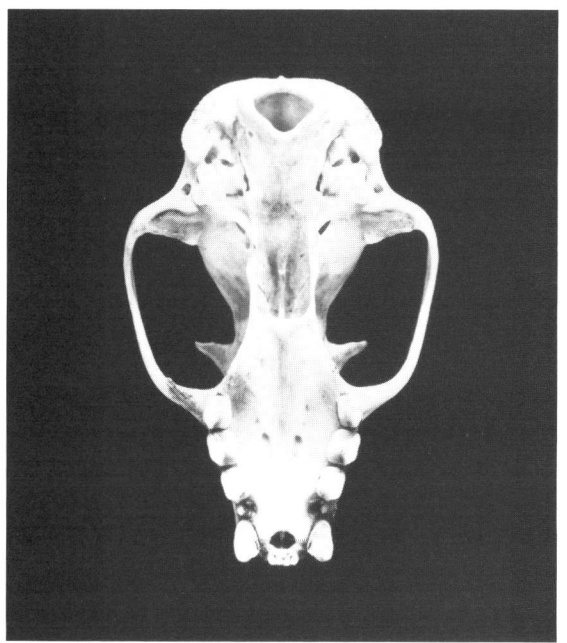

Fig. 46. *Thoopterus nigrescens* – AMNH 222773 – ♂ (Celebes)

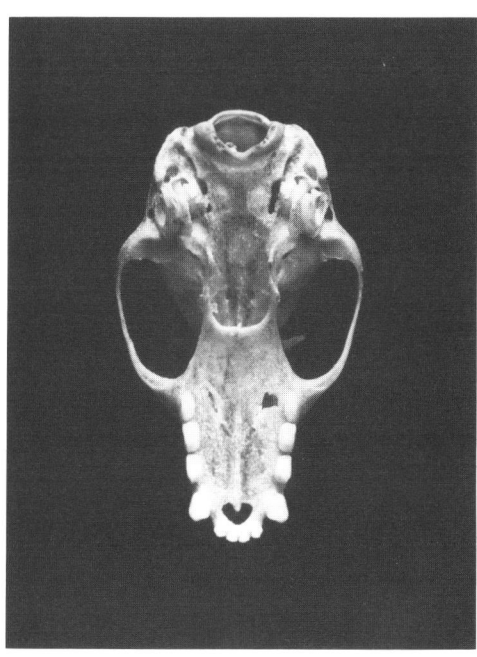

Fig. 47. *Sphaerias blanfordi* – AMNH 240004 – ♂ (Thailand)

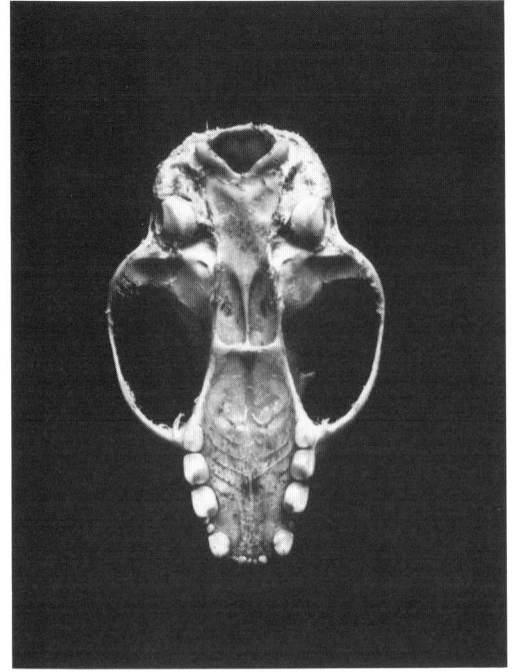
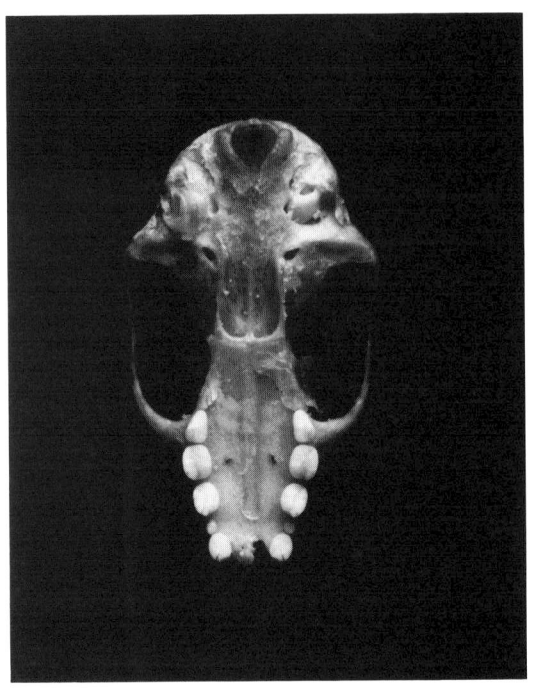

Fig. 48. *Aethalops alecto* – AMNH 216757 – ♀ (Malaya)

Fig. 49. *Penthetor lucasi* – AMNH 106824 – ♀ (Borneo)

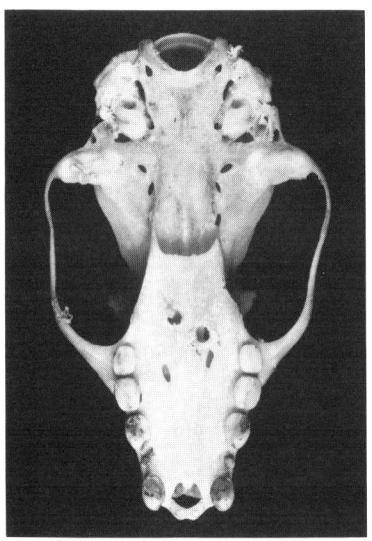
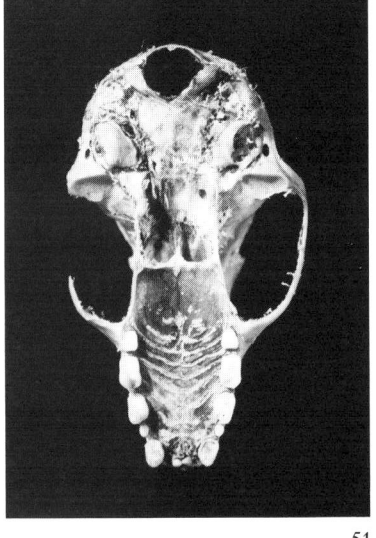

Fig. 50. *Latidens salimalii* – BNHM 1563 – ?♂ (India: Madras – type)

Fig. 51. *Alionycteris paucidentata* – BM 73.1790 – ♂ (Philippines: Mindanao)

Fig. 52. *Otopteropus cartilaginodus* – UMMZ 156972 – ♀ (Philippines: Luzon)

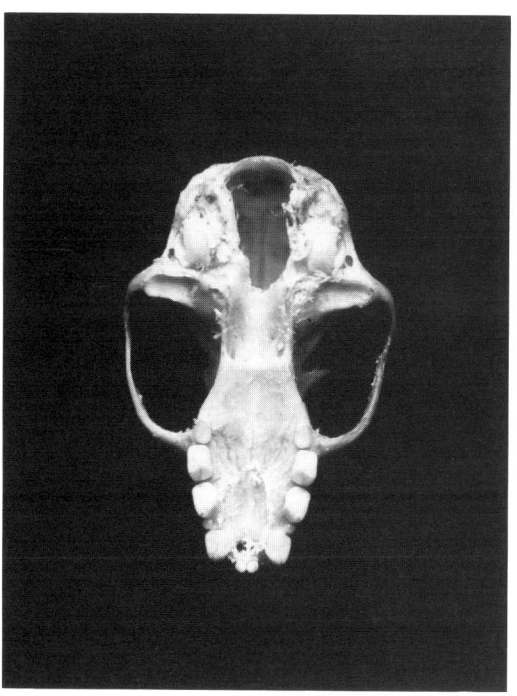

Fig. 53. *Haplonycteris fischeri* – AMNH 187088 – ♂ (Philippines: Luzon)

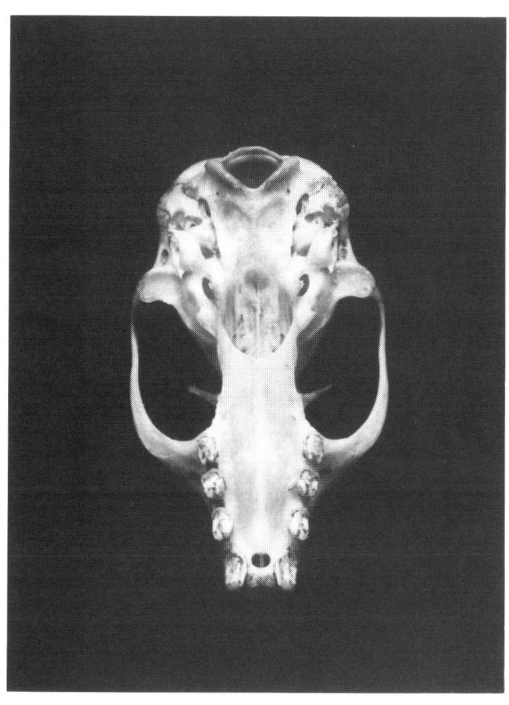

Fig. 54. *Paranyctimene raptor* – AMNH 194854 – ♂ (Papua New Guinea)

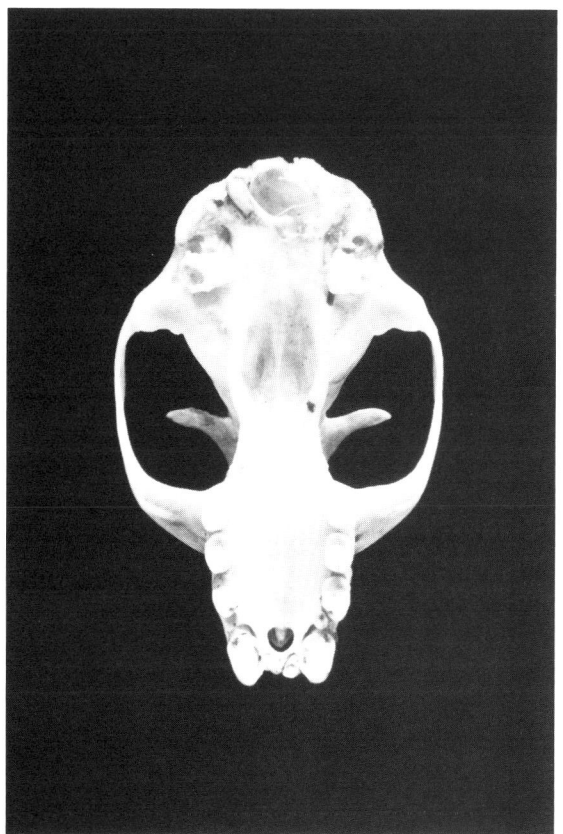

Fig. 55. *Nyctimene aello* – AMNH 105100 – ♂ (Papua New Guinea)

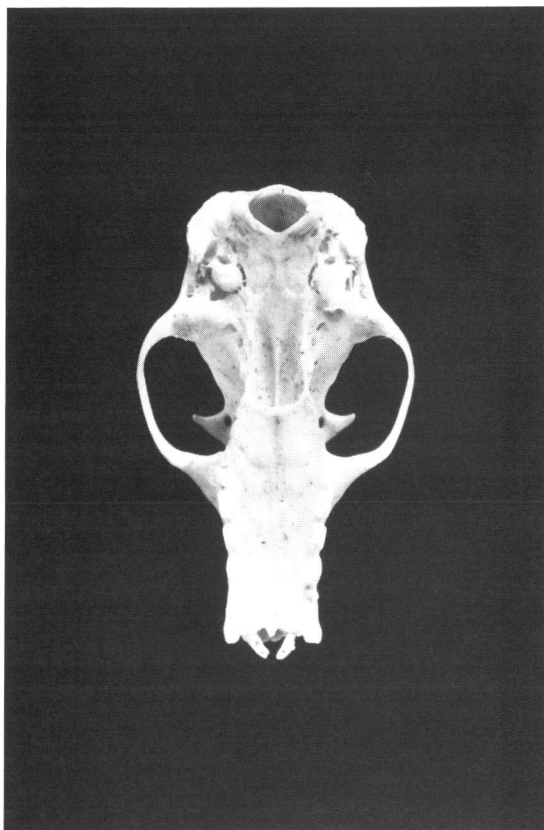

Fig. 56. *Eonycteris spelaea* – AMNH 241775 – unsexed (Philippines: Luzon)

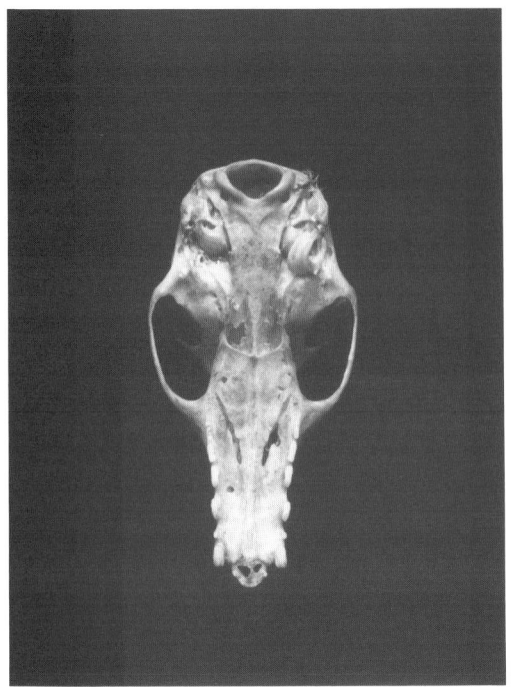

Fig. 57. *Megaloglossus woermanni* – AMNH 236283 – ♂ (Cameroon)

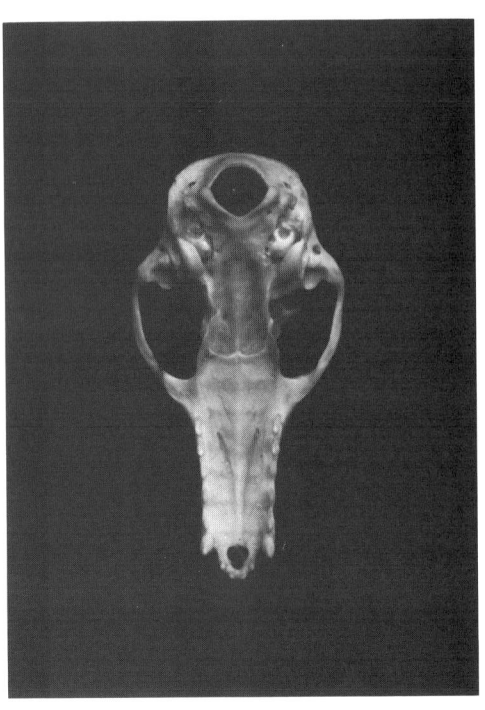

Fig. 58. *Macroglossus sobrinus* – AMNH 103208 – ♂ (Mentawai is.)

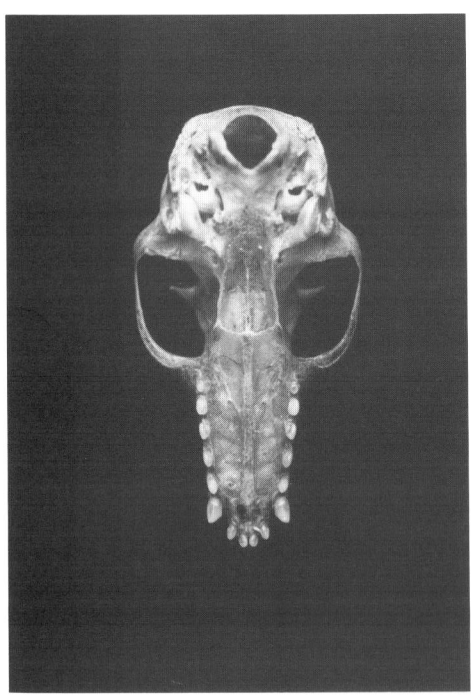

Fig. 59. *Syconycteris australis* – AMNH 157375 – ♀ (Papua New Guinea)

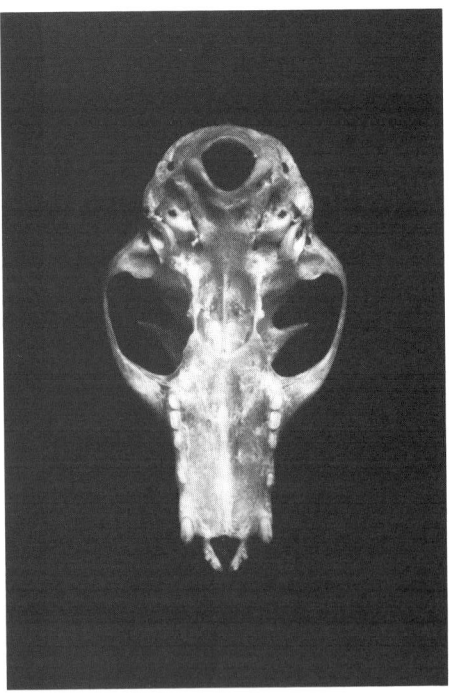

Fig. 60. *Melonycteris melanops* – AMNH 237312 – ♀ (Papua New Guinea: Tolokiwa)

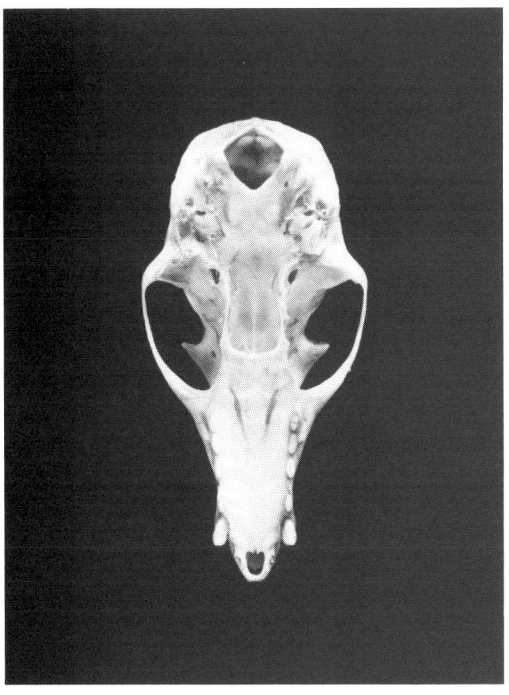

Fig. 61. *Notopteris macdonaldi* – AMNH 119453 – ♀ (Fiji)

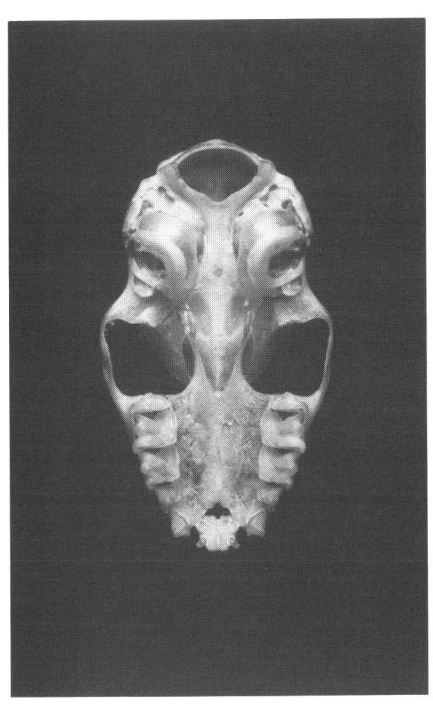

Fig. 62. *Rhinopoma hardwickei* – AMNH 208126 – ♀ (India: Uttar Pradesh)

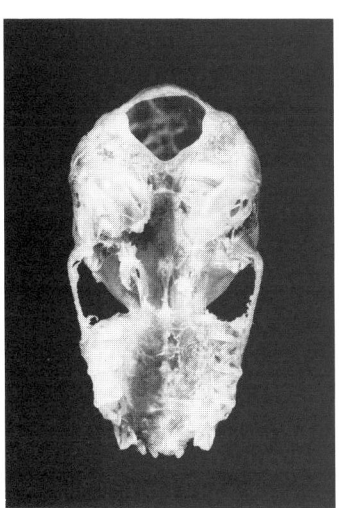

Fig. 63. *Craseonycteris thonglongyai* – BM 77.2993 – ♂ (Thailand)

Fig. 64. *Emballonura furax* – AMNH 221958 – ♂ (New Guinea: Irian Jaya; Japen)

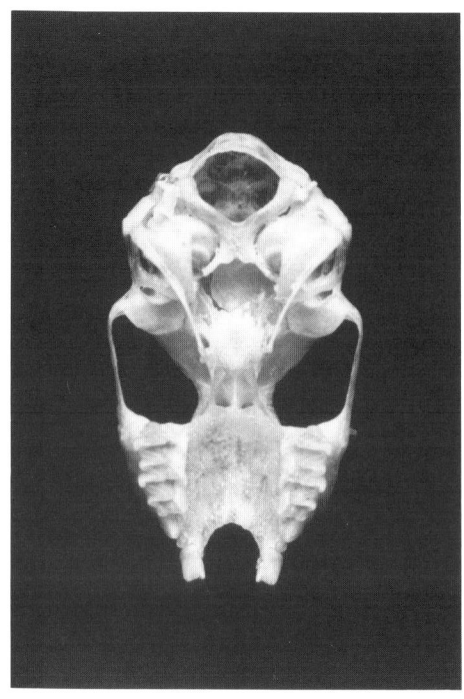

Fig. 65. *Coleura afra* – AMNH 187350 – ♂ (Kenya)

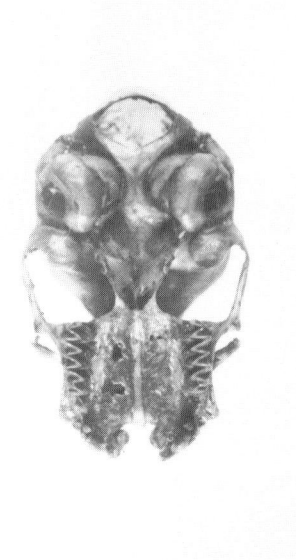

Fig. 66. *Rhynchonycteris naso* – AMNH 7439 – ♀ (Trinidad)

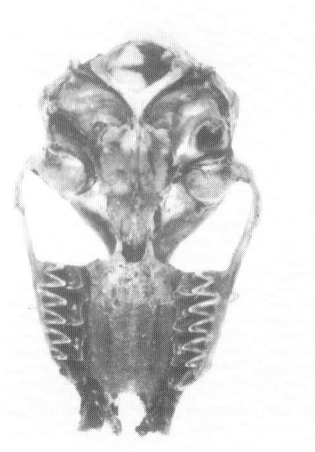

Fig. 67. *Saccopteryx bilineata* – AMNH 7508 – ♀ (Trinidad)

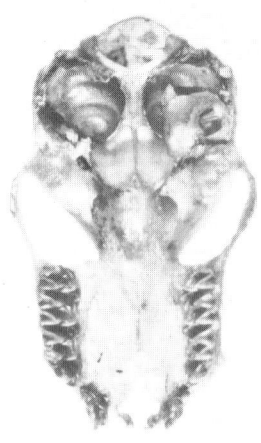

Fig. 68. *Centronycteris maximiliani* – KU 32088 – unsexed (Mexico)

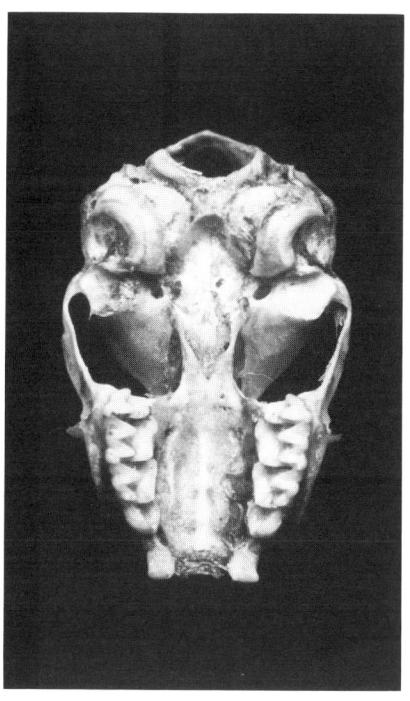

Fig. 69. *Peropteryx macrotis* – USNM 101931 – ♀ (Trinidad)

Fig. 70. *Cormura brevirostris* – AMNH 79546 – ♂ (Brazil: Amazonas)

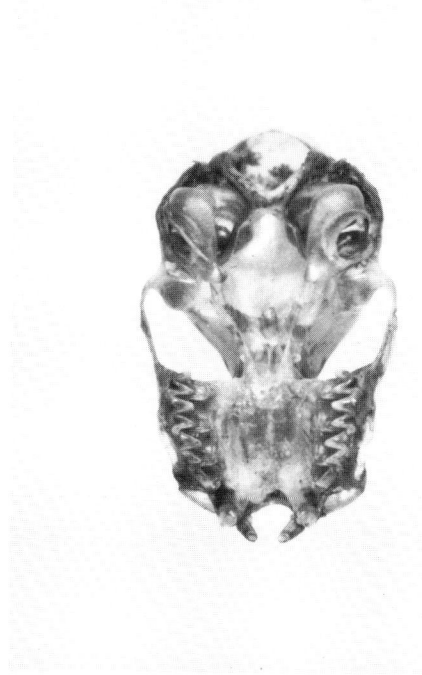
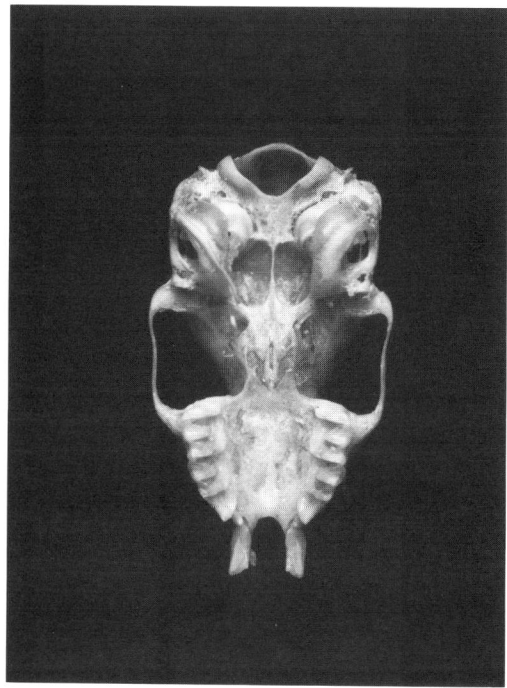

Fig. 71. *Balantiopteryx plicata* – AMNH 189571 – ♀ (Mexico)

Fig. 72. *Taphozous australis* – AMNH 154714 – ♂ (Australia: Queensland)

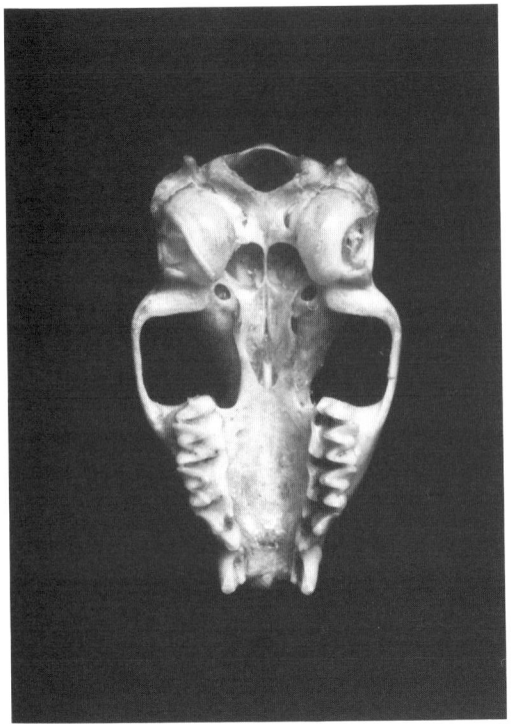

Fig. 73. *Saccolaimus flaviventris* – AMNH 107755 – ♀ (Australia: Queensland)

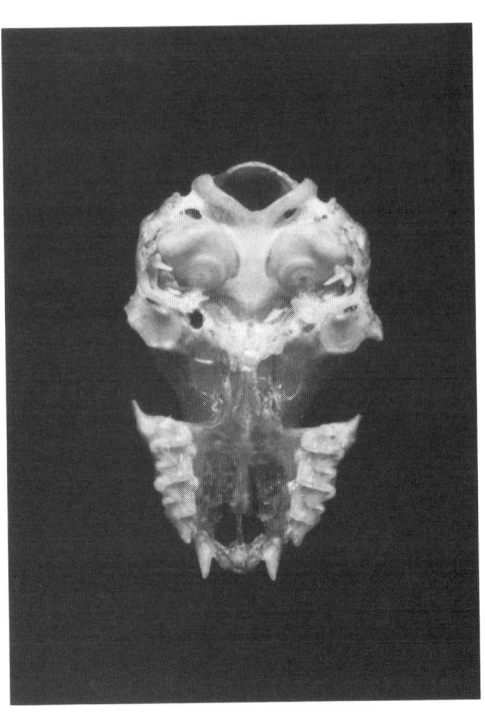

Fig. 74. *Cyttarops alecto* – LACM 26625 – ♂ (Costa Rica)

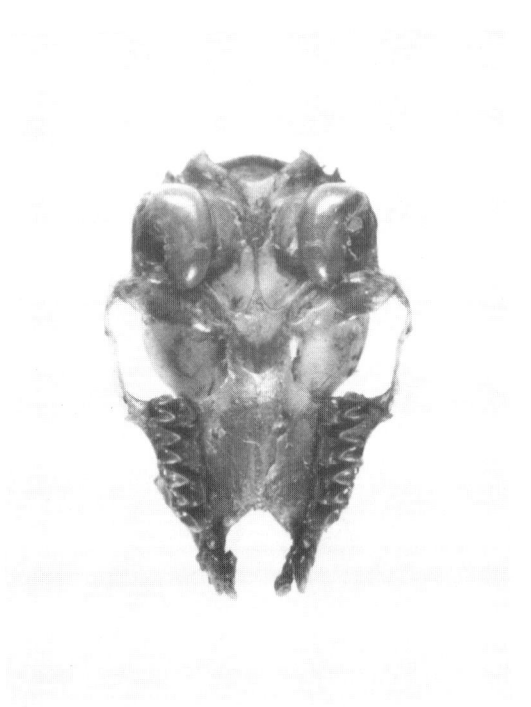

Fig. 75. *Diclidurus albus* – AMNH 149167 – ♀ (Colombia)

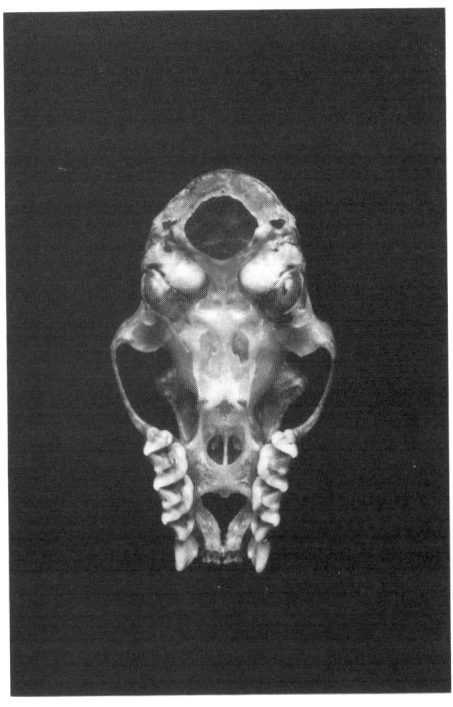

Fig. 76. *Nycteris tragata* – AMNH 216802 – ♂ (Malaya)

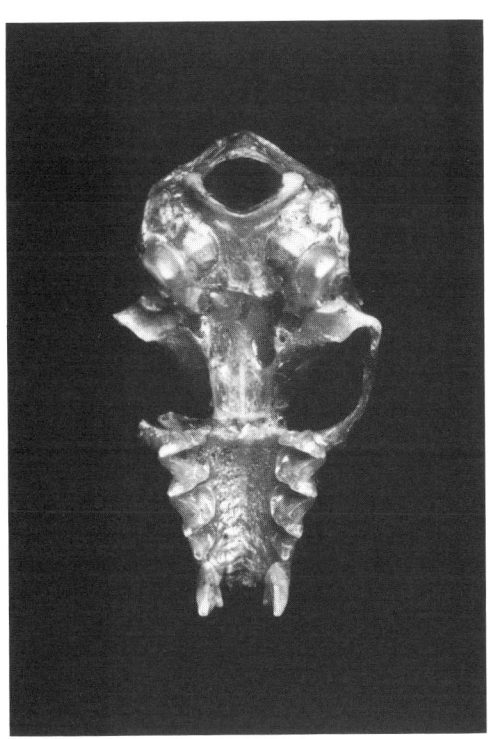

Fig. 77. *Megaderma lyra* – AMNH 208820 – ♂ (India: Maharashtra)

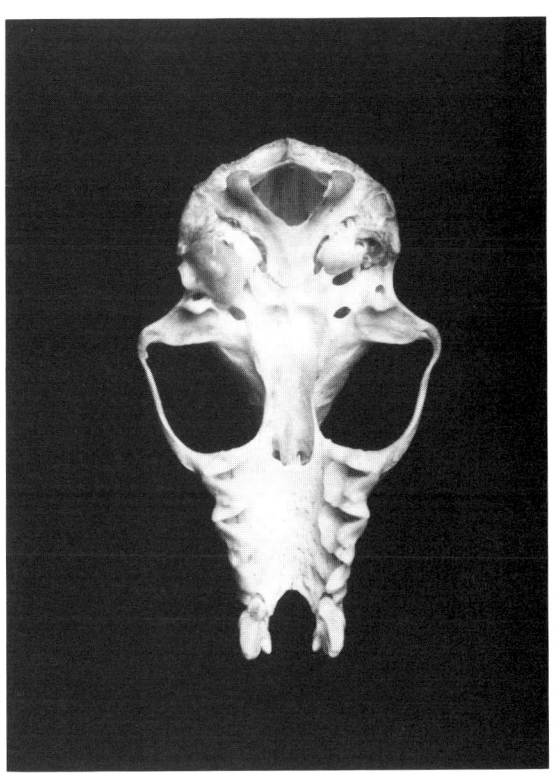

Fig. 78. *Macroderma gigas* – AMNH 162699 – ♂ (Australia: Queensland)

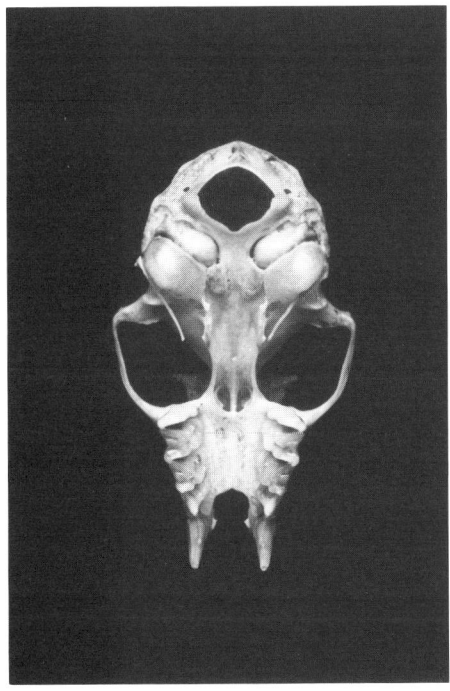

Fig. 79. *Cardioderma cor* – AMNH 187337 – ♀ (Kenya)

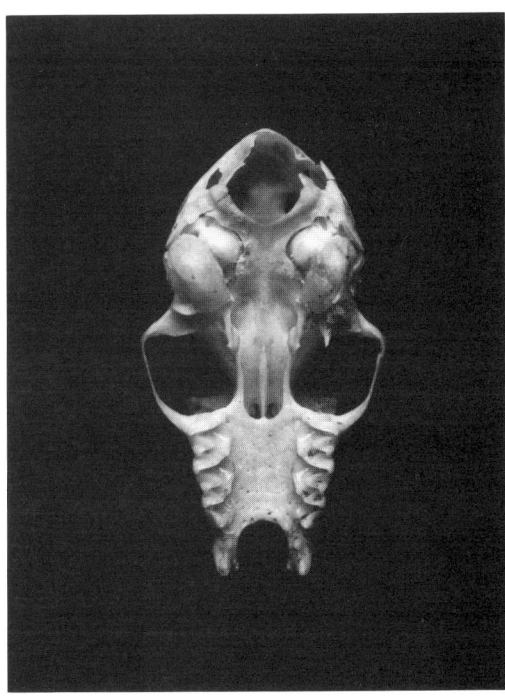

Fig. 80. *Lavia frons* – AMNH 165817 ♂ (Chad)

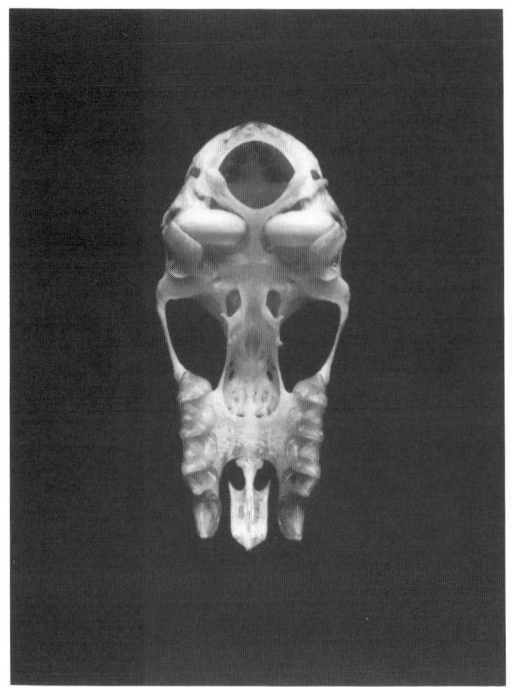

Fig. 81. *Rhinolophus affinis* – AMNH 216 821 – ♀ (Malaya)

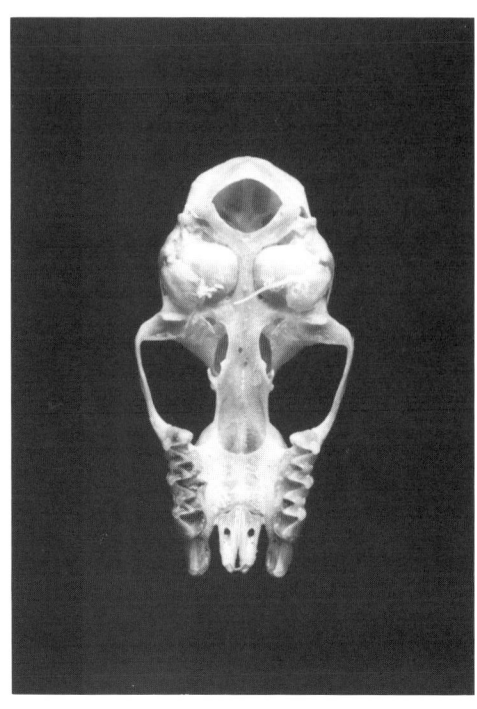

Fig. 82. *Hipposideros cyclops* – AMNH 239 398 – ♂ (Ivory Coast)

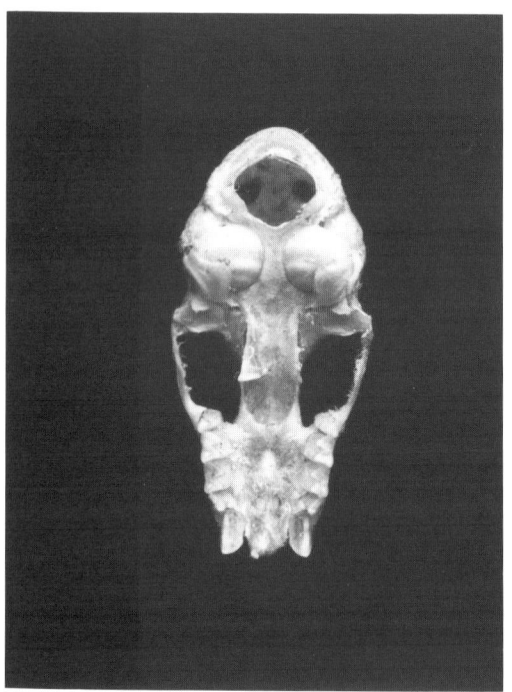

Fig. 83. *Anthops ornatus* – AMNH 99 908 – ♂ (Solomons: Choiseul)

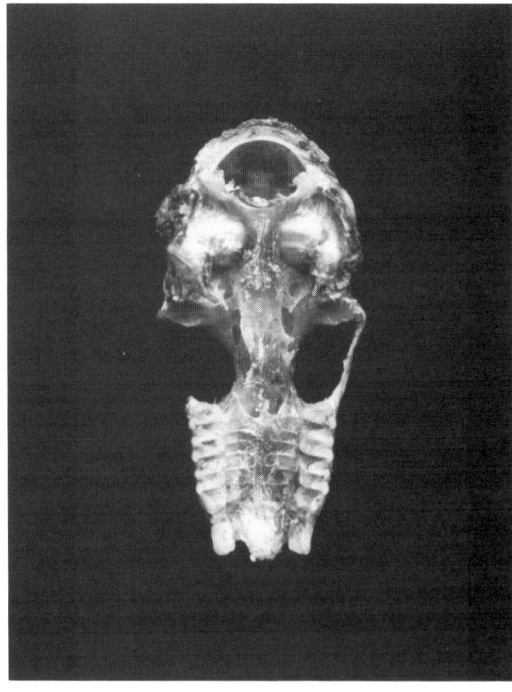

Fig. 84. *Aselliscus tricuspidatus* – AMNH 221 998 – ♀ (New Guinea: Irian Jaya; Japen)

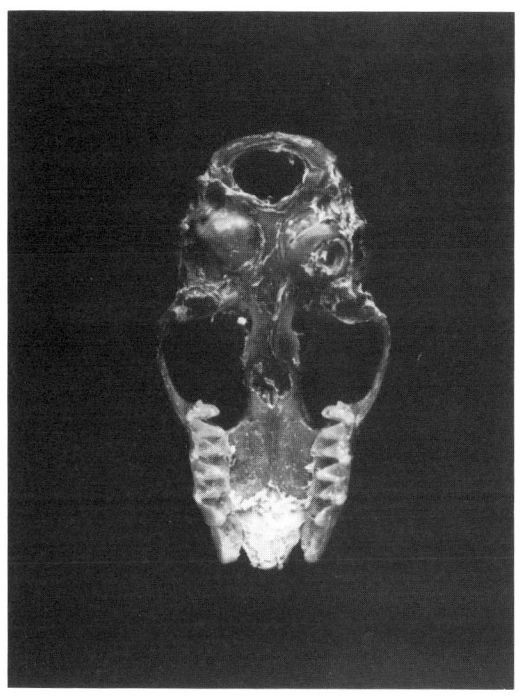

Fig. 85. *Asellia tridens* – AMNH 175963 – unsexed (Afghanistan)

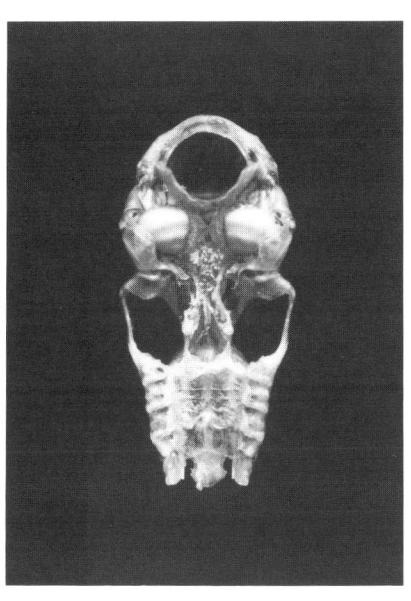

Fig. 86. *Rhinonycteris aurantius* – AMNH 197215 – ♂ (Australia: W. Australia)

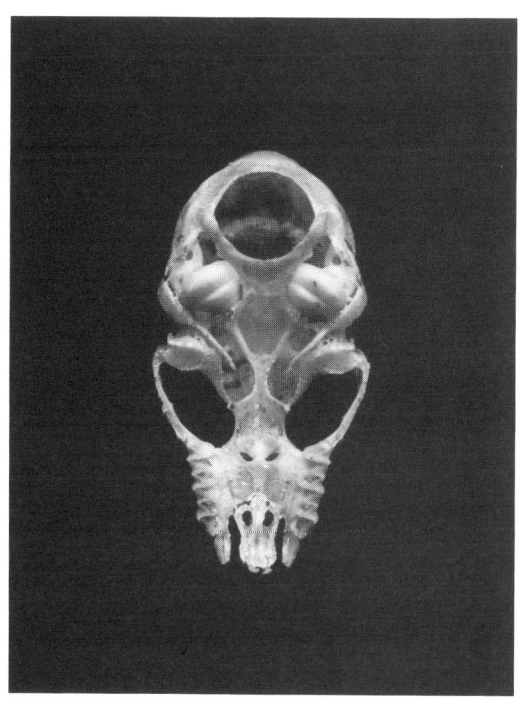

Fig. 87. *Cloeotis percivali* – AMNH 168160 – ♀ (Botswana)

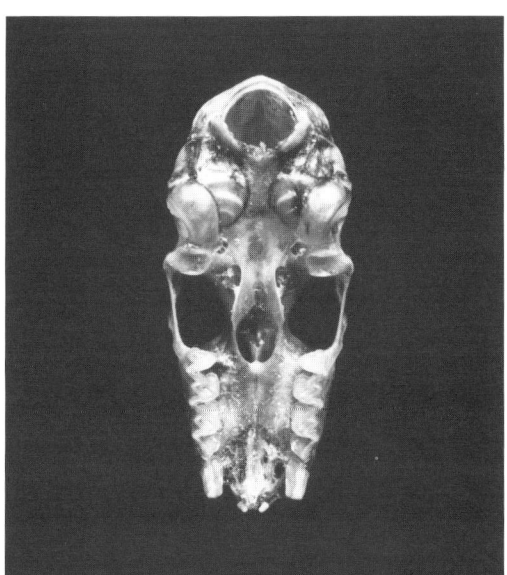

Fig. 88. *Triaenops persicus* – AMNH 207083 – ♂ (Tanzania)

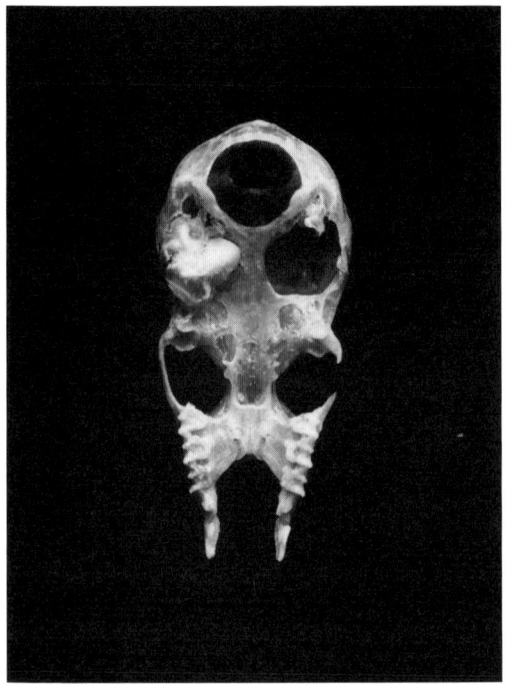
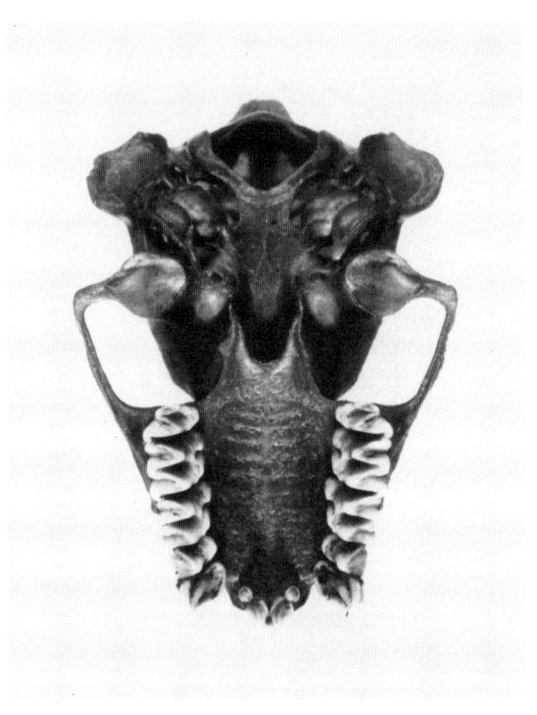

Fig. 89. *Coelops frithi* – AMNH 107 508 – ♂ (Bali)
[*Paracoelops megalotis* – no photograph available]

Fig. 90. *Noctilio leporinus* – AMNH 180 266 – ♂ (Trinidad)

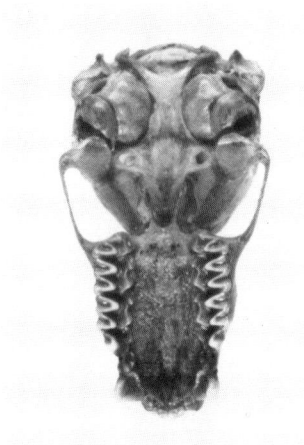
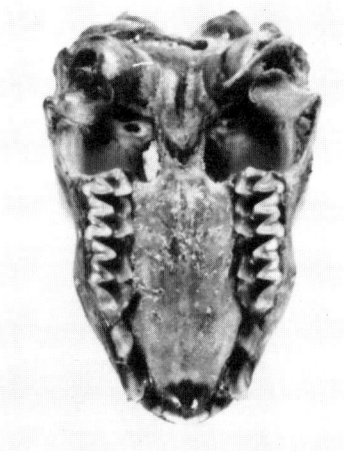
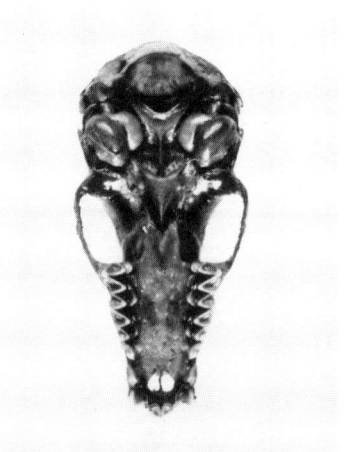

Fig. 91. *Pteronotus parnellii* – AMNH 182 693 – ♀ (Trinidad)

Fig. 92. *Mormoops megalophylla* – AMNH 175 176 – ♀ (Trinidad)

Fig. 93. *Micronycteris megalotis* – AMNH 175 877 – ♂ (Trinidad)

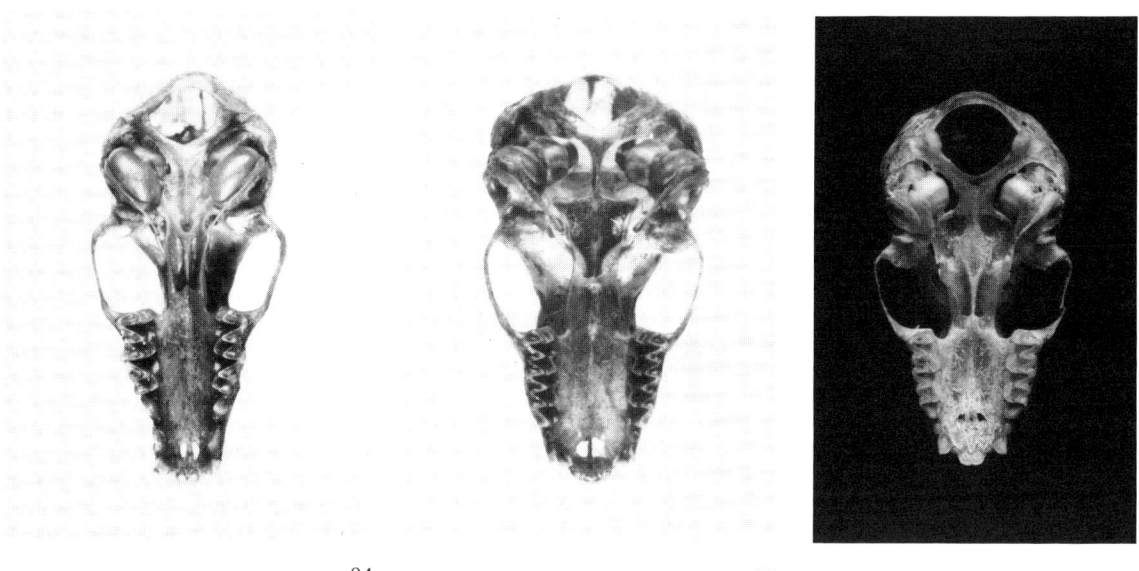

94   95   96

Fig. 94. *Macrotus waterhousii* – AMNH 182158 – ♂ (Mexico)

Fig. 95. *Lonchorhina aurita* – AMNH 184701 – ♀ (Trinidad)

Fig. 96. *Macrophyllum macrophyllum* – AMNH 177671 – ♀ (Nicaragua)

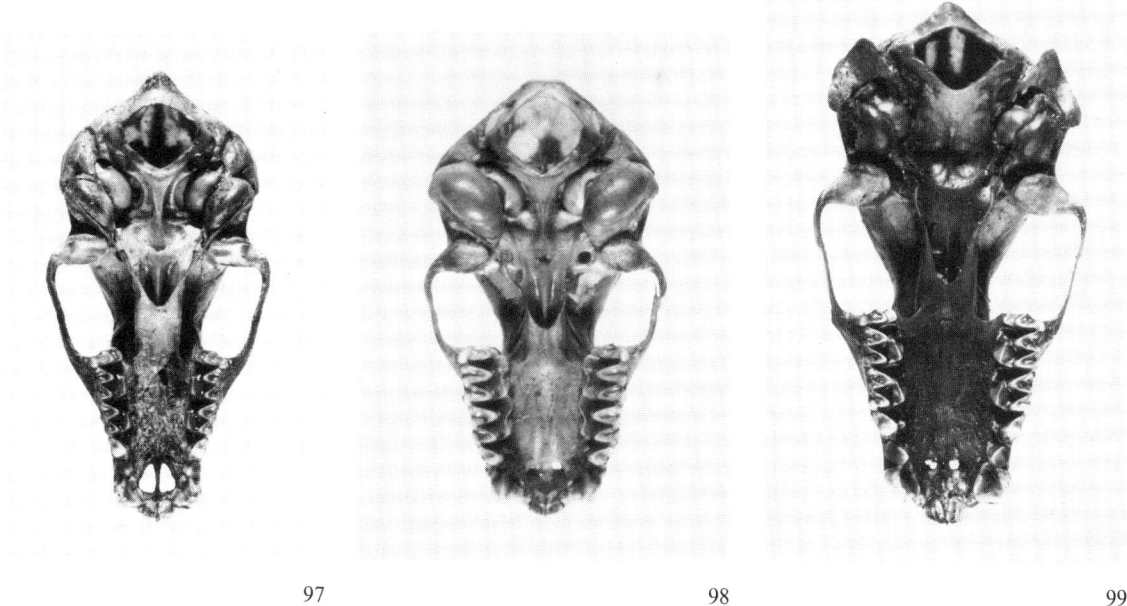

97   98   99

Fig. 97. *Tonatia bidens* – AMNH 180261 – ♂ (Trinidad)

Fig. 98. *Mimon crenulatum* – AMNH 175586 – ♀ (Trinidad)

Fig. 99. *Phyllostomus hastatus* – AMNH 24140 – ♀ (Trinidad)

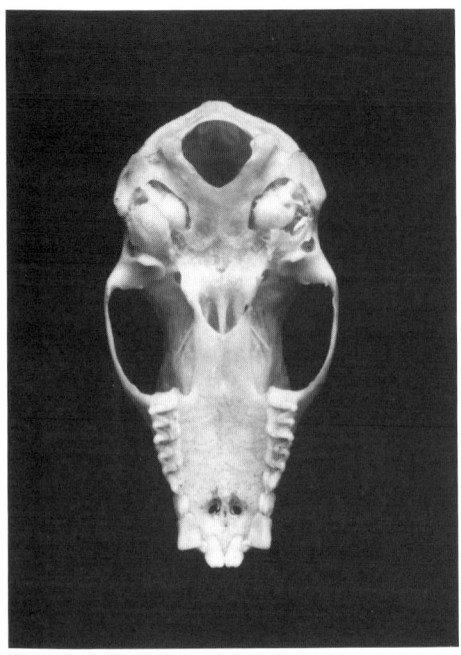

Fig. 100. *Phylloderma stenops* – AMNH 126 869 – ♀ (Honduras)

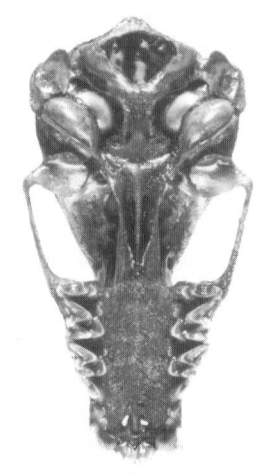

Fig. 101. *Trachops cirrhosus* – AMNH 175 603 – ♀ (Trinidad)

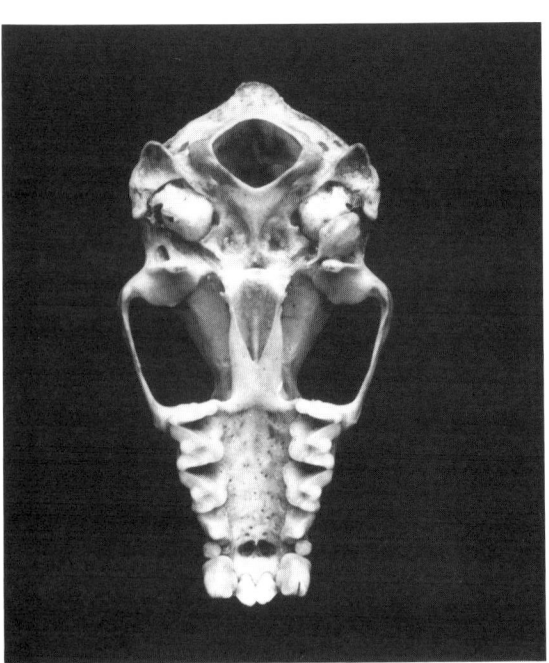

Fig. 102. *Chrotopterus auritus* – AMNH 36 988 – ♂ (Brazil: Mato Grosso)

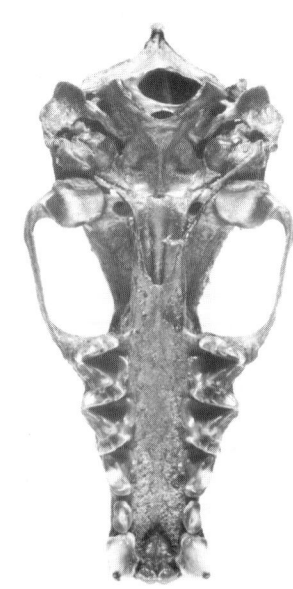

Fig. 103. *Vampyrum spectrum* – AMNH 17 517 – ♂ (Trinidad)

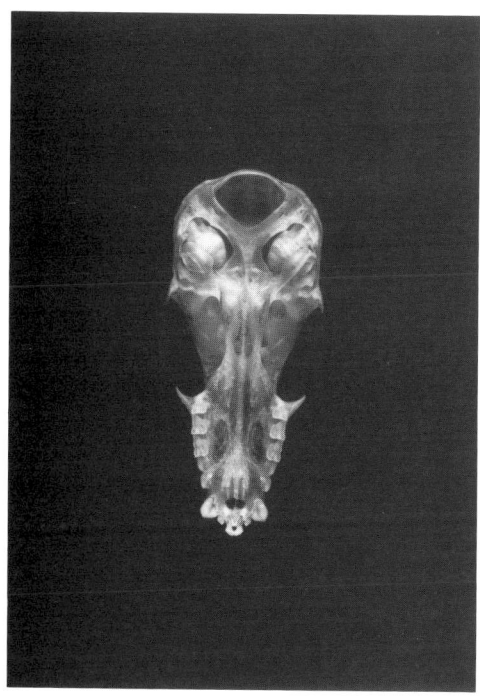

Fig. 104. *Lionycteris spurrelli* – AMNH 260004 – ♀ (Venezuela)

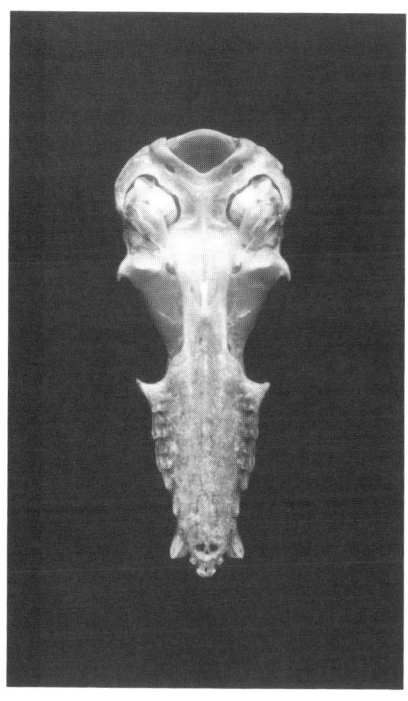

Fig. 105. *Lonchophylla handleyi* – AMNH 230315 – ♂ (Peru)

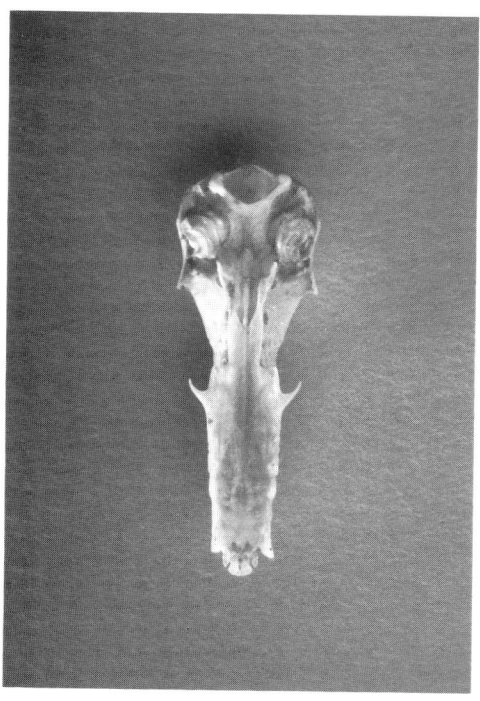

Fig. 106. *Platalina genovensium* – MCZ 32955 – ♂ (Peru)

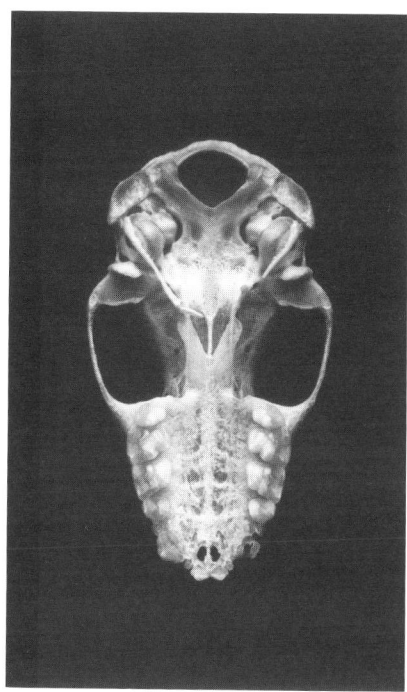

Fig. 107. *Brachyphylla cavernarum* – AMNH 214255 – ♀ (Virgin Is.: St. Croix)

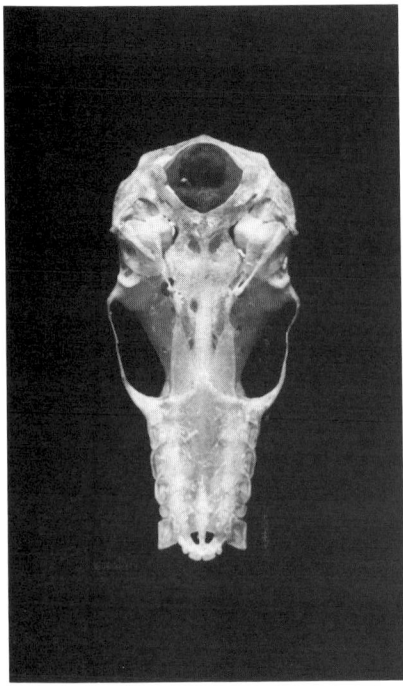
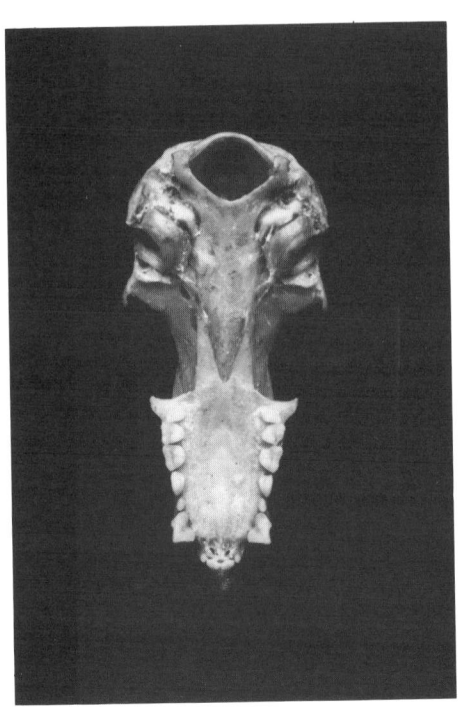

Fig. 108. *Erophylla bombifrons* – AMNH 212998 – ♀ (Dominican Republic)

Fig. 109. *Phyllonycteris poeyi* – AMNH 23759 – ♂ (Cuba)

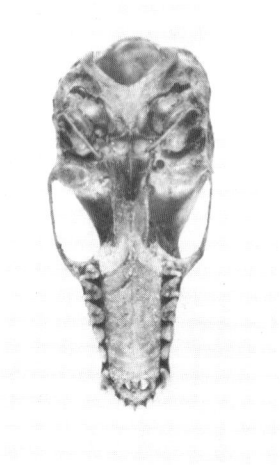
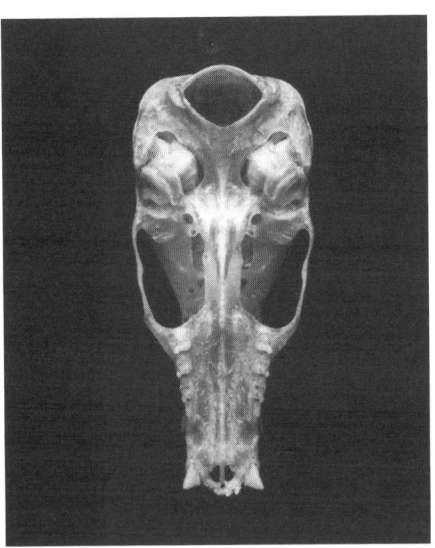

Fig. 110. *Glossophaga soricina* – AMNH 176581 – ♀ (Trinidad)

Fig. 111. *Monophyllus redmani* – AMNH 212997 – ♂ (Dominican Republic)

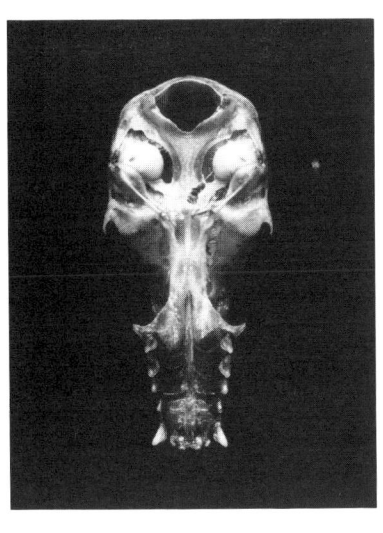 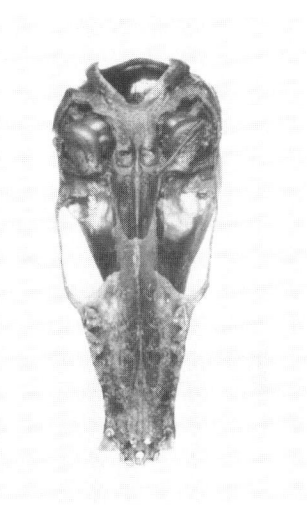 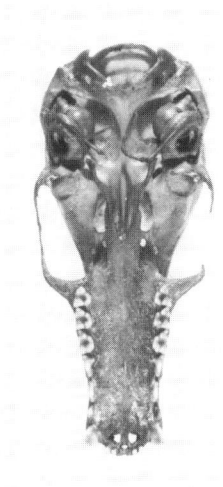

112 113 114

Fig. 112. *Lichonycteris obscurus* – AMNH 244621 – ♂ (Bolivia)

Fig. 113. *Leptonycteris nivalis* – AMNH 180348 – ♀ (Mexico)

Fig. 114. *Anoura geoffroyi* – AMNH 175827 – ♂ (Trinidad)

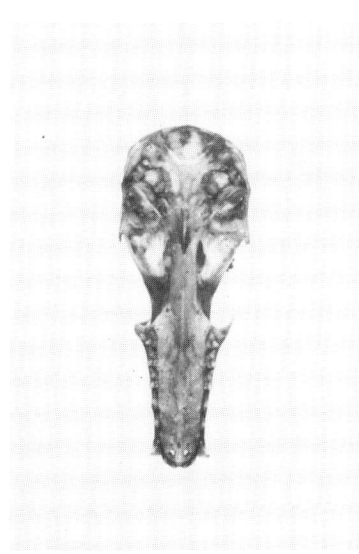 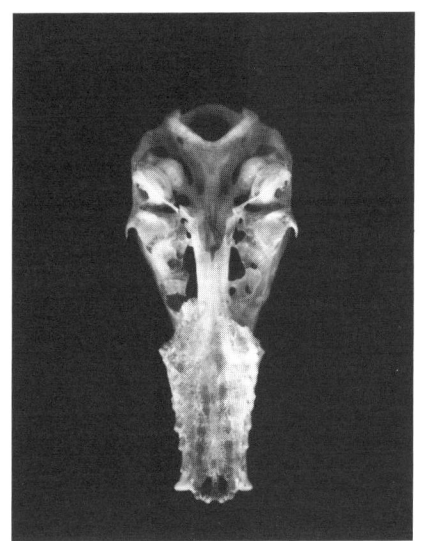 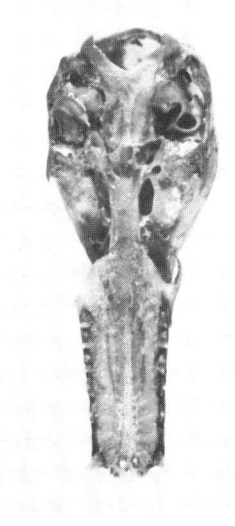

115 116 117

Fig. 115. *Hylonycteris underwoodi* – AMNH 189687 – ♂ (Mexico)

Fig. 116. *Sceronycteris ega* – USNM 407889 – ♂ (Venezuela)

Fig. 117. *Choeroniscus intermedius* – AMNH 6072 – ♀ (Trinidad)

118

119

120

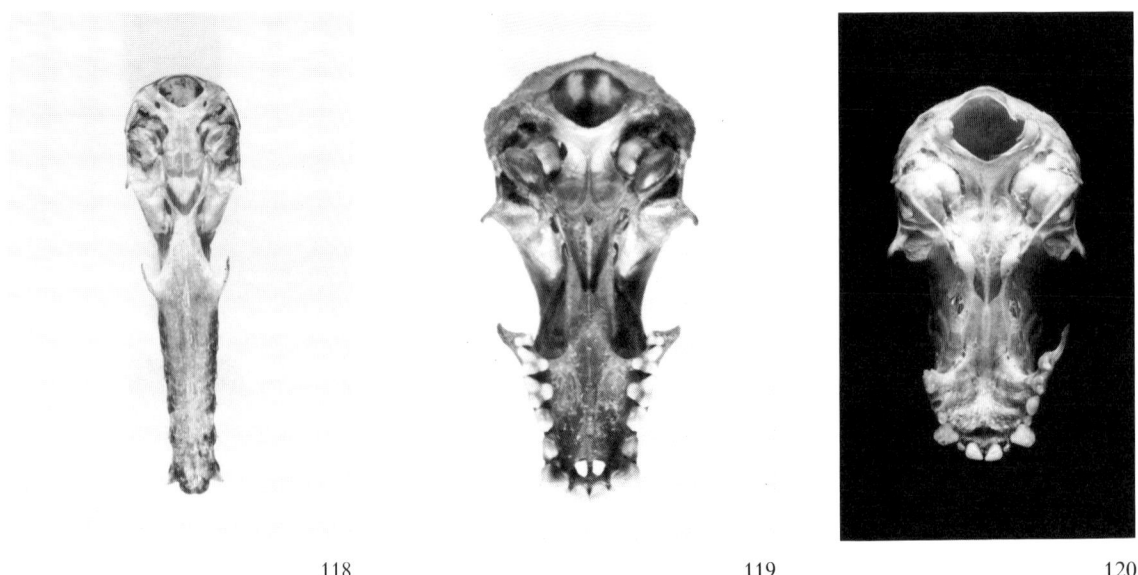

Fig. 118. *Choeronycteris harrisoni* – UMMZ 110524 – ♂ (Mexico)

Fig. 119. *Carollia perspicillata* – AMNH 184730 – ♂ (Trinidad)

Fig. 120. *Rhinophylla fischerae* – AMNH 230496 – ♀ (Peru)

121

122

123

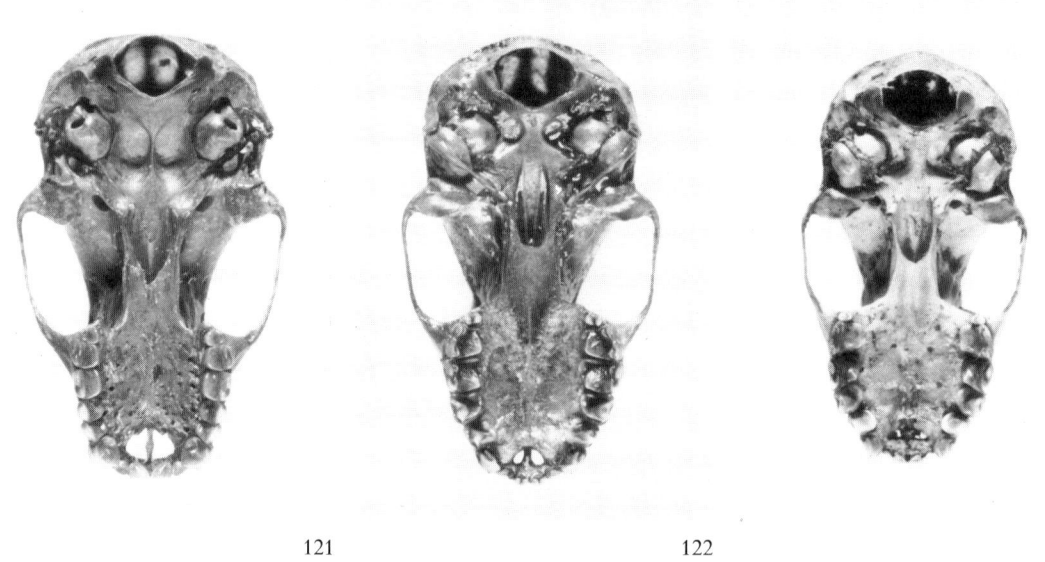

Fig. 121. *Sturnira tildae* – AMNH 149625 – ♂ (Trinidad)

Fig. 122. *Uroderma bilobatum* – AMNH 175649 – ♀ (Trinidad)

Fig. 123. *Vampyrops helleri* – AMNH 149624 – ♀ (Trinidad)

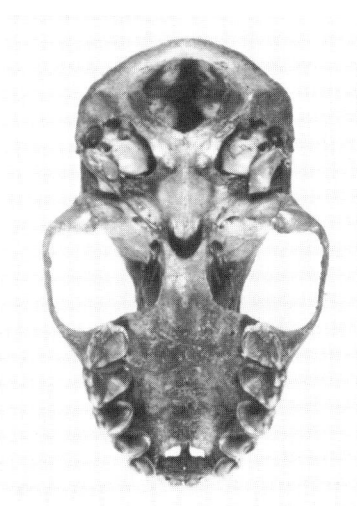
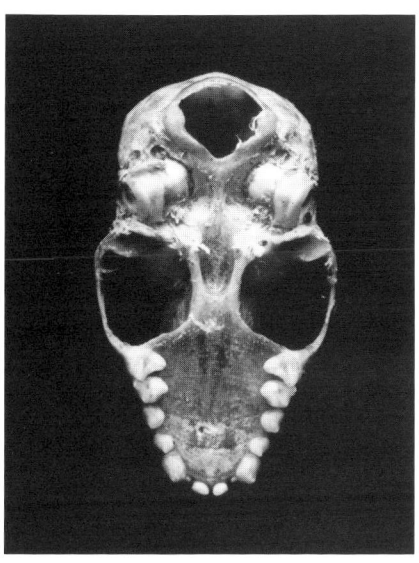

Fig. 124. *Vampyrodes caraccioloi* – AMNH 175642 – ♀ (Trinidad)

Fig. 125. *Vampyressa pusilla* – AMNH 233193 – ♂ (Colombia)

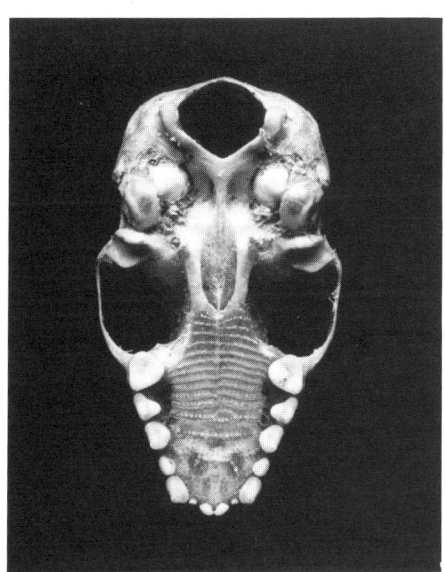

Fig. 126. *Chiroderma villosum* – AMNH 175599 – ♀ (Trinidad)

Fig. 127. *Mesophylla macconnelli* – AMNH 233746 – ♀ (Peru)
[*Ectophylla alba* – no photograph available]

Fig. 128. *Artibeus jamaicensis* – AMNH 184697 – ♀ (Trinidad)

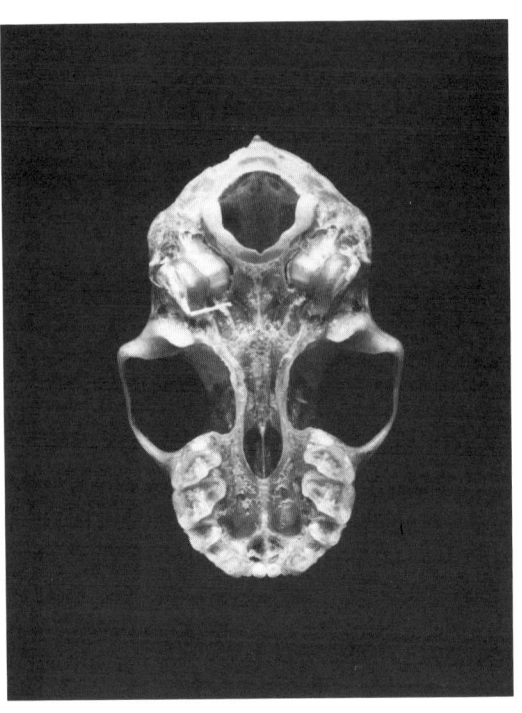

Fig. 129. *Ardops nichollsi* – AMNH 213954 – ♂ (Lesser Antilles: Martinique)

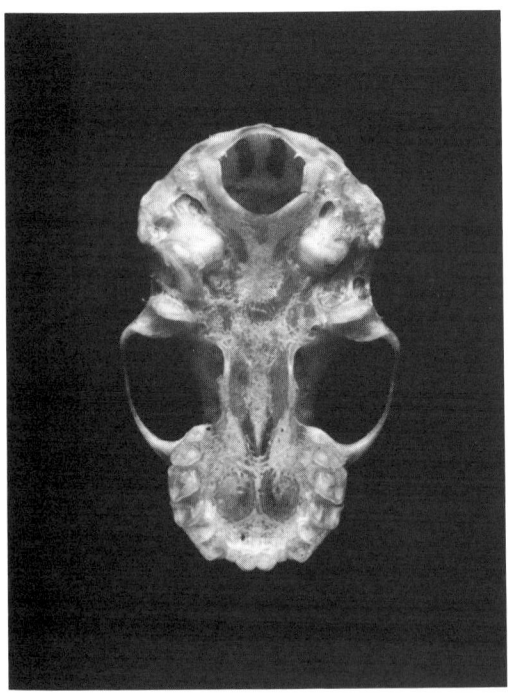

Fig. 130. *Phyllops falcatus* – AMNH 143662 – ♀ (?Hispaniola)

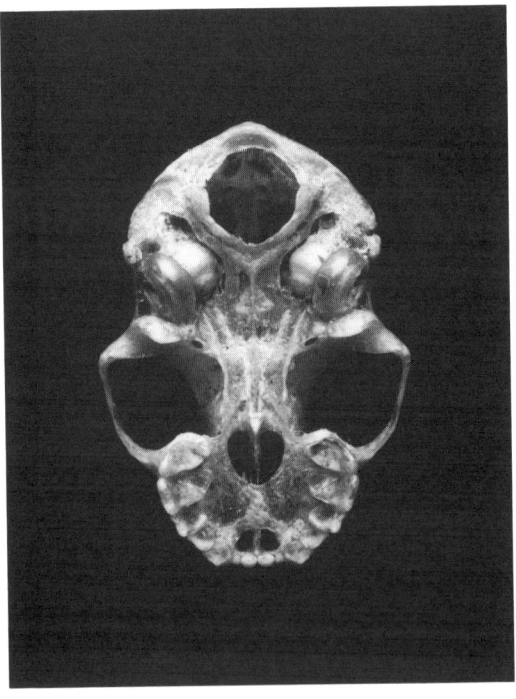

Fig. 131. *Ariteus flavescens* – AMNH 4121 – ♂ (Jamaica)

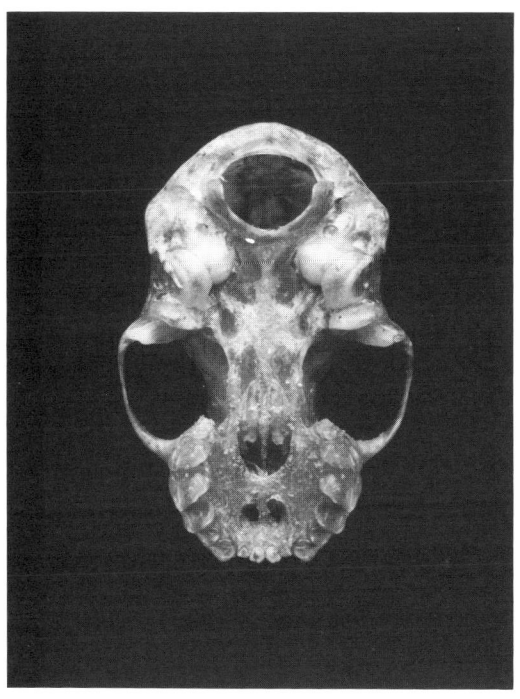

Fig. 132. *Stenoderma rufum* – AMNH 208 982 – ♂ (Puerto Rico)

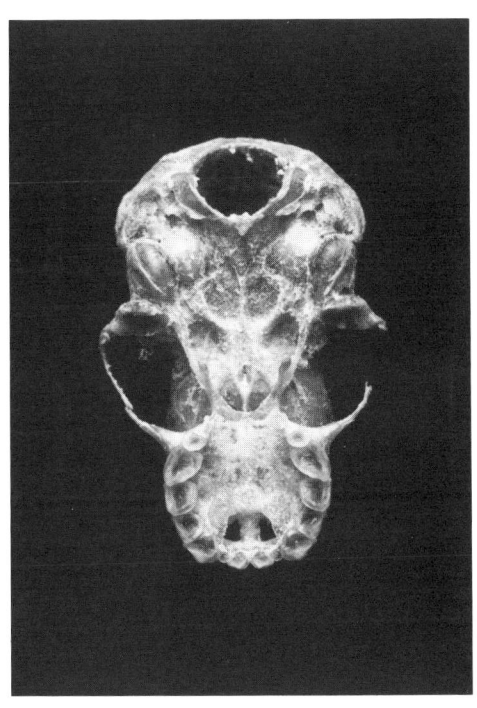

Fig. 133. *Pygoderma bilabiatum* – AMNH 246 398 – ♀ (Bolivia)

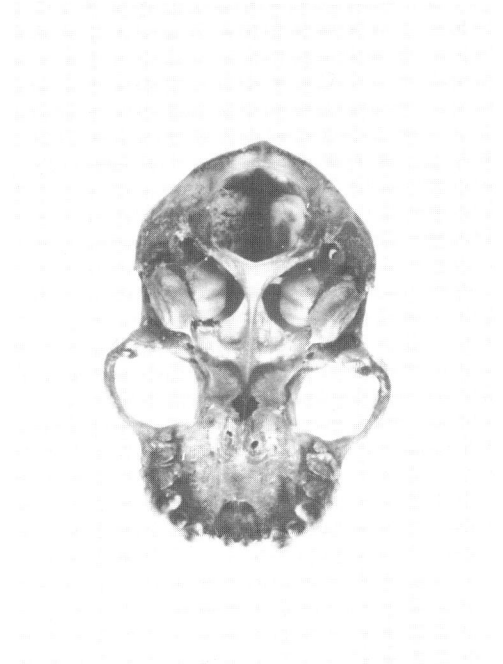

Fig. 134. *Ametrida centurio* – AMNH 142 613 – ♀ (Venezuela)

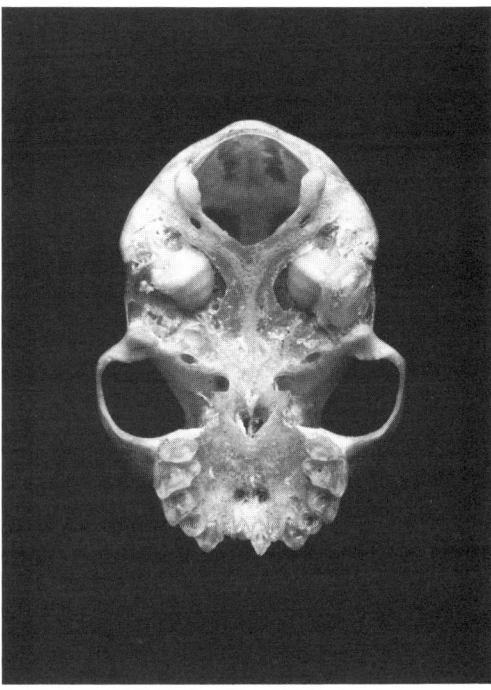

Fig. 135. *Sphaeronycteris toxophyllum* – AMNH 21 344 – ♀ (Venezuela)

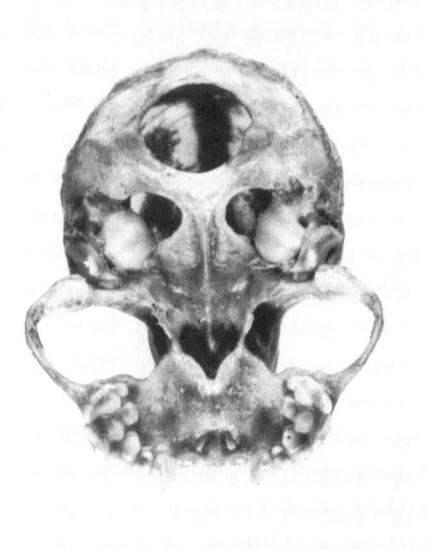

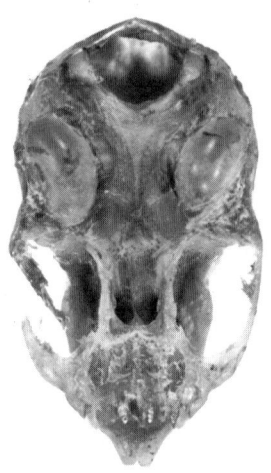

Fig. 136. *Centurio senex* – AMNH 183 862 – ♂ (Trinidad)

Fig. 137. *Diphylla ecaudata* – AMNH 165 640 – ♂ (Mexico)

138

139

140

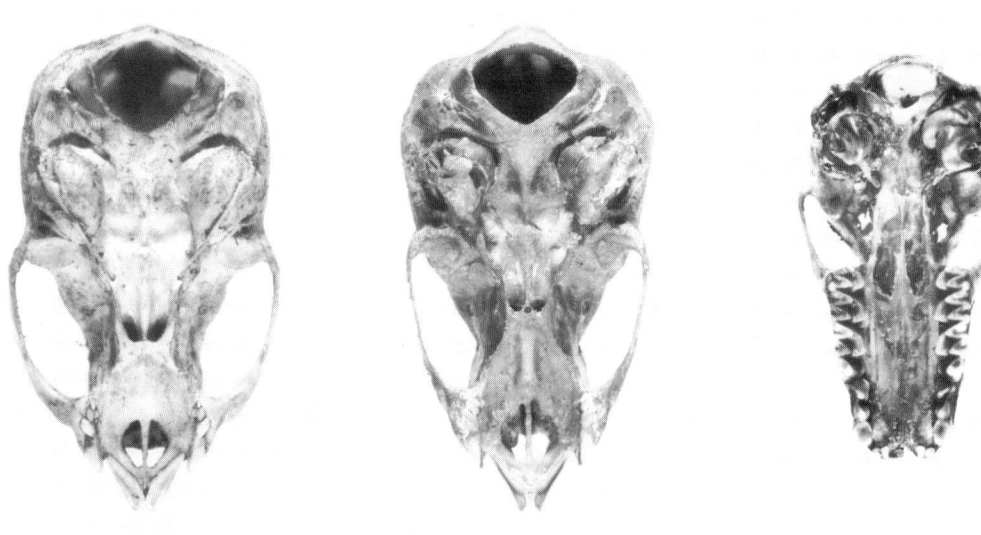

Fig. 138. *Diaemus youngi* – AMNH 175 654 – ♀ (Trinidad)

Fig. 139. *Desmodus rotundus* – AMNH 175 683 – ♀ (Trinidad)

Fig. 140. *Natalus tumidirostris* – AMNH 176 500 – ♂ (Trinidad)

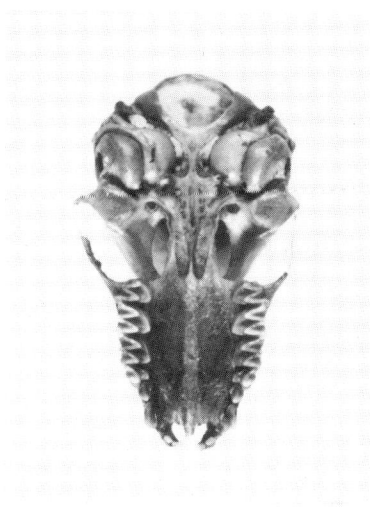
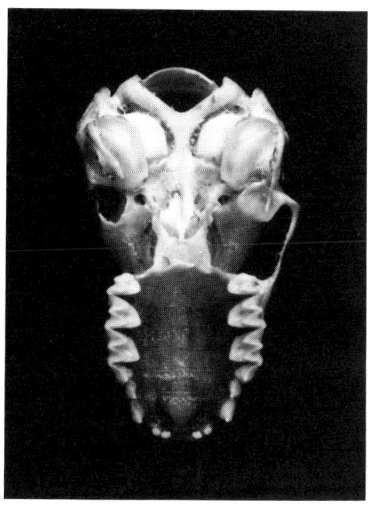

Fig. 141. *Furipterus horrens* – AMNH 142903 – ♂ (Guyana)

Fig. 142. *Amorphochilus schnablii* – AMNH 28601 – ♂ (Peru)

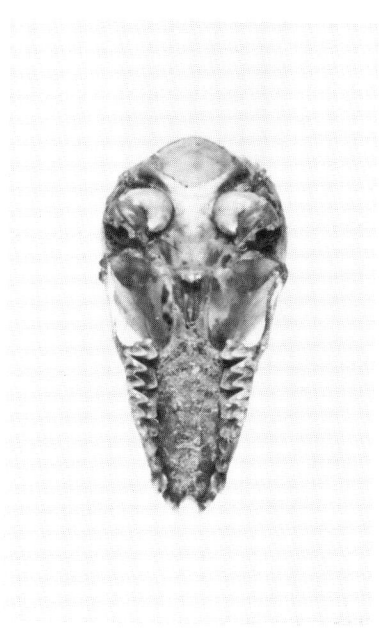
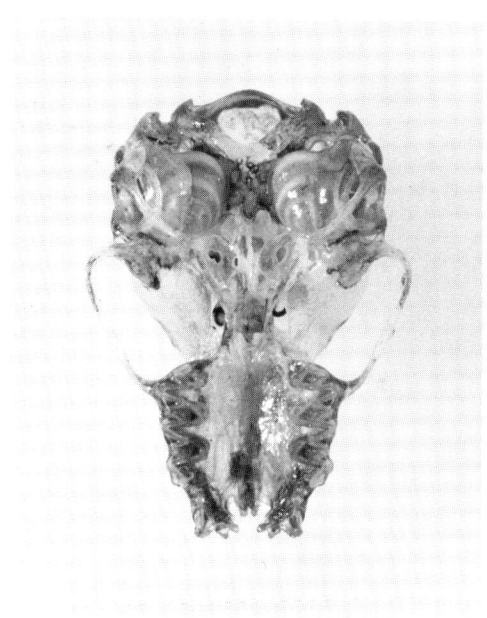

Fig. 143. *Thyroptera tricolor* – AMNH 183860 – ♂ (Trinidad)

Fig. 144. *Myzopoda aurita* – ROM 46926 – ♂ (Madagascar)

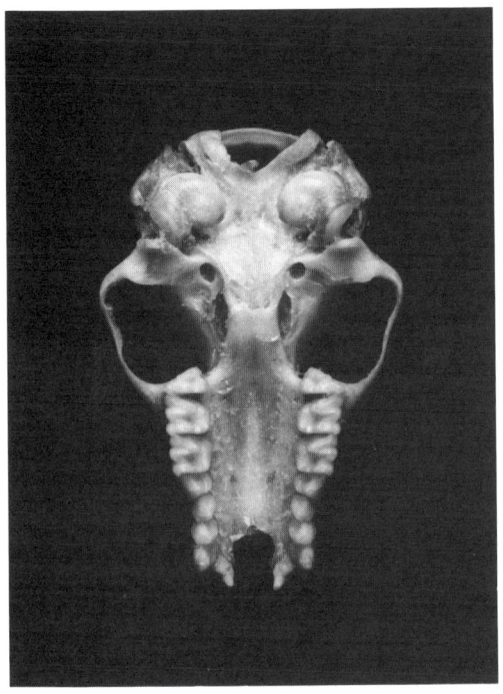

Fig. 145. *Kerivoula papillosa* – AMNH 103784 – ♀ (Borneo)

Fig. 146. *Myotis nigricans* – AMNH 175725 – ♀ (Trinidad)

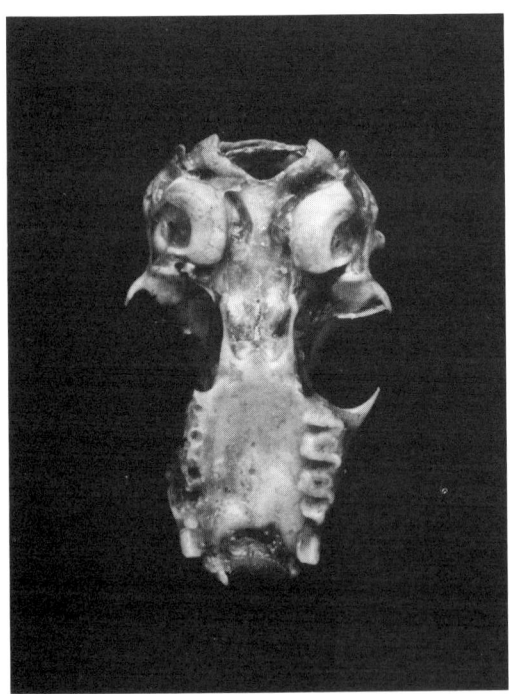

Fig. 147. *Lasionycteris noctivagans* – AMNH 40679 – ♂ (U. S. A.: Oregon)

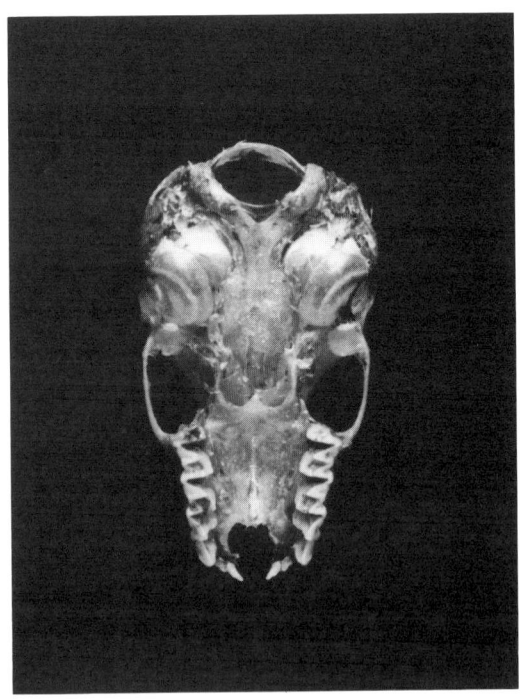

Fig. 148. *Barbastella leucomelas* – AMNH 245382 – ♀ (U. S. S. R.: Uzbekistan)

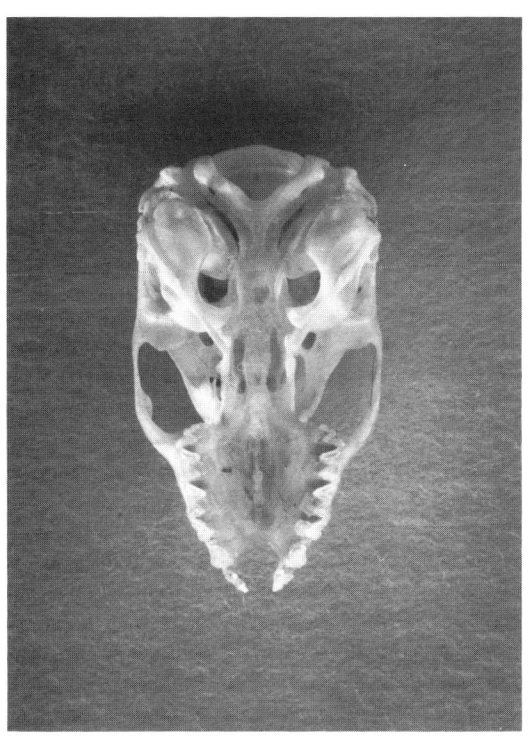

Fig. 149. *Euderma maculatum* – CM 42752 – ♀ (U. S. A.: Utah)

Fig. 150. *Plecotus townsendii* – AMNH 166984 – ♀ (Mexico)

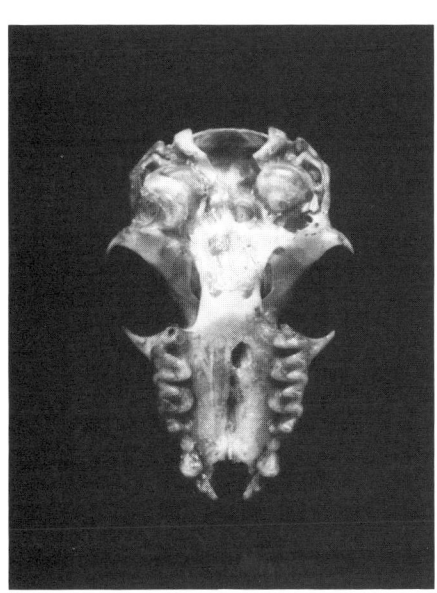

Fig. 151. *Eudiscopus denticulus* – AMNH 54789 – ♀ (Burma)

Fig. 152. *Pipistrellus subflavus* – KU 29882 – ♂ (Mexico)

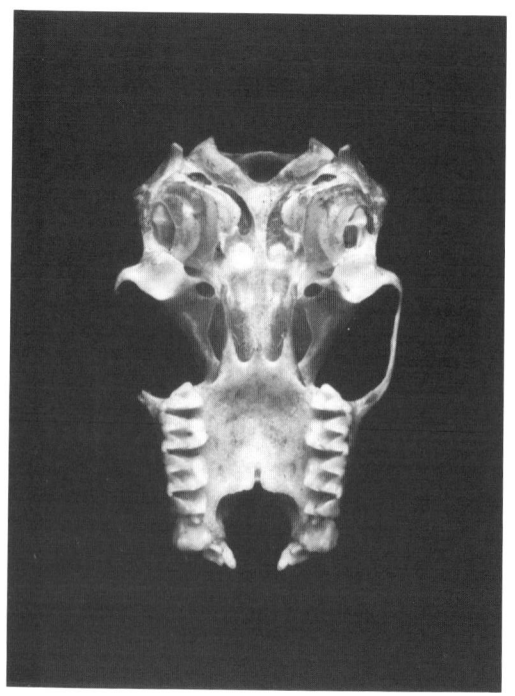

Fig. 153. *Nyctalus noctula* – AMNH ?33640 – ?♂ (England)

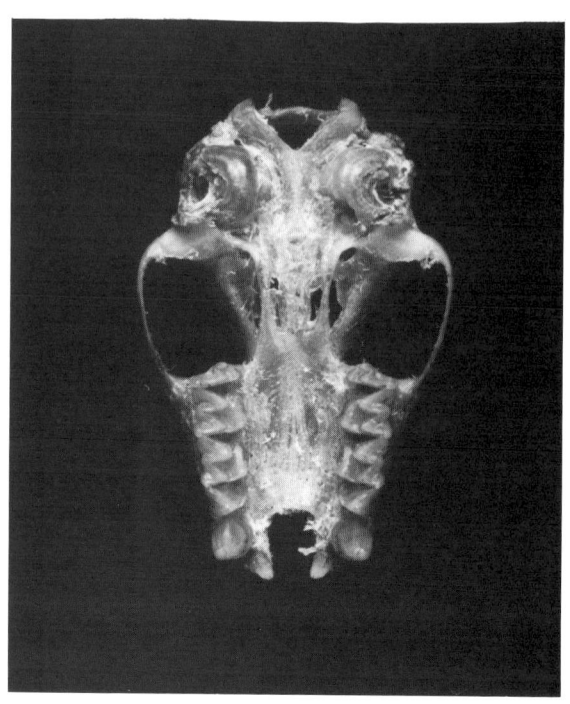

Fig. 154. *Ia io* – AMNH 113693 – unsexed (China: Sichuan)

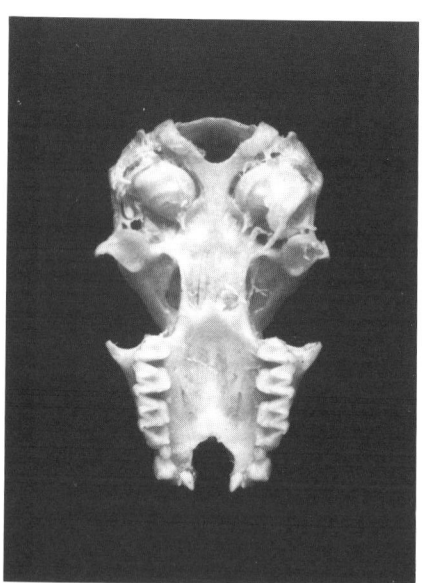

Fig. 155. *Glischropus tylopus* – AMNH 103803 – ♀ (Borneo)

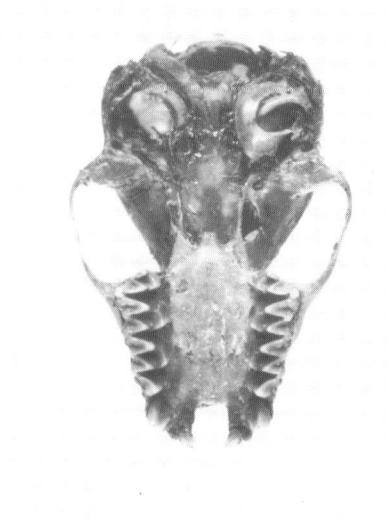

Fig. 156. *Eptesicus brasiliensis* – MCZ 11267 – ♂ (Tobago)

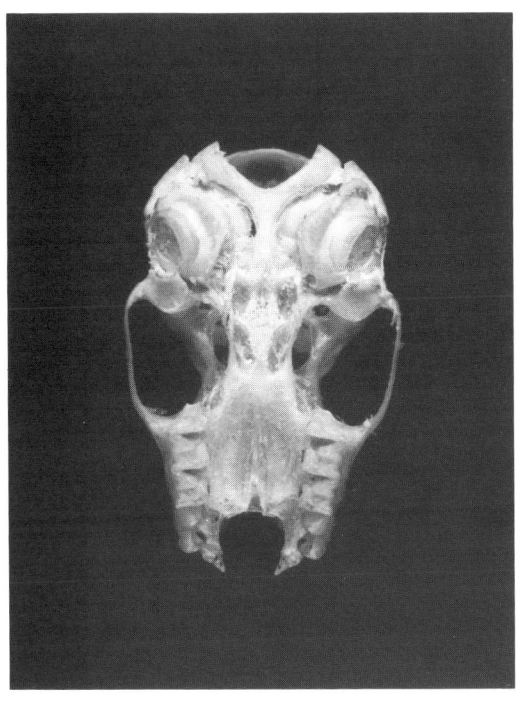

Fig. 157. *Vespertilio murinus* – AMNH 217031 – ♂ (Switzerland)

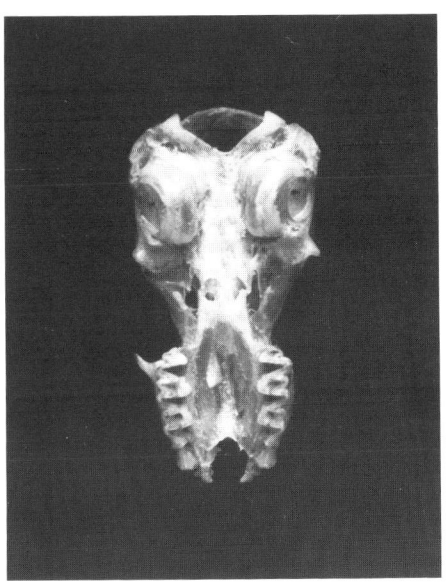

Fig. 158. *Laephotis angolensis* – AMNH 87244 – ♂ (Angola)

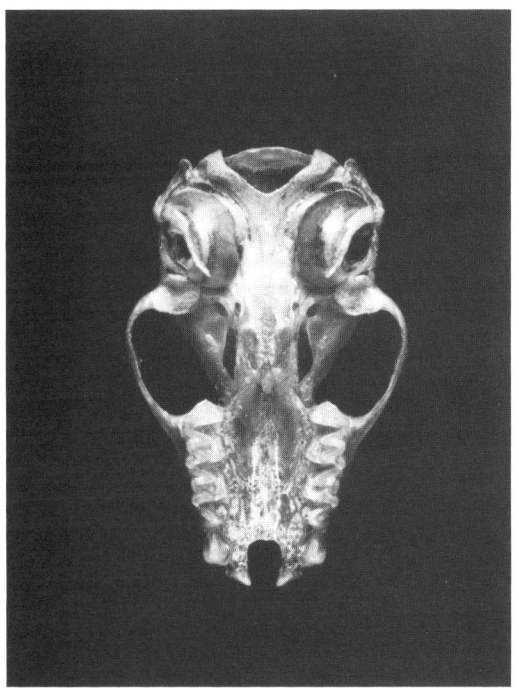

Fig. 159. *Histiotus montanus* – AMNH 188781 – ♀ (Uruguay)

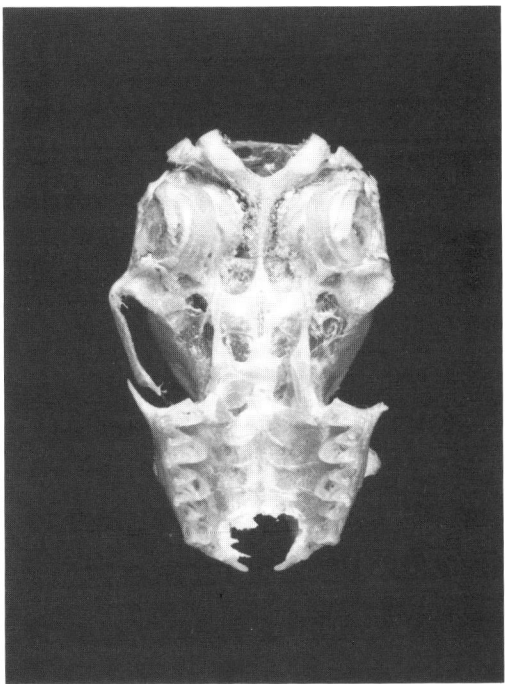

Fig. 160. *Philetor brachypterus* – AMNH 247525 – ♀ (Malaya)

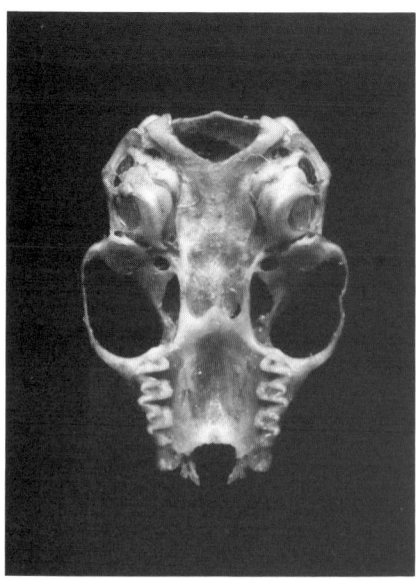

Fig. 161. *Tylonycteris robustula* – AMNH 216980 – ♂ (Malaya)

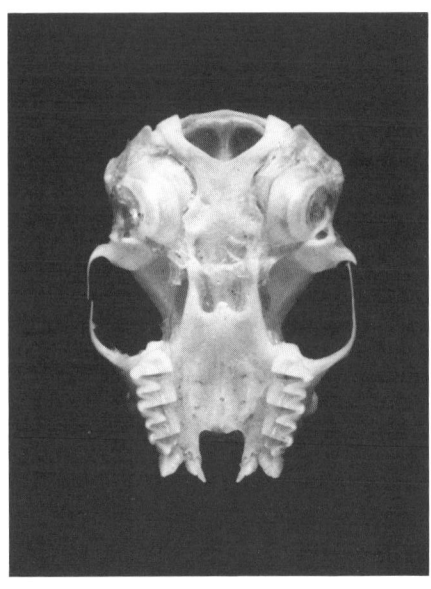

Fig. 162. *Mimetillus moloneyi* – AMNH 115851 – ♂ (Zambia)

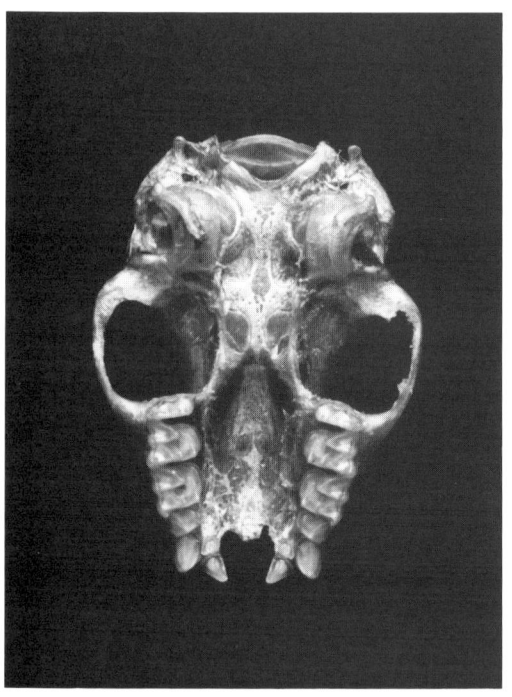

Fig. 163. *Hesperoptenus gaskelli* – AMNH 226785 – ♂ (Celebes)

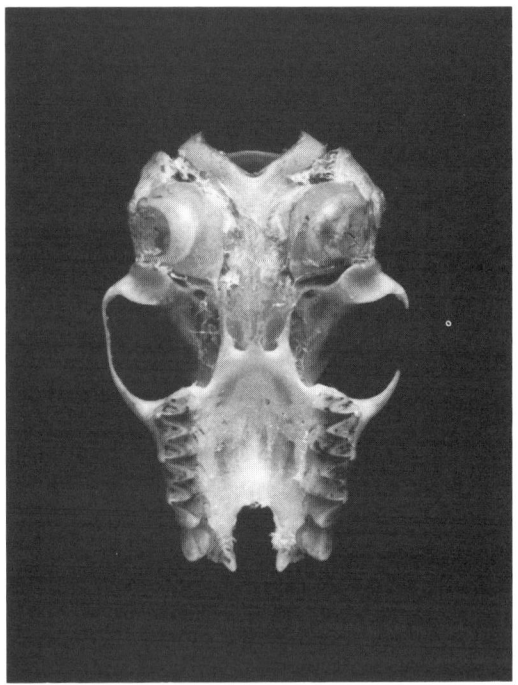

Fig. 164. *Chalinolobus gouldii* – AMNH 160283 – ♀ (Australia: S. Australia)

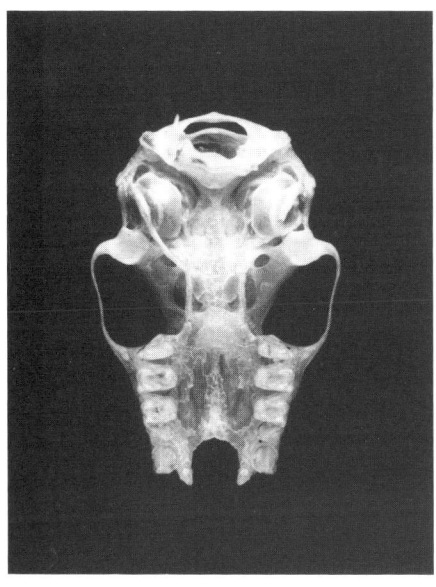 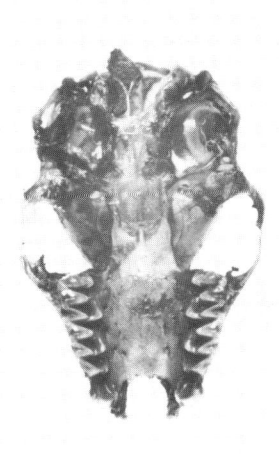

Fig. 165. *Nycticeius humeralis* – AMNH 219890 – ♀ (U. S. A.: Florida)

Fig. 166. *Rhogeesa tumida* – AMNH 69968 – ♂ (Venezuela)

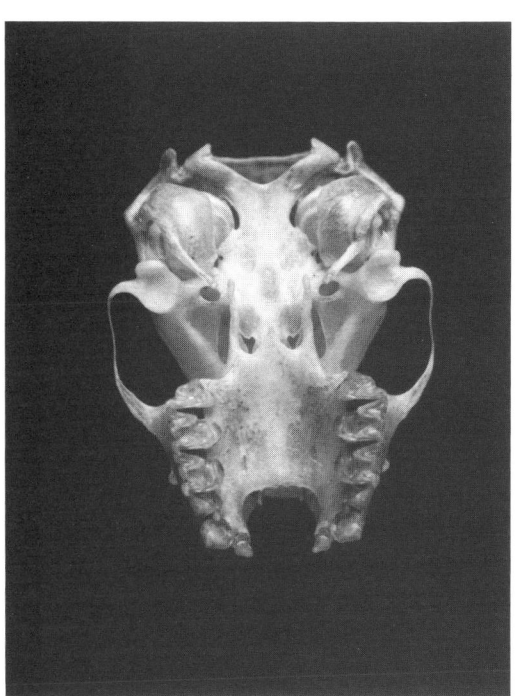 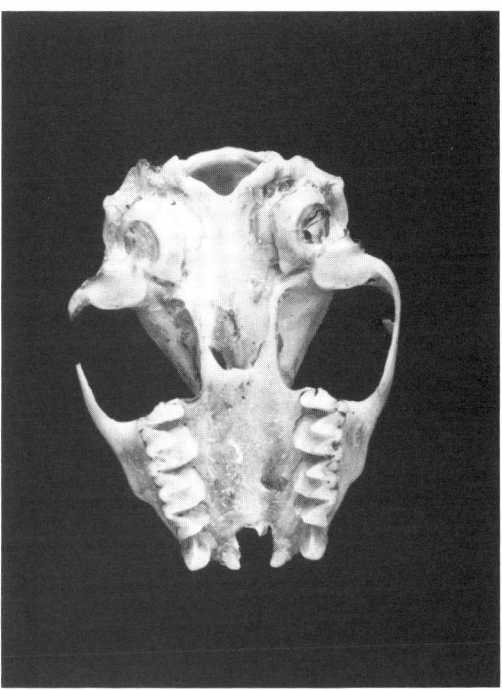

Fig. 167. *Scotoecus hirundo* – AMNH 241055 – ♀ (Cameroon)

Fig. 168. *Scotomanes ornatus* – AMNH 119481 – unsexed (China: Sichuan)

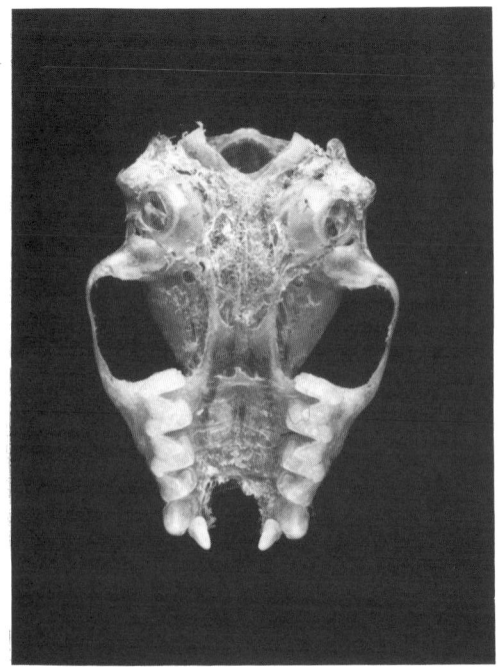
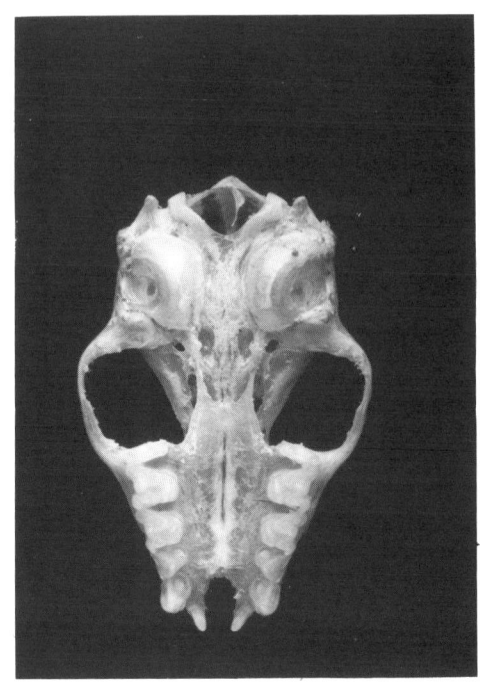

Fig. 169. *Scotophilus kuhlii* – AMNH 242305 – ♂ (Celebes)

Fig. 170. *Otonycteris hemprichi* – AMNH 212071 – ♂ (Pakistan)

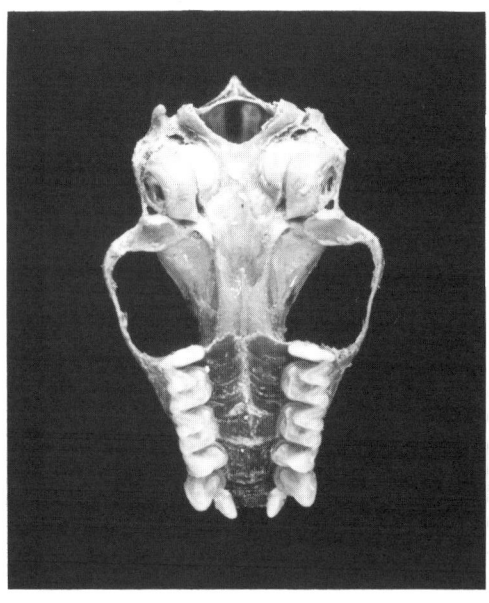

Fig. 171. *Lasiurus borealis* – AMNH 175719 – ♂ (Trinidad)

Fig. 172. *Bauerus dubiaquercus* – AMNH 256833 – ♂ (Belize)

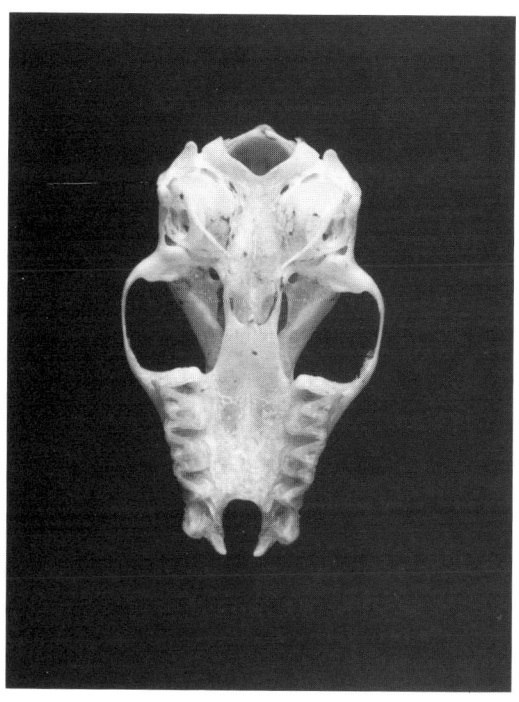

Fig. 173. *Antrozous pallidus* – AMNH 138 324 – ♀ (U. S. A.: California)

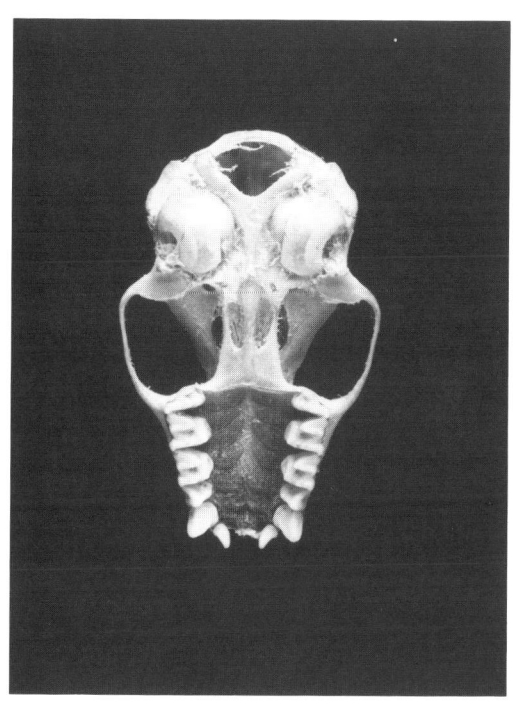

Fig. 174. *Nyctophilus gouldi* – AMNH 108 862 – ♀ (Papua New Guinea)

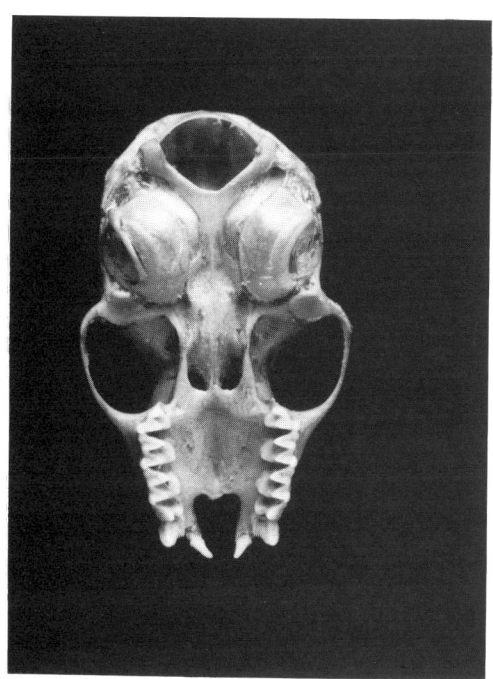

Fig. 175. *Pharotis imogene* – AMNH 160 266 – ♀ (Papua New Guinea)

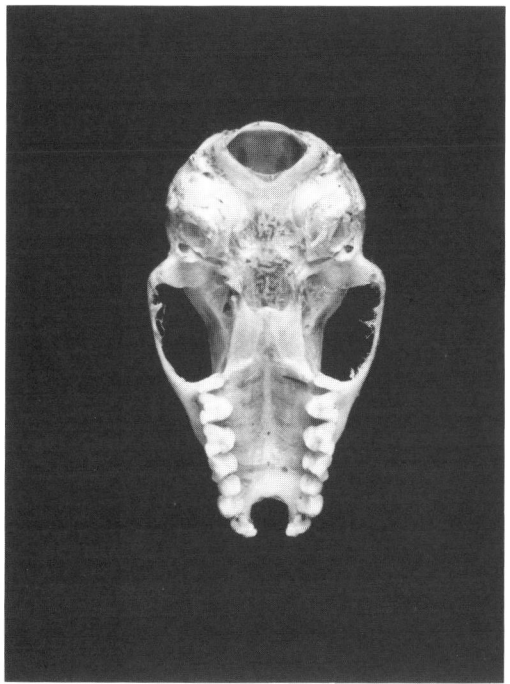

Fig. 176. *Murina leucogaster* – AMNH 244 272 – unsexed (U. S. S. R.: Primorsk)

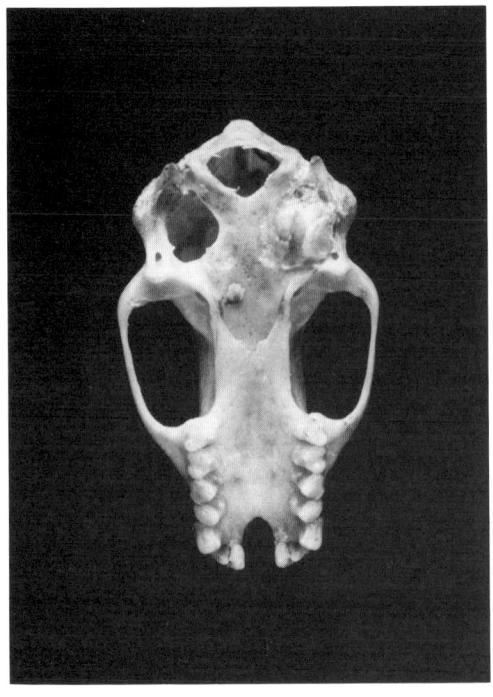

Fig. 177. *Harpiocephalus harpia* – AMNH 208656 – ♂ (India: Assam)

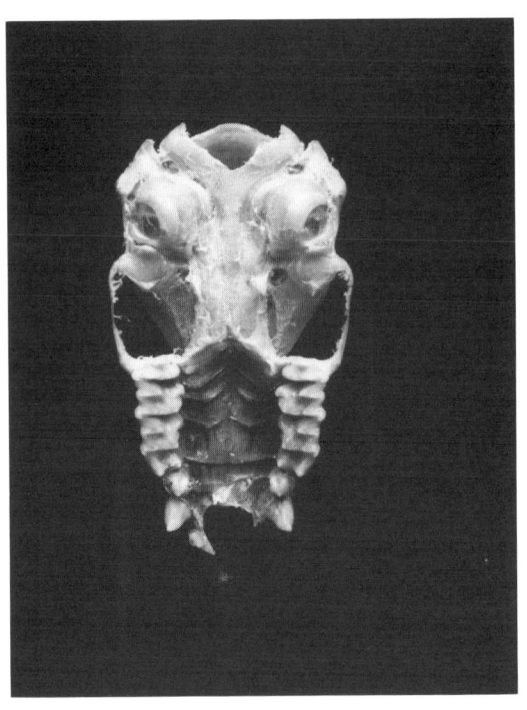

Fig. 178. *Miniopterus magnater* – AMNH 192956 – ♂ (Papua New Guinea)

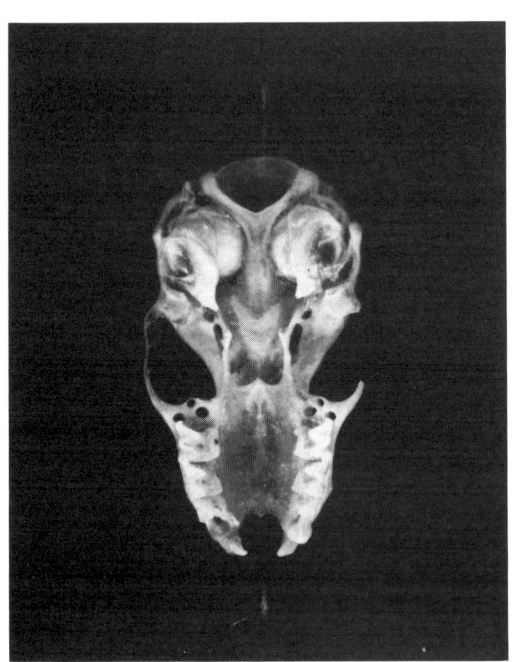

Fig. 179. *Tomopeas ravus* – USNM 103930 – ♀ (Peru)

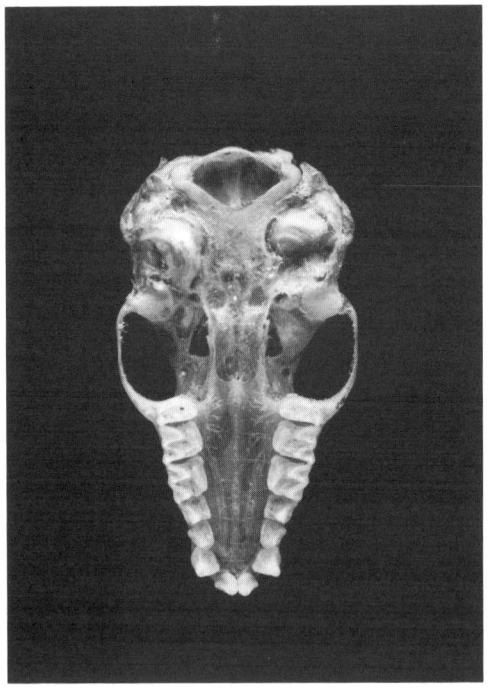

Fig. 180. *Mystacina robusta* – AMNH 214243 – ♂ (New Zealand)

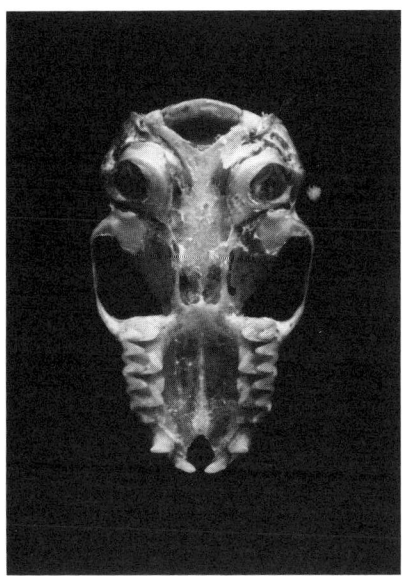

Fig. 181. *Mormopterus kalinowskii* – AMNH 165625 – ♂ (Peru)

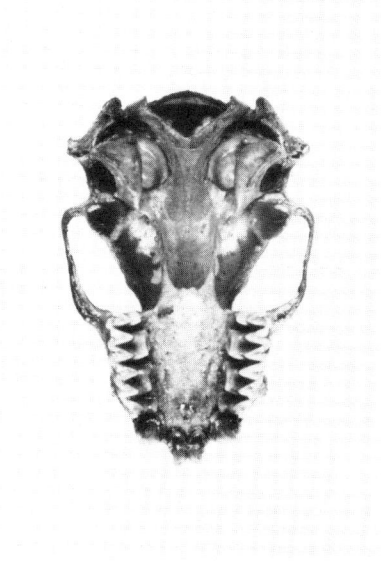

Fig. 182. *Molossops greenhalli* – AMNH 175326 – ♂ (Trinidad)

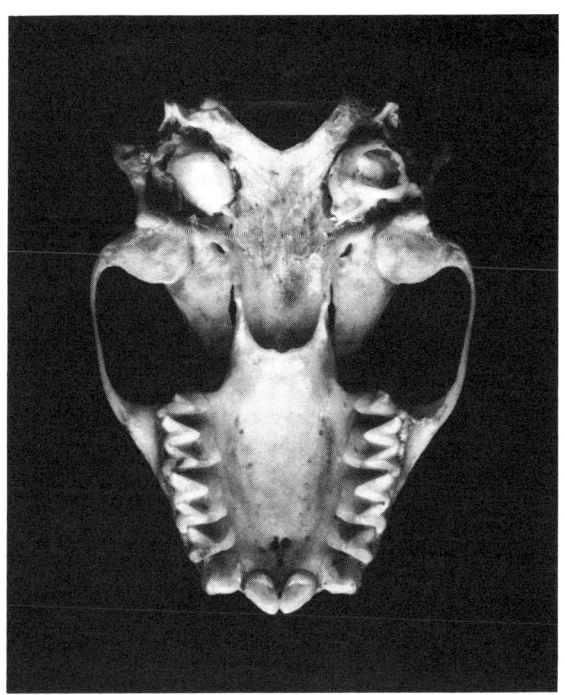

Fig. 183. *Cheiromeles torquatus* – AMNH 103922 – ♀ (Borneo)

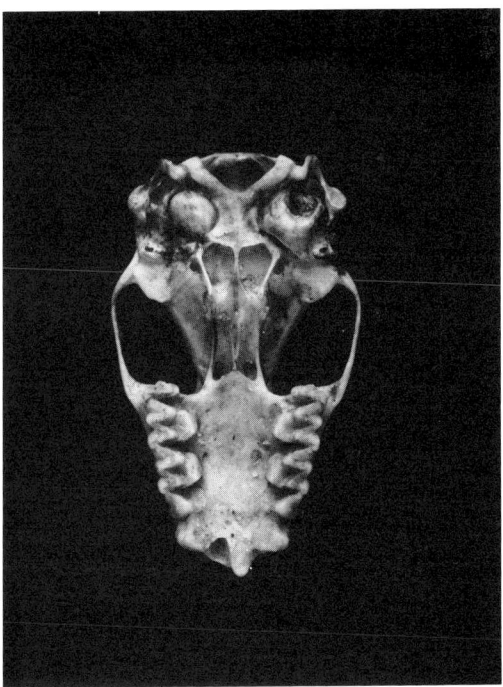

Fig. 184. *Myopterus daubentonii* – AMNH 48854 – ♀ (Zaire)

Fig. 185. *Tadarida brasiliensis* – USNM 102073 – ♂ ("Tobago")

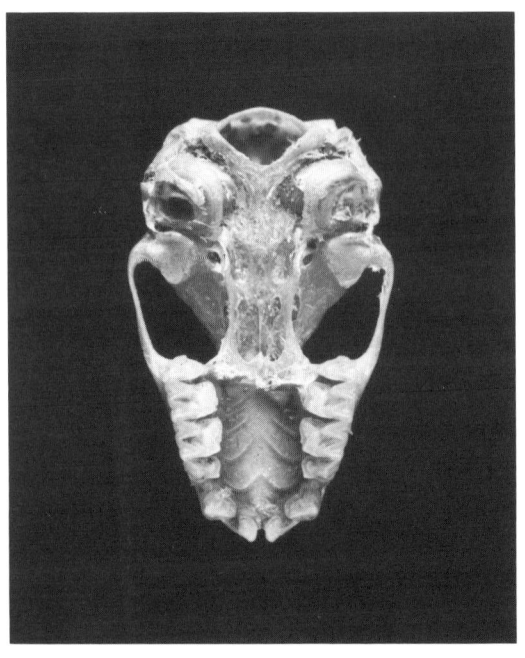

Fig. 186. *Chaerephon jobensis* – AMNH 216689 – ♂ (Australia: W. Australia)

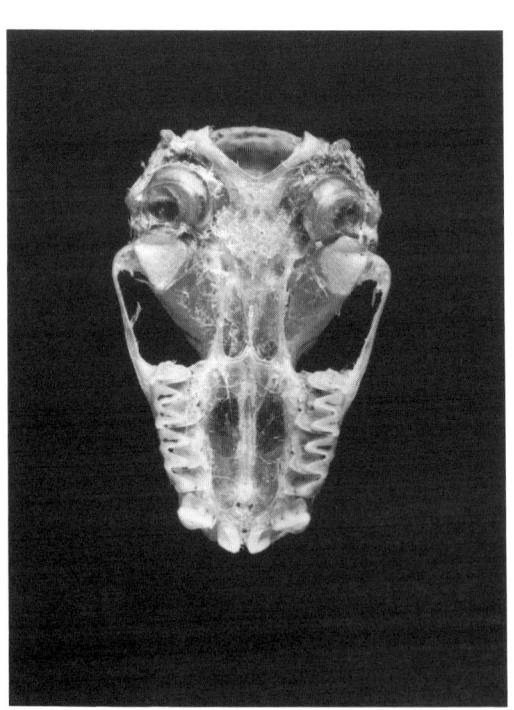

Fig. 187. *Mops condylurus* – AMNH 219805 – ♂ (Uganda)

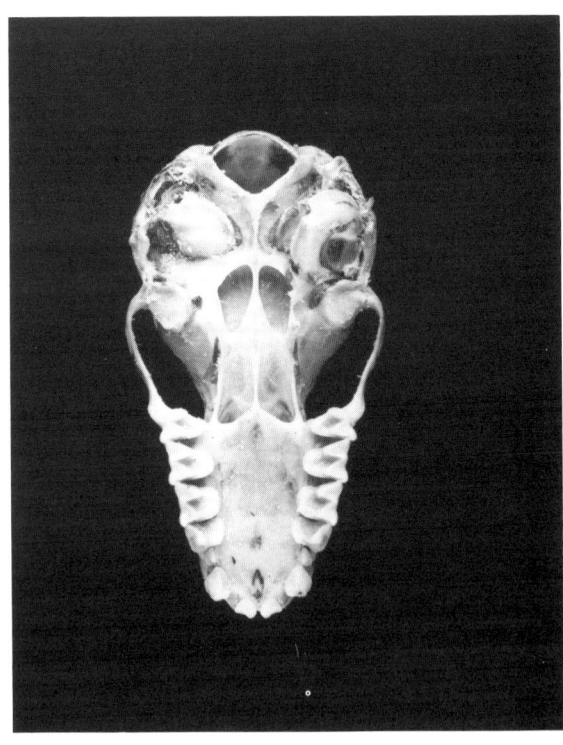

Fig. 188. *Otomops martiensseni* – AMNH 172858 – ♂ (S. Africa: Natal)

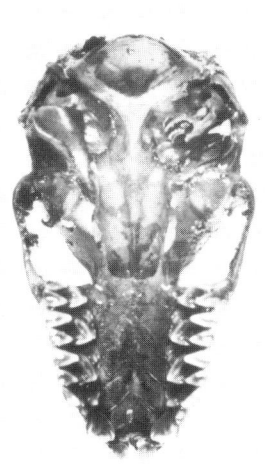

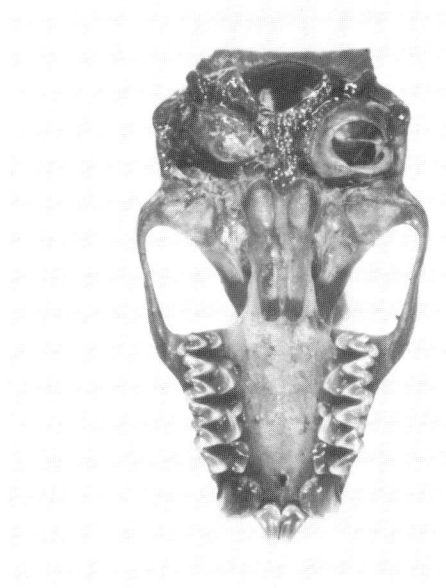

Fig. 189. *Nyctinomops laticaudatus* – AMNH 179963 – ♀ (Trinidad)

Fig. 190. *Eumops underwoodi* – AMNH 126862 – ♂ (Mexico)

Fig. 191. *Promops centralis* – AMNH 175652 – ♀ (Trinidad)

Fig. 192. *Molossus molossus* – AMNH 179987 – ♂ (Trinidad)

I am indebted to several museums for photographs of skulls or loan of skulls to have photographs made by our Photography Studio. These are (with their acronyms and the names of those who gave me assistance): British Museum (Natural History) (BM – J. E. Hill), Carnegie Museum (CM – D. A. Schlitter and S. B. McLaren), Los Angeles County Museum (LACM – L. J. Barkley), Museum of Comparative Zoology (MCZ – M. Rutzmoser and M. Massaro), National Museum of Natural History (USNM – D. E. Wilson and R. D. Fisher), Royal Ontario Museum (ROM – J. Eger), University of Michigan (UMMZ – P. Myers).

I especially want to thank Mr. John E. Hill, who not only arranged to have photographs made of the skulls of several British Museum specimens, but also provided a photograph of the type of *Latidens salimalii* which had been borrowed from the Bombay Natural History Society. I am also indebted to him for much stimulating discussion of bat systematics, which cleared up many obscure points. Dr. Guy Musser and the entire staff of the Mammal Department here at the American Museum of Natural History (AMNH) have, as usual, been very supportive.

Figs. 1, 3–6 are from Hill and Smith (1984). Figs. 2, 7–19, are from Anderson and Jones, 1984, Orders and Families of Recent Mammals of the World, John Wiley and Sons, New York, a copyrighted work, and is used with permission. Fig. 2 (in part) originally appeared in Slaughter and Walton (eds), 1970, About Bats: A Chiropteran Symposium, Southern Methodist University Press, Dallas, Texas, and is used with permission.

Table 1. Characters of Families

	A	B	C	D	E	F	G	H	I	J	K	L	M	N	O	P	Q
1	−	−	−	+	−	−	−	+	±	±	−	−	−	−	±	−	+
2	−	−	+	+	+	+	+	+	+	+	+	+	+	+	+	+	+
3	−	−	−	−	−	−	−	−	−	−	−	+	−	−	−	−	−
4	±	+	+	+	+	+	+	+	+	+	+	+	+	+	+	+	+
5	±	−	+	+	−	+	+	+	+	±	−	−	−	−	−	+	−
6	−	−	−	−	+	+	+	+	+	±	+	+	+	+	+	+	−
7	−	−	+	−	−	+	−	−	+	+	+	+	+	+	+	+	+
8	−	−	+	−	+	+	+	−	−	−	−	−	+	−	−	−	+
9	+	−	−	+	+	+	−	−	−	−	−	−	−	−	−	−	−
10	−	−	−	+	+	−	+	−	−	−	−	−	−	−	−	−	−
11	+	−	−	−	−	+	−	−	−	−	−	−	−	−	+	−	±
12	±	−	−	±	−	−	+	+	+	+	+	+	+	+	+	+	+
13	±	+	+	±	+	+	+	−	−	−	−	−	−	−	−	−	−
14	−	+	+	−	+	+	+	+	+	+	+	+	+	+	+	+	+
15	+	+	+	±	−	+	+	+	+	+	−	−	−	−	±	+	+
16	±	+	+	+	+	+	+	+	+	±	−	+	−	−	±	+	+
17	+	−	−	−	−	−	−	−	±	−	−	−	−	−	−	−	−
18	+	−	−	−	−	−	−	−	±	−	−	−	−	−	−	−	−
19	±	+	−	−	+	+	+	−	+	+	−	+	−	−	±	−	−
20	−	−	−	−	−	−	−	±	−	+	+	+	+	±	−	−
21	+	−	−	−	−	−	+	−	−	−	−	−	−	−	−	−	−

A (Pteropodidae)
B (Rhinopomatidae)
C (Craseonycteridae)
D (Emballonuridae)
E (Nycteridae)
F (Megadermatidae)
G (Rhinolophidae)
H (Noctilionidae)
I (Mormoopidae)
J (Phyllostomidae)
K (Natalidae)
L (Furipteridae)
M (Thyropteridae)
N (Myzopodidae)
O (Vespertilionidae)
P (Mystacinidae)
Q (Molossidae)

1 (wings lengthened)
2 (second digit of wing reduced)
3 (claw on 1st manal digit lost)
4 (claw on 2nd manal digit lost)
5 (tail reduced or lost)
6 (interfemoral membrane extensive)
7 (trochiter markedly enlarged)
8 (significant rib and/or vertebral fusion)
9 (postorbital processes present)
10 (nasal branch of premaxillary lost)
11 (palatal branch of premaxillary lost)
12 (premaxillaries fused to maxillaries)
13 (premaxillaries loose and movable)
14 (periotic bone free from remainder of skull)
15 (incisor number reduced)
16 (premolar number reduced)
17 (molar number reduced)
18 (molar pattern modified as an adaptation for fruit eating)
19 (rhinarium modified to form some kind of noseleaf-like structure)
20 (ear pinna more or less funnel-shaped)
21 (tragus absent).

TABLE 2

Dental formulae of genera of bats. Only "normal" formulae are given, so some individual specimens may exhibit formulae not listed for their genera. If, however, a variant is common in at least one species of a genus, that genus is listed for both formulae with the addition of "(part)".

i2/3 c1/1 p3/3 m3/3 x 2 = 38 – *Natalus; Thyroptera; Myzopoda; Kerivoula, Myotis* (part)

i2/3 c1/1 p2/3 m3/3 x 2 = 36 – *Furipterus, Amorphochilus; Lasionycteris, Plecotus, Eudiscopus, Miniopterus*

i2/2 c1/1 p3/3 m2/3 x 2 = 34 – *Eidolon, Rousettus, Myonycteris* (part), *Pteropus, Acerodon, Pteralopex, Eonycteris* (part), *Megaloglossus, Macroglossus, Syconycteris, Melonycteris* (part)

i2/3 c1/1 p2/2 m3/3 x 2 = 34 – *Emballonura; Myotis* (part) *Barbastella, Euderma, Pipistrellus* (part), *Nyctalus, Io, Glischropus, Chalinolobus* (part), *Murina, Harpiocephalus*

i2/2 c1/1 p2/3 m3/3 x 2 = 34 – *Pteronotus, Mormoops; Micronycteris* (part), *Macrotus, Lonchorhina, Macrophyllum, Phylloderma, Trachops, Vampyrum, Lionycteris, Lonchophylla, Platalina, Glossophaga, Monophyllus*

i2/2 c1/1 p3/3 m2/2 x 2 = 32 – *Myonycteris* (part), *Eonycteris* (part)

i1/2 c1/1 p3/3 m2/3 x 2 = 32 – *Boneia*

i2/2 c1/1 p2/3 m2/3 x 2 = 32 – *Neopteryx*

i2/2 c1/1 p3/3 m1/3 x 2 = 32 – *Plerotes*

i2/1 c1/1 p3/3 m2/3 x 2 = 32 – *Melonycteris* (part)

i1/3 c1/1 p2/2 m3/3 x 2 = 32 – *Coleura, Rhynchonycteris, Saccopteryx, Centronycteris, Peropteryx, Cormura, Balantiopteryx, Cyttarops, Diclidurus; Pipistrellus* (part), *Scotoecus* (part), *Lasiurus* (part); *Tadarida* (part)

i2/3 c1/1 p1/2 m3/3 x 2 = 32 – *Nycteris; Pipistrellus* (part), *Eptesicus, Vespertilio, Laephotis, Histiotus, Philetor, Tylonycteris, Mimetillus, Hesperoptenus, Chalinolobus* (part)

i1/2 c1/1 p2/3 m3/3 x 2 = 32 – *Rhinolophus*

i2/1 c1/1 p2/3 m3/3 X 2 = 32 – *Micronycteris* (part), *Tonatia, Chrotopterus*

i2/2 c1/1 p2/2 m3/3 x 2 = 32 – *Phyllostomus, Brachyphylla, Erophylla, Phyllonycteris, Carollia, Rhinophylla, Sturnira* (part), *Uroderma, Vampyrops, Artibeus* (part), *Ardops, Phyllops, Stenoderma, Ametrida, Sphaeronycteris*

i2/0 c1/1 p3/3 m3/3 x 2 = 32 – *Anoura*

i2/1 c1/1 p3/3 m2/2 x 2 = 30 – *Styloctenium, Balionycteris*

i1/1 c1/1 p3/3 m2/3 x 2 = 30 – *Harpyionycteris*

i2/2 c1/1 p3/3 m1/2 X 2 = 30 – *Cynopterus, Chironax, Thoopterus, Sphaerias*

i1/2 c1/1 p2/2 m3/3 x 2 = 30 – *Taphozous, Saccolaimus; Hipposideros, Anthops, Aselliscus, Rhinonycteris, Cloeotis, Triaenops, Coelops, Paracoelops; Mormopterus* (part), *Molossops* (part), *Tadarida* (part), *Chaerephon, Mops* (part), *Otomops, Nyctinomops, Eumops* (part), *Promops*

i2/1 c1/1 p2/2 m3/3 x 2 = 30 – *Mimon, Sturnira* (part)

i2/2 c1/1 p2/3 m2/2 x 2 = 30 – *Leptonycteris*

i2/0 c1/1 p2/3 m3/3 x 2 = 30 – *Hylonycteris, Scleronycteris, Choeroniscus, Choeronycteris*

i2/2 c1/1 p2/2 m2/3 x 2 = 30 – *Vampyrodes, Vampyressa* (part), *Mesophylla, Artibeus* (part), *Ariteus*

i1/3 c1/1 p1/2 m3/3 x 2 = 30 – *Nycticeius, Rhogeesa, Scotoecus* (part), *Scotomanes, Scotophilus, Otonycteris, Lasiurus* (part), *Bauerus, Nyctophilus, Pharotis; Mormopterus* (part)

i1/1 c1/1 p2/3 m2/3 x 2 = 28 – *Dobsonia*

i2/2 c1/1 p2/3 m1/2 x 2 = 28 – *Hypsignathus, Epomops, Epomophorus, Micropteropus, Nanonycteris, Scotonycteris, Casinycteris, Dyacopterus*

i2/1 c1/1 p3/3 m1/2 x 2 = 28 – *Ptenochirus, Megaerops, Aethalops, Penthetor, Notopteris*

i1/2 c1/1 p1/2 m3/3 x 2 = 28 – *Rhinopoma; Craseonycteris; Asellia; Antrozous, Tomopeas; Mormopterus* (part), *Molossops* (part), *Mops* (part), *Eumops* (part)

i0/2 c1/1 p2/2 m3/3 x 2 = 28 – *Megaderma*

i2/1 c1/1 p1/2 m3/3 x 2 = 28 – *Noctilio*

i2/2 c1/1 p2/2 m2/2 x 2 = 28 – *Vampyressa* (part), *Chiroderma, Ectophylla, Artibeus* (part), *Pygoderma, Centurio*

i1/1 c1/1 p2/2 m3/3 x 2 = 28 – *Mystacina; Mops* (part)

i1/0 c1/1 p2/3 m2/3 x 2 = 26 – ? *Aproteles*

i1/1 c1/1 p3/3 m1/2 x 2 = 26 – *Latidens, Alionycteris*

i0/2 c1/1 p1/2 m3/3 x 2 = 26 – *Macroderma, Cardioderma, Lavia*

i2/0 c1/1 p2/3 m2/2 x 2 = 26 – *Lichonycteris*

i2/1 c1/1 p2/2 m2/2 x 2 = 26 – *Vampyressa* (part)

i2/2 c1/1 p1/2 m2/2 x 2 = 26 – *Diphylla*

i1/1 c1/1 p1/2 m3/3 x 2 = 26 – *Molossops* (part), *Myopterus, Cheiromeles, Molossus*

i1/1 c1/1 p3/3 m1/1 x 2 = 24 – *Otopteropus, Haplonycteris*

i1/0 c1/1 p3/3 m1/2 x 2 = 24 – *Paranyctimene, Nyctimene*

i1/2 c1/1 p1/2 m2/1 x 2 = 22 – *Diaemus*

i1/2 c1/1 p1/2 m1/1 x 2 = 20 – *Desmodus*

TABLE 3

Forearm lengths for the genera of bats arranged according to size classes.

21–25 mm. – *Craseonycteris; Ametrida; Kerivoula; Myotis, Pipistrellus, Tylonycteris, Hesperoptenus, Rhogeesa; Molossops*

26–30 mm. – *Craseonycteris; Cloeotis; Lichonycteris, Vampyressa, Ectophylla, Mesophylla, Ametrida; Natalus; Kerivoula, Myotis, Pipistrellus, Glischropus, Eptesicus, Tylonycteris, Mimetillus, Hesperoptenus, Nycticeius, Rhogeesa, Scotoecus, Nyctophilus, Murina; Mormopterus, Molossops, Mops*

31–35 mm. – *Emballonura, Rhynchonycteris, Saccopteryx, Balantiopteryx; Nycteris; Rhinolophus, Hipposideros, Cloeotis, Coelops; Pteronotus; Micronycteris, Macrophyllum, Tonatia, Lionycteris, Lonchophylla, Glossophaga, Monophyllus, Lichonycteris, Anoura, Hylonycteris, Scleronycteris, Choeroniscus, Carollia, Rhinophylla, Sturnira, Vampyrops, Vampyressa, Mesophylla, Artibeus, Ametrida; Natalus; Furipterus, Amorphochilus; Thyroptera; Kerivoula, Myotis, Plecotus, Eudiscopus, Pipistrellus, Glischropus, Eptesicus, Laephotis, Philetor, Tylonycteris, Chalinolobus, Nycticeius, Rhogeesa, Scotoecus, Nyctophilus, Murina, Miniopterus, Tomopeas; Mormopterus, Molossops, Myopterus, Tadarida, Chaerophon, Mops, Molossus*

36–40 mm. – *Balionycteris, Megaloglossus, Macroglossus, Syconycteris, Emballonura, Rhynchonycteris, Saccopteryx, Peropteryx, Balantiopteryx; Nycteris; Rhinolophus, Hipposideros, Aselliscus, Asellia, Cloeotis, Coelops; Pteronotus; Micronycteris, Macrophyllum, Tonatia, Lionycteris, Lonchophylla, Glossophaga, Monophyllus, Lichonycteris, Anoura, Hylonycteris, Choeroniscus, Choeronycteris, Carollia, Sturnira, Uroderma, Vampyrops, Vampyressa, Chiroderma, Artibeus, Phyllops, Ariteus, Pygoderma, Sphaeronycteris; Natalus; Furipterus, Amorphochilus; Thyroptera; Kerivoula, Myotis, Lasionycteris, Barbastella, Plecotus, Eudiscopus, Pipistrellus, Nyctalus, Eptesicus, Vespertilio, Laephotis, Phi-*

letor, Hesperoptenus, Chalinolobus, Nycticeius, Scotoecus, Lasiurus, Nyctophilus, Pharotis, Murina, Miniopterus; Mystacina; Mormopterus, Molossops, Myopterus, Tadarida, Chaerephon, Mops, Nyctinomops, Eumops, Molossus

41–45 mm. – *Nanonycteris, Scotonycteris, Balionycteris, Chironax, Aethalops, Alionycteris, Magaloglossus, Macroglossus, Syconycteris, Melonycteris; Rhinopoma; Emballonura, Coleura, Rhynchonycteris, Saccopteryx, Centronycteris, Peropteryx, Cormura, Balantiopteryx; Nycteris; Rhinolophus, Hipposideros, Aselliscus, Asellia, Rhinonycteris, Triaenops, Coelops, Paracoelops; Pteronotus, Mormoops; Micronycteris, Lonchorhina, Tonatia, Mimon, Lonchophylla, Erophylla, Phyllonycteris, Glossophaga, Monophyllus, Anoura, Choeroniscus, Carollia, Sturnira, Uroderma, Vampyrops, Chiroderma, Artibeus, Ardops, Phyllops, Ariteus, Pygoderma, Sphaeronycteris, Centurio; Natalus; Kerivoula, Myotis, Lasionycteris, Barbastella, Plecotus, Pipistrellus, Nyctalus, Eptesicus, Vespertilio, Laephotis, Histiotus, Hesperoptenus, Chilinolobus, Scotophilus, Lasiurus, Antrozous, Nyctophilus, Murina, Harpiocephalus, Miniopterus; Mystacina; Mormopterus, Molossops, Myopterus, Tadarida, Chaerephon, Mops, Nyctinomops, Eumops, Promops, Molossus*

46–50 mm. – *Micropteropus, Nanonycteris, Scotonycteris, Casinycteris, Megaerops, Chironax, Sphaerias, Aethalops, Alionycteris, Otopteropus, Haplonycteris, Paranyctimene, Nyctimene, Megaloglossus, Macroglossus, Syconycteris, Melonycteris; Rhinopoma; Emballonura, Coleura, Saccopteryx, Centronycteris, Peropteryx, Cormura, Balantiopteryx, Cyttarops; Nycteris; Lavia; Rhinolophus, Hipposideros, Anthops, Asellia, Rhinonycteris, Triaenops, Coelops; Pteronotus, Mormoops; Micronycteris, Macrotus, Lonchorhina, Tonatia, Mimon, Lonchophylla, Platalina, Erophylla, Phyllonycteris, Monophyllus, Leptonycteris, Anoura, Choeronycteris, Carollia, Sturnira, Uroderma, Vampyrops, Vampyrodes, Chiroderma, Artibeus, Ardops, Stenoderma, Centurio, Diaemus; Natalus; Myzopoda; Myotis, Euderma, Plecotus, Pipistrellus, Nyctalus, Eptesicus, Vespertilio, Histiotus, Hesperoptenus, Chalinolobus, Nycticeius, Scotomanes, Scotophilus, Lasiurus, Bauerus, Antrozous, Nyctophilus, Harpiocephalus, Miniopterus; Mystacina; Molossops, Myopterus, Tadarida, Chaerephon, Mops, Otomops, Nyctinomops, Eumops, Promops, Molossus*

51–55 mm. – *Myonycteris, Plerotes, Micropteropus, Nanonycteris, Scotonycteris, Casinycteris, Cynopterus, Megaerops, Sphaerias, Aethalops, Paranyctimene, Nyctimene, Macroglossus, Melonycteris; Rhinopoma; Emballonura, Coleura, Saccopteryx, Peropteryx, Taphozous, Diclidurus; Nycteris, Megaderma, Cardioderma, Lavia; Rhinolophus, Hipposideros, Anthops, Asellia, Triaenops; Noctilio; Pteronotus, Mormoops; Micronycteris, Macrotus, Lonchorhina, Tonatia, Mimon, Leptonycteris, Sturnira, Vampyrops, Vampyrodes, Chiroderma, Artibeus, Ardops, Stenoderma, Diphylla, Diaemus, Desmodus; Myotis, Euderma, Pipistrellus, Nyctalus, Eptesicus, Vespertilio, Histiotus, Hesperoptenus, Nycticeius, Scotomanes, Scotophilus, Otonycteris, Lasiurus, Bauerus, Antrozons, Harpiocephalus, Miniopterus; Myopterus, Tadarida, Chaerephon, Mops, Otomops, Nyctinomops, Eumops, Promops, Molossus*

56–60 mm. – *Myonycteris, Epomophorus, Micropteropus, Casinycteris, Cynopterus, Megaerops, Penthetor, Paranyctimene, Nyctimene, Syconycteris, Melonycteris, Notopteris; Rhinopoma; Coleura, Taphozous, Diclidurus; Nycteris; Megaderma, Cardioderma, Lavia; Rhinolophus, Hipposideros, Triaenops; Noctilio; Pteronotus, Mormoops; Micronycteris, Lonchorhina, Tonatia, Mimon, Phyllostomus, Trachops, Brachyphylla, Leptonycteris, Sturnira, Vampyrops, Chiroderma, Artibeus, Diphylla, Diaemus, Desmodus; Myotis, Nyctalus, Eptesicus, Hes-*

peroptenus, Nycticeius, Scotomanes, Scotophilus, Otonycteris, Lasiurus, Bauerus, Antrozous, Miniopterus; Myopterus, Tadarida, Mops, Otomops, Nyctinomops, Eumops, Promops

61–65 mm. – *Rousettus, Myonycteris, Epomophorus, Micropteropus, Casinycteris, Cynopterus, Ptenochirus, Penthetor, Nyctimene, Eonycteris, Melonycteris, Notopteris; Rhinopoma; Taphozous, Saccolaimus, Diclidurus; Nycteris; Megaderma, Lavia; Rhinolophus, Hipposideros, Triaenops; Noctilio; Pteronotus, Mormoops, Phyllostomus, Phylloderma, Trachops, Brachyphylla, Sturnira, Vampyrops, Artibeus, Desmodus; Myotis, Nyctalus, Hesperoptenus, Scotomanes, Scotophilus, Otonycteris, Lasiurus, Antrozous; Tadarida, Mops, Otomops, Nyctinomops, Eumops*

66–70 mm. – *Rousettus, Myonycteris, Epomophorus, Micropteropus, Cynopterus, Ptenochirus, Latidens, Nyctimene, Eonycteris, Notopteris; Rhinopoma; Taphozous, Saccolaimus, Diclidurus; Nycteris; Megaderma; Rhinolophus, Hipposideros; Noctilio; Pteronotus; Phyllostomus, Phylloderma, Brachyphylla, Artibeus; Myotis, Nyctalus, Scotophilus, Otonycteris; Tadarida, Mops, Otomops, Eumops*

71–75 mm. – *Rousettus, Myonycteris, Dobsonia, Epomophorus, Scotonycteris, Cynopterus, Ptenochirus, Thoopterus, Nyctimene, Eonycteris, Notopteris; Rhinopoma; Taphozous, Saccolaimus, Diclidurus; Megaderma; Rhinolophus, Hipposideros; Noctilio; Phyllostomus, Phylloderma, Artibeus; Myotis, Io, Scotophilus; Cheiromeles, Otomops, Eumops*

76–80 mm. – *Rousettus, Dobsonia, Epomops, Epomophorus, Scotonycteris, Cynopterus, Ptenochirus, Dyacopterus, Thoopterus, Nyctimene, Eonycteris, Taphozous, Saccolaimus; Rhinolophus, Hipposideros; Noctilio; Phyllostomus, Phylloderma, Chrotopterus; Io, Scotophilus; Cheiromeles, Eumops*

81–85 mm. – *Rousettus, Dobsonia, Harpyionycteris, Epomops, Epomophorus, Cynopterus, Ptenochirus, Dyacopterus, Nyctimene, Eonycteris, Saccolaimus; Hipposideros; Noctilio; Phyllostomus, Phylloderma, Chrotopterus; Scotophilus; Cheiromeles, Eumops*

86–90 mm. – *Rousettus, Pteropus, Styloctenium, Dobsonia, Harpyionycteris, Epomops, Epomophorus, Cynopterus, Ptenochirus, Dyacopterus, Nyctimene; Saccolaimus; Hipposideros; Noctilio; Phyllostomus; Scotophilus; Cheiromeles*

91–95 mm. – *Rousettus, Boneia, Pteropus, Styloctenium, Harpyionycteris, Epomops, Epomophorus, Dyacopterus; Saccolaimus; Hipposideros; Noctilio; Phyllostomus*

96–100 mm. – *Rousettus, Boneia, Pteropus, Styloctenium, Dobsonia, Epomops, Epomophorus, Hipposideros; Phyllostomus*

101–105 mm. – *Eidolon, Rousettus, Boneia, Pteropus, Dobsonia, Epomops; Macroderma; Hipposideros; Vampyrum*

106–110 mm. – *Eidolon, Pteropus, Neopteryx, Dobsonia; Macroderma; Hipposideros; Vampyrum*

111–115 mm. – *Eidolon, Pteropus, Dobsonia; Macroderma; Hipposideros; Vampyrum*

116–120 mm. – *Eidolon, Pteropus, Pteralopex, Dobsonia, Hypsignathus*

121–135 mm. – *Eidolon, Pteropus, Acerodon, Dobsonia, Hypsignathus*

136–140 mm. – *Pteropus, Acerodon, Pteralopex, Dobsonia, Hypsignathus*

141–145 mm. – *Pteropus, Acerodon, Pteralopex, Dobsonia*

146–155 mm. – *Pteropus, Acerodon, Dobsonia*

156–160 mm. – *Pteropus, Acerodon, Pteralopex, Dobsonia*

161–165 mm. – *Pteropus, Acerodon*

166–180 mm. – *Pteropus*

181–210 mm – *Pteropus, Accrodon*

211–220 mm. – *Pteropus*

Literature

AELLEN, V. (1952): Contribution à l'Etude des Chiroptères du Cameroun. – Mém. Soc. neuchateloise Sci. Nat., **8**: 1–121.
- (1954): Déscription d'un nouvel *Hipposideros* (Chiroptera) de la Côte d'Ivoire. – Rev. suisse. Zool., **61**: 473–483.
- (1957): Les Chiroptères africaines du Musée zoologique de Strassbourg. – Rev. suisse Zool., **64**: 189–214.
- (1959): Chiroptères nouveaux d'Afrique. – Arch. Sci. Genève, **12**: 217–235.
- (1966): Notes sur *Tadarida teniotis* (RAF.) (Mammalia, Chiroptera) I. Systématique, paléontologie et peuplement, répartition géographique. – Rev. suisse zool., **73**: 119–159.
- & A. BROSSET (1968): Chiroptères du sud du Congo (Brazzaville). – Rev. Suisse Zool., **75**: 435–458.

ANDERSEN, K. (1905a): On some Bats of the Genus *Rhinolophus*, with Remarks on their Mutual Affinities and Descriptions of Twenty-six new Forms. – Proc. Zool. Soc. London, **2**: 75–145.
- (1905b): On the Bats of the *Rhinolophus philippinensis* Group, with Descriptions of Five new Species. – Ann. Mag. nat. Hist., (7) **16**: 243–257.
- (1905c): On the Bats of the *Rhinolophus macrotis* Group, with Descriptions of Two new Forms. – Ann. Mag. nat. Hist., (7) **16**: 289–292.
- (1905d): A List of the Species and Subspecies of the Genus *Rhinolophus*, with some Notes on their Geographical Distribution. – Ann. Mag. nat. Hist., (7) **16**: 648–662.
- (1907): Chiropteran Notes. – Ann. Mus. civico Stor. Nat. Genova, (3) **3**: 1–41.
- (1912): Catalogue of the Chiroptera in the Collection of the British Museum, 2nd ed., vol. 1: Megachiroptera. – Trustees Brit. Mus., London.
- (1918): Diagnoses of new Bats of the Families Rhinolophidae and Megadermatidae. – Ann. Mag. nat. Hist., (9) **2**: 371–384.

BAKER, R. J. (1984): A sympatric cryptic species of mammal: A new species of *Rhogeesa* (Chiroptera: Vespertilionidae). – Syst. Zool., **33**: 178–183.
- & H. H. GENOWAYS (1976): A new species of *Chiroderma* from Guadeloupe, West Indies (Chiroptera: Phyllostomatidae). – Occ. Pap. Mus. Texas, Tech Univ., **39**: 1–9.

BAUD, J. F. (1979): *Myotis aelleni*, nov. spec., chauve-souris nouvelle d'Argentine (Chiroptera. Vespertilionidae). – Rev. suisse Zool., **86**: 267–278.

BEAUX, O. DE (1923): Mammiferi della Somalia Italiana. – Atti. Soc. Ital. Sci. nat., **62**: 247–316.

BERGMANS, W. (1975): A new species of *Dobsonia* PALMER 1898 (Mammalia, Megachiroptera) from Waigeo, with notes on other members of the genus. – Beaufortia, **23**: 1–13.
- (1975): On the differences between sympatric *Epomops franqueti* (TOMES 1860) and *Epomops buettikoferi* (MATSCHIE 1899), with additional notes on the latter species (Mammalia, Megachiroptera). – Beaufortia, **23**: 141–152.
- (1976): A revision of the African genus *Myonycteris* MATSCHIE 1899 (Mammalia, Megachiroptera). – Beaufortia, **24**: 189–216.
- (1978): On *Dobsonia* PALMER 1898 from the Lesser Sunda Islands (Mammalia: Megachiroptera). – Senckenbergiana biol., **59**: 1–18.
- (1979): Taxonomy and zoogeography of *Dobsonia* PALMER 1898, from the Louisiade Archipelago, the D'Entrecasteaux Group, Trobriand Island and Woodlark Island (Mammalia, Megachiroptera). – Beaufortia, **29**: 199–214.
- (1980): A new fruit bat of the genus *Myonycteris* MATSCHIE 1899, from eastern Kenya and Tanzania (Mammalia, Megachiroptera). – Zool. Mededelingen, **55**: 171–181.
- & J. E. HILL (1980): On a new species of *Rousettus* GRAY 1821, from Sumatra and Borneo (Mammalia: Megachiroptera). – Bull. Bril. Mus. nat. Hist., (Zool.) **38**: 95–104.
- & S. SARBINI (1985): Fruit bats of the genus *Dobsonia* PALMER 1898 from the islands of Biak, Owii, Numfoor, and Yapen, Irian Jaya (Mammalia, Megachiroptera). – Beaufortia, **34**: 181–189.

BHATNAGAR, K. P. (1980): The chiropteran vomeronasal organ: its relevance to the phylogeny of bats. – In D. E. WILSON & A. L. GARDNER (eds.), Proceedings Fifth International Bat Research Conference. – Texas Tech Press, Lubbock.

BÖHME, W. & R. HUTTERER (1979): Kommentierte Liste einer Säugetier-Aufsammlung aus dem Senegal. – Bonner zool. Beitr., **29**: 303–323.

BROSSET, A. (1984): Chiroptères d'altitude du Mont Nimba (Guinée). Déscription d'un espèce nouvelle, *Hipposideros lamottei*. – Mammalia, **48**: 545–555.

CABRERA, A. (1909): Un nuevo "*Rhinolophus*" filipino. – Bol. Hist. Nat., **9**: 304–306.
- (1958): Catalogo de los Mamiferos de America del Sur, I (Metatheria-Unguiculata-Carnivora). – Rev. Mus. Argentino Cienc. nat. "Bernardino Rivadavia", Cienc. zool, **4**: 1–307.

CAKENBERGHE, V. VAN & F. DE VREE (1985): Systematics of African *Nycteris*. – Proc. Internation. Symp. African Vertebr., Bonn: 53–90.

CHASEN, F. N. (1940): A handlist of Malaysian Mammals. – Bull. Raffles Mus., Singapore, **15**: 1–209.

CORBET, G. B. (1978): The Mammals of the Palearctic Region: a taxonomic review. – Trustees Brit. Mus., London.
- (1984): The mammals of the Palearctic Region: a taxonomic review. – Supplement, Trustees Brit. Mus., London.

DAVIS, W. B. (1966): Review of the South American Bats of the genus *Eptesicus*. – Southwestern Naturalist, **11**: 245–274.
- (1968): Review of the Genus *Uroderma* (Chiroptera). – J. Mammal., **49**: 676–698.
- (1969, 1970): A Review of the Small Fruit Bats (Genus *Artibeus*) of Middle America. – Southwestern Naturalist, **14**: 15–29, 389–402.
- (1973): Geographic Variation in the Fishing Bat, *Noctilio leporinus*. – J. Mammal., **54**: 862–874.
- (1976): Geographic Variation in the Lesser Noctilio, *Noctilio albiventris* (Chiroptera). – J. Mammal., **57**: 687–707.
- (1980): New *Sturnira* (Chiroptera: Phyllostomidae) from Central and South America, with Key to Currently Recognized Species. – Occ. Pap. Mus. Texas Tech Univ., **70**: 1–5.

- (1984): Review of the Large Fruit-eating Bats of the *Artibeus "lituratus"* complex (Chiroptera: Phyllostomidae) in Middle America. - Occ. Pap. Mus. Texas Tech Univ., **93**: 1-16.
- & D. C. CARTER (1962): Review of the Genus *Leptonycteris* (Mammalia: Chiroptera). - Proc. biol. Soc. Washington, **75**: 193-197.
- (1978): A Review of the Round-eared Bats of the *Tonatia silvicola* complex, with Descriptions of Three New Taxa. - Occ. Pap. Mus. Texas Tech Univ., **53**: 1-12.
DEBLASE, A. F. (1980): The Bats of Iran: Systematics, Distribution, Ecology. - Fieldiana Zool., (n. s.) **4**: 1-424.
DOBSON, G. E. (1912): Catalogue of the Chiroptera in the Collection of the British Museum. - Trustees Brit. Mus., London.
DORST, J. (1947): Une nouvelle Chauve Souris de l'Indochine française *Paracoelops megalotis*. - Bull. Mus. Paris, (2e ser.) **19**: 436-437.
EGER, J. L. (1977): Systematics of the Genus *Eumops* (Chiroptera: Molossidae). - Life Sci. Contrib. Royal Ontario Mus., **110**: 1-69.
- & R. L. PETERSON (1979): Distribution and systematic relationship of *Tadarida bivittata* and *Tadarida ansorgei* (Chiroptera: Molossidae). - Canadian J. Zool., **57**: 1887-1895.
ELLERMAN, J. R. & T. C. S. MORRISON-SCOTT (1951): Checklist of Palearctic and Indian Mammals 1758-1946. - Trustees Brit. Mus., London.
EL-RAYAH, M. A. (1981): A New Species of Bat of the Genus *Tadarida* (Family Molossidae) from West Africa. - Royal Ontario Mus. Life Sci. Occ. Pap., **36**: 1-10.
ENGSTROM, M. D. & D. E. WILSON (1981): Systematics of *Antrozous dubiaquercus* (Chiroptera: Vespertilionidae), with Comments on the Status of *Bauerus* VAN GELDER. - Ann. Carnegie Mus., **50**: 371-383.
FEILER, A. (1984): Über die Säugetiere der Insel São Tomé (Mammalia). - Zool. Abhand. Staatl. Mus. Tierkunde Dresden, **40**: 75-78.
FELTEN, H. (1964a): Zur Taxonomie indo-australischer Fledermäuse der Gattung *Tadarida* (Mammalia, Chiroptera). - Senckenbergiana biol., **45**: 1-13.
- (1964b): Flughunde der Gattung *Pteropus* von Neukaledonian und den Loyalty-Inseln (Mammalia, Chiroptera). - Senckenbergiana biol., **45**: 671-683.
- & D. KOCK (1972): Weitere Flughunde der Gattung *Pteropus* von den Neuen Hebriden, sowie den Banks- und Torresinseln, Pazifischer Ozean (Mammalia: Chiroptera). - Senckenbergiana biol., **53**: 179-188.
FENTON, M. B. & R. L. PETERSON (1972): Further notes on *Tadarida aloysiisabaudiae* and *Tadarida russata* (Chiroptera: Molossidae - Africa). - Canadian J. Zool., **50**: 19-24.
FINDLEY, J. S. (1972): Phenetic Relationships among Bats of the Genus *Myotis*. - Syst. Zool., **21**: 31-52.
FRANCIS, C. M. & J. E. HILL (1986): A review of the Bornean *Pipistrellus* (Mammalia: Chiroptera). - Mammalia, **50**: 43-55.
FREEMAN, P. W. (1981): A Multivariate Study of the Family Molossidae (Mammalia: Chiroptera): Morphology, Ecology, Evolution. - Fieldiana Zool., (n. s.) **7**: 1-173.
GAISLER, J. (1970): The Bats (Chiroptera) collected in Afghanistan by the Czechoslovak Expeditions of 1965-1967. - Acta Sci. Nat. Acad. Sci. Bohemo-Slovacae Brno, n. s. **4**: (6), 1-56.
- (1971): Systematic Review and Distinguishing Characters of the Bats (Chiroptera) Hitherto Recorded in Afghanistan. - Zool. Listy, **20**: 97-101.
GARDNER, A. L. & D. C. CARTER (1972): A review of the Peruvian species of *Vampyrops* (Chiroptera, Phyllostomatidae). - J. Mammal., **53**: 72-82.
GENOWAYS, H. H. & R. J. BAKER (1975): A new species of *Eptesicus* from Guadeloupe, Lesser Antilles (Chiroptera: Vespertilionidae). - Occ. Pap. Mus. Texas Tech Univ., **34**: 1-7.
- & S. L. WILLIAMS (1979): Notes on Bats (Mammalia: Chiroptera) from Bonaire and Curaçao, Dutch West Indies. - Ann. Carnegie Mus., **48**: 311-321.
- & S. L. WILLIAMS (1980): Results of the Alcoa Foundation-Suriname Expeditions. I. A new species of Bat of the genus *Tonatia* (Mammalia: Phyllostomatidae). - Ann. Carnegie Mus., **49**: 203-211.
GOODWIN, G. G. (1942): A Summary of Recognizable Species of *Tonatia* with Descriptions of two new Species. - J. Mammal., **23**: 204-209.
- (1959): Bats of the Subgenus *Natalus*. - Amer. Mus. Novitates, **1977**: 1-22.
- & A. M. GREENHALL (1961): A Review of the Bats of Trinidad and Tobago. - Bull. Amer. Mus. nat. Hist., **122**: 187-302.
- (1962): Two New Bats from Trinidad, with Comments on the Status of the Genus *Mesophylla*. - Amer. Mus. Novitates, **2080**: 1-18.
GOODWIN, R. E. (1979): The Bats of Timor: Systematics and Ecology. - Bull. Amer. Mus. nat. Hist., **163**: 73-122.
GRANDIDIER, G. (1937): Mamifères nouveaux de la région de Diego-Suarez (Madagascar). - Bull. Mus. Nation. Hist. nat., (2e ser.) **9**: 347-353.
GRIFFITHS, T. A. (1982): Systematics of the New World Nectar-Feeding Bats (Mammalia, Phyllostomidae), Based on Morphology of the Hyoid and Lingual Regions. - Amer. Mus. Novitates, **2742**: 1-45.
HALL, E. R. (1981): The Mammals of North America, **1**. - John Wiley & Sons, New York.
HANÁK, V. & J. GAISLER (1969): Notes on the Taxonomy and Ecology of *Myotis longipes* (DOBSON 1873). - Zool. Listy, **18**: 195-206.
- & I. HORÁČEK (1983-1984): Some Comments on the Taxonomy of *Myotis daubentoni* (KUHL 1819) (Chiroptera, Mammalia). - Myotis, **21-22**: 7-19.
HANDLEY, C. O. (1956): A new Species of Free-tailed Bat (Genus *Mormopterus*) from Peru. - Proc. biol. Soc. Washington, **69**: 197-202.
- (1959): A Revision of American Bats of the Genera *Euderma* and *Plecotus*. - Proc. U. S. nation. Mus., **110**: 95-246.
- (1960): Descriptions of New Bats from Panama. - Proc. U. S. nation. Mus., **112**: 459-479.
- (1966): Descriptions of New Bats (*Choeroniscus* and *Rhinophylla*) from Colombia. - Proc. biol. Soc. Washington, **79**: 83-88.
- (1984): New Species of Mammals from Northern South America: a Long-tongued Bat, Genus *Anoura* GRAY. - Proc. biol. Soc. Washington, **97**: 512-513.
- (1987): New species of Mammals from Northern South America: Fruit-Eating Bats, Genus *Artibeus* LEACH. - Fieldiana, Zool., (n. s.) **39**: 163-172.
- & K. C. FERRIS (1972): Descriptions of New Bats of the genus *Vampyrops*. - Proc. biol. Soc. Washington, **84**: 519-524.
HARRISON, D. L. (1959): A New Subspecies of Lesser Long-Winged Bat *Miniopterus minor* PETERS 1867, from the Comoro Islands. - Durban Mus. Novitates, **5**: 191-196.
- (1965): Remarks on some Trident Leaf-Nosed Bats (Genus *Asellia* GRAY 1838), Obtained by the Israel South Red Sea Expedition, 1962. - Sea Fisheries Research Station, Haifa, Bull., **38**: 3-5.

– (1975): A New Species of African Free-Tailed Bat (Chiroptera: Molossidae), Obtained by the Zaire River Expedition. – Mammalia, **39**: 313–318.

– (1979): A New Species of Pipistrelle Bat (*Pipistrellus*: Vespertilionidae) from Oman, Arabia. – Mammalia, **43**: 573–576.

– (1982): Observations on some rare Arabian *Pipistrellus* (Chiroptera: Vespertilionidae) with special reference to the external male genitalia. – Bonner zool. Beitr., **33**: 187–190.

HAYMAN, R. W. (1945): A new Genus of Fruit-Bat and a new Squirrel from Celebes. – Ann. Mag. nat. Hist., (11) **12**: 569–578.

– & J. E. HILL (1971): Order Chiroptera. – In J. MEESTER & H. W. SETZER (eds.), The Mammals of Africa, an Identification Manual. – Smithsonian Institution Press, Washington.

HEANEY, L. R. & R. L. PETERSON (1984): A New Species of Tube-Nosed Fruit Bat (*Nyctimene*) from Negros Island, Philippines (Mammalia: Pteropodidae). – Occ. Pap. Mus. Zool. Univ. Michigan, **708**: 1–16.

HERNANDEZ-CAMACHO, J. & A. CADENA-G. (1978): Notas para la Revision des Genero *Lonchorhina* (*Chiroptera, Phyllostomidae*). – Caldasia, **12**: 201–251.

HILL, J. E. (1956): The Mammals of Rennell Island. – Nat. Hist. Rennell Island, British Solomon Islands, **1**: 73–84.

– (1961): Indo-Australian Bats of the Genus *Tadarida*. – Mammalia, **25**: 29–56.

– (1962): Notes on some Insectivores and Bats from Upper Burma. – Proc. zool. Soc. London, **139** (1): 110–137.

– (1963a): A Revision of the Genus *Hipposideros*. – Bull. Brit. Mus. nat. Hist. (Zool.), **11**: 1–129.

– (1963b): Notes on some Tube-Nosed Bats, Genus *Murina*, from Southeastern Asia, with Descriptions of a New Species and a New Subspecies. – Federation Mus. J., (n. s.) **8**: 48–59.

– (1964): Notes on Bats from British Guiana, with the Description of a New Genus and Species of *Phyllostomidae*. – Mammalia, **28**: 553–572.

– (1965): Asiatic Bats of the Genera *Kerivoula* and *Phoniscus* (Vespertilionidae), with a Note on *Kerivoula aerosa* TOMES. – Mammalia, **29**: 524–556.

– (1966): A Review of the Genus *Philetor* (Chiroptera: Vespertilionidae). – Bull. Brit. Mus. nat. Hist., (Zool.) **14**: 371–387.

– (1969): The Status of *Myopterus senegalensis* OKEN 1816 (Chiroptera: Molossidae). – Mammalia, **33**: 727–729.

– (1971): The bats of Aldabra Atoll, western Indian Ocean. – Phil. Trans. Roy. Soc. London, (B) **260**: 573–576.

– (1972a): A Note on *Rhinolophus rex* ALLEN 1923 and *Rhinomegalophus paradoxolophus* BOURRET 1951 (Chiroptera: Rhinolophidae). – Mammalia, **36**: 428–434.

– (1972b): The Gunong Benom Expedition, 1967. 4. New Records of Malayan Bats, with Taxonomic Notes and the Description of a New *Pipistrellus*. – Bull. Brit. Mus. nat. Hist., (Zool.) **23**: 21–42.

– (1974a): A Review of *Laephotis* THOMAS 1901 (Chiroptera: Vespertilionidae). – Bull. Brit. Mus. nat. Hist., (Zool.) **27**: 73–82.

– (1974b): New Records of Bats from Southeastern Asia, with Taxonomic Notes. – Bull. Brit. Mus. nat. Hist., (Zool.) **27**: 127–138.

– (1974c): A Review of *Scotoecus* THOMAS 1901 (Chiroptera: Vespertilionidae). – Bull. Brit. Mus. nat. Hist., (Zool.) **27**: 167–188.

– (1974d): A New Family, Genus and Species of Bat (Mammalia: Chiroptera) from Thailand. – Bull. Brit. Mus. nat. Hist., (Zool.) **27**: 301–336.

– (1976): Bats Referred to *Hesperoptenus* PETERS, 1869 (Chiroptera: Vespertilionidae) with the Description of a New Subgenus. – Bull. Brit. Mus. nat. Hist., (Zool.) **30**: 1–28.

– (1977a): A Review of the Rhinopomatidae (Mammalia: Chiroptera). – Bull. Brit. Mus. nat. Hist., (Zool.) **32**: 29–43.

– (1977b): African bats allied to *Kerivoula lanosa* (A. SMITH, 1847) (*Chiroptera: Vespertilionidae*). – Rev. Zool. Afr., **91**: 623–633.

– (1980a): The Status of *Vespertilio borbonicus* E. GEOFFROY, 1803 (Chiroptera: Vespertilionidae). – Zool. Mededelingen, **55**: 287–295.

– (1980b): A note on *Lonchophylla* (Chiroptera: Phyllostomatidae) from Ecuador and Peru, with the description of a new species. – Bull. Brit. Mus. nat. Hist., (Zool.) **38**: 233–236.

– (1982): A review of the leaf-nosed bats *Rhinonycteris, Cloeotis* and *Triaenops* (Chiroptera: Hipposideridae). – Bonner zool. Beitr., **33**: 165–186.

– (1983): Bats (Mammalia: Chiroptera) from Indo-Australia. – Bull. Brit. Mus. nat. Hist., (Zool.) **45**: 103–208.

– (1985a): Records of bats (Chiroptera) from New Guinea, with the description of a new *Hipposideros* (Hipposideridae). – Mammalia, **49**: 525–535.

– (1985b): The status of *Lichonycteris degener* MILLER 1931 (Chiroptera: Phyllostomidae). – Mammalia, **49**: 579–582.

– (1986): A note on *Balantiopteryx infusca* (THOMAS 1897) (Chiroptera: Emballonuridae). – Mammalia, **50**: 558–560.

– & W. N. BECKON (1978): A new species of *Pteralopex* THOMAS 1888 (Chiroptera: Pteropodidae) from the Fiji Islands. – Bull. Brit. Mus. nat. Hist., (Zool.) **34**: 65–82.

– & M. J. DANIEL (1985): Systematics of the New Zealand short-tailed bat *Mystacina* GRAY 1843 (Chiroptera: Mystacinidae). – Bull. Brit. Mus. nat. Hist., (Zool.) **48**: 279–300.

– & C. M. FRANCIS (1984): New bats (Mammalia: Chiroptera) and new records of bats from Borneo and Malaya. – Bull. Brit. Mus. nat. Hist., (Zool.) **47**: 305–329.

– & D. L. HARRISON (1987): The baculum in the Vespertilioninae (Chiroptera: Vespertilionidae) with a systematic review, a synopsis of *Pipistrellus* and *Eptesicus*, and the descriptions of a new genus and subgenus. – Bull. Brit. Mus. nat. Hist., (Zool.) **52**: 225–305.

– & K. F. KOOPMAN (1981): The status of *Lamingtona lophorhina* MCKEAN & CALABY 1968 (Chiroptera, Vespertilionidae). – Bull. Brit. Mus. nat. Hist., (Zool.) **41**: 275–278.

– & P. MORRIS (1971): Bats from Ethiopia Collected by the Great Abbai Expedition, 1968. – Bull. Brit. Mus. nat. Hist., (Zool.) **21**: 25–49.

– & D. A. SCHLITTER (1982): A Record of *Rhinolophus arcuatus* (Chiroptera: Rhinolophidae) from New Guinea, with the Description of a New Subspecies. – Ann. Carnegie Mus., **51**: 455–464.

– & J. D. SMITH (1984): Bats. A Natural History. – Univ. Texas Press, Austin.

– & K. THONGLONGYA (1972): Bats from Thailand and Cambodia. – Bull. Brit. Mus. nat. Hist., (Zool.) **22**: 171–196.

– & G. TOPAL (1973): The Affinities of *Pipistrellus ridleyi* THOMAS 1898 and *Glischropus rosseti* OEY 1951 (Chiroptera: Vespertilionidae). – Bull. Brit. Mus. nat. Hist., (Zool.) **24**: 447–454.

– & S. YENBUTRA (1984): A new species of the *Hipposideros bicolor* group (Chiroptera: Hipposideridae) from Thailand. – Bull. Brit. Mus. nat. Hist., (Zool.) **47**: 77–82.

- & M. Yoshiyuki (1980): A New Species of *Rhinolophus* (Chiroptera, Rhinolophidae) from Iriomote Island, Ryukyu Islands, with Notes on the Asiatic Members of the *Rhinolophus pusillus* Group. – Bull. Nation. Sci. Mus., (ser. A, Zool.), **6**: 179–189.
- A. Zubaid, & G. W. H. Davidson (1986): The taxonomy of the *Hipposideros bicolor* group (Chiroptera: Hipposideridae) from southeastern Asia. – Mammalia, **50**: 535–540.

Honacki, J. H., K. E. Kinman, & J. W. Koeppl (eds.) (1982): Mammal Species of the World. – Allen Press and Association of Systematics Collections, Lawrence, Kansas.

Horáček, I. & V. Hanák (1983–1984): Comments on the Systematics and Phylogeny of *Myotis nattereri* (Kuhl 1818). – Myotis, **21–22**: 20–29.
- (1985–1986): Generic Status of *Pipistrellus savii* and Comments on Classification of the Genus *Pipistrellus* (Chiroptera, Vespertilionidae). – Myotis, **23–24**: 9–16.

Hyndman, D. & J. I. Menzies (1980): *Aproteles bulmerae* (Chiroptera: Pteropodidae) of New Guinea is not extinct. – J. Mammal., **61**: 159–160.

Ibáñez, C. (1980): Descripcion de un Nuevo Género de Quiroptero Neotropical de la Familia Molossidae. – Doñana Acta Vertebrata, **7**: 104–111.

Imaizumi, Y (1970): The Handbook of Japanese Land Mammals, 1. – Shin-Shicho-Sha Co., Tokyo.

Jenkins, P. & J. E. Hill (1980): The status of *Hipposideros galeritus* Cantor 1846 and *Hipposideros cervinus* (Gould 1854) (Chiroptera: Hipposideridae). – Bull. Brit. Mus. nat. Hist. (Zool.), **41**: 279–294.

Jones, J. K. & D. C. Carter (1976): Annotated Checklist, with Keys to Subfamilies and Genera. *In* R. J. Baker, J. K. Jones, & D. C. Carter (eds.), Biology of Bats of the New World Family Phyllostomatidae, Part I. – Special Publ. Mus. Texas Tech Univ., **10**: 7–47.

Jong, N. De & W. Bergmans (1981): A Revision of the Fruit Bats of the Genus *Dobsonia* Palmer 1898 from Sulawesi and Some Nearby Islands (Mammalia, Megachiroptera, Pteropodinae). – Zool. Abh. Staatl. Mus. Tierkunde Dresden, **37**: 209–224.

Kitchener, D. J. & N. Caputi (1985): Systematic Revision of Australian *Scoteanax* and *Scotorepens* (Chiroptera: Vespertilionidae), with Remarks on Relationships to other Nycticeini. – Rec. West. Austral. Mus., **12**: 85–146.
- & B. Jones (1986): Revision of Australo-Papuan *Pipistrellus* and of *Falsistrellus* (Microchiroptera: Vespertilionidae). – Rec. West. Austral. Mus., **12**: 435–495.
- B. Jones, & N. Caputti (1987): Revision of Australian *Eptesicus* (Microchiroptera: Vespertilionidae). – Rec. West. Austral. Mus., **13**: 427–500.

Klingener, D. & G. K. Creighton (1984): On small bats of the Genus *Pteropus* from the Philippines. – Proc. biol. Soc. Washington, **97**: 395–403.

Kock, D. (1969a): Die Fledermaus-Fauna des Sudan. – Abhandl. Senckenbergischen naturforsch. Ges., **521**: 1–238.
- (1969b): Eine neue Gattung und Art cynopteriner Flughunde von Mindanao, Philippinen (Mammalia, Chiroptera). – Senckenbergiana biol., **50**: 319–327.
- (1969c): Eine bemerkenswerte neue Gattung und Art Flughund von Luzon, Philippinen (Mammalia, Chiroptera). – Senckenbergiana biol., **50**: 329–338.
- (1975): Ein Originalexemplar von *Nyctinomus ventralis* Heuglin 1861 (Mammalia: Chiroptera: Molossidae). – Stuttgarter Beitr. Naturk., (ser. A, Biol.), **272**: 1–9.

Koopman, K. F. (1971): Taxonomic Notes on *Chalinolobus* and *Glauconycteris* (Chiroptera, Vespertilionidae). – Amer. Mus. Novitates, **2451**: 1–10.
- (1975): Bats of the Sudan. – Bull. Amer. Mus. nat. Hist., **154**: 353–444.
- (1978): Zoogeography of Peruvian Bats With Special Emphasis on the Role of the Andes. – Amer. Mus. Novitates, **2651**: 1–33.
- (1979): Zoogeography of Mammals from Islands off the Northeastern Coast of New Guinea. – Amer. Mus. Novitates, **2590**: 1–17.
- (1982): Results of the Archbold Expeditions No. 109. Bats from Eastern Papua and the East Papuan Islands. – Amer. Mus. Novitates, **2747**: 1–34.
- (1982): Biogeography of the Bats of South America. – *In* M. A. Mares & H. H. Genoways (eds.), Mammalian Biology in South America. – Special Publ. Ser., Pymatuming Lab. Ecology, Univ. Pittsburgh, **6**: 273–302.
- (1983): A Significant Range Extension for *Philetor* (Chiroptera, Vespertilionidae) with Remarks on Geographical Variation. – J. Mammal., **64**: 525–526.
- (1984a): Taxonomic and Distributional Notes on Tropical Australian Bats. – Amer. Mus. Novitates, **2778**: 1–48.
- (1984b): Bats. – *In* S. Anderson & J. K. Jones (eds.), Orders and Families of Recent Mammals of the World. – John Wiley and Sons, New York: 145–186.
- (1984c): A Progress Report on the Systematics of African *Scotophilus* (Vespertilionidae). – Proc. 6th Internation. Bat Res. Conf., Ile-Ife, Nigeria: 102–113.
- & G. T. Macintyre (1980): Phylogenetic analysis of chiropteran dentition. – *In* D. E. Wilson & A. L. Gardner (eds.). – Proceedings, Fifth International Bat Research Conference. – Texas Tech Press, Lubbock.
- R. E. Mumford & J. E. Heisterberg (1978): Bat Records from Upper Volta, West Africa. – Amer. Mus. Novitates, **2643**: 1–6.

Kuroda, N (1938): A List of the Japanese Mammals. – Tokyo.

Laurie, E. M. O. & J. E. Hill (1954): List of Land Mammals of New Guinea, Celebes, and Adjacent Islands 1758–1952. – Trustees Brit. Mus., London.

Laval, R. K. (1973): A Revision of the Neotropical Bats of the Genus *Myotis*. – Nat. Hist. Mus., Los Angeles Co., Sci. Bull, **15**: 1–54.
- (1973): The Systematics of the Genus *Rhogeesa*. – Occ. Pap. Mus. nat. Hist. Univ. Kansas, Lawrence, **19**: 1–47.

Lawrence, B. (1939): Collections from the Philippine Islands. Mammals. – Bull. Mus. comparative Zool., **86**: 28–73.

Legendre, S. (1984): Etude odontologique des représentants actuels du groupe *Tadarida* (Chiroptera, Molossidae). Implications phylogéniques, systématiques et zoogéographiques. – Rev. suisse. Zool., **91**: 399–442.

Lekagul, B. & J. A. Mc Neely (1977): Mammals of Thailand. – Assoc. for the Conservation of Wildlife, Bangkok.

Linares, O. J. (1971): A new subspecies of Funnel-eared bat (*Natalus stramineus*) from western Venezuela. – Bull. S. California Acad. Sci., **70**: 81–84.

Luckett, W. P. (1980): The use of fetal membrane data in assessing chiropteran phylogeny. – *In* D. E. Wilson & A. L. Gardner (eds.). – Proceedings, Fifth International Bat Research Conference. Texas Tech Press, Lubbock.

Maeda, K. (1980): Review on the Classification of Little tube-nosed bats, *Murina aurata* group. – Mammalia, **44**: 531–551.
- (1982): Studies on the Classification of *Miniopterus* in Eurasia, Australia, and Melanesia. – Honyurui Kagaku (Mammalian Science), suppl., **1**: 1–176.

MARTIN, C. O. & D. J. SCHMIDLY (1982): Taxonomic Review of the Pallid Bat, *Antrozous pallidus* (LE CONTE). – Spec. Publ. Mus. Texas Tech Univ., **18**: 1–48.

MEDWAY, LORD (1977): Mammals of Borneo. – Monogr. Malaysian Branch Royal Asiatic Soc., **7**: 1–172.

MENZIES, J. I. (1977): Fossil and Subfossil Fruit Bats from the Mountains of New Guinea. – Austral. J. Zool., **25**: 329–336.

MILLER, G. S. (1907): The Families and Genera of Bats. – U. S. Nation. Mus. Bull., **57**: 1–282.

MÖHRES, F. P. (1953): Über die Ultraschallorientierung der Hufeisennasen (Chiroptera-Rhinolophinae). Z. vergleichende Physiologie, **34**: 547–588.

MORI, T. (1928): Four new species of bats (Vespertilionidae) from Korea. – Annotationes zoologicae Japonenses, **11**: 389–395.

MUSSER, G. G., K. F. KOOPMAN & D. CALIFIA (1982): The Sulawesan *Pteropus arquatus* and *P. argentatus* are *Acerodon celebensis*; the Philippine *P. leucotis* is an *Acerodon*. – J. Mammal., **63**: 319–328.

OCHOA, J. & C. IBANEZ (1982): Nuevo Murcielago del Genero *Lonchorhina* (Chiroptera: Phyllostomidae). – Mem. Soc. Cienc. nat. La Salle, **42**: 145–159.

OJASTI, J. & O. J. LINARES (1971): Adiciones a la Fauna de Murcielagos de Venezuela con notas sobre los Especies del Genero *Diclidurus* (Chiroptera). – Acta biol. Venez., **7**: 421–441.

OTTENWALDER, J. A. & H. H. GENOWAYS (1982): Systematic Review of the Antillean Bats of the *Natalus micropus*-complex (Chiroptera: Natalidae). – Ann. Carnegie Mus., **51**: 17–38.

OWEN, R. D. (1987): Phylogenetic Analyses of the Bat Subfamily Stenodermatinae (Mammalia: Chiroptera). – Spec. Publ. Mus. Texas Tech Univ., **26**: 1–65.

– & W. D. WEBSTER (1983): Morphological variation in the ipanema bat, *Pygoderma bilabiatum*, with description of a new subspecies. – J. Mammal., **64**: 146–149.

PETERSON, R. L. (1965a): A review of the flat-headed bats of the family *Molossidae* from South America and Africa. – Life Sci. Contrib. Royal Ontario Mus., **64**: 1–32.

– (1965b): A Review of the Bats of the Genus *Ametrida*, Family Phyllostomidae. – Life Sci. Contrib. Royal Ontario Mus., **65**: 1–13.

– (1968): A New Bat of the Genus *Vampyressa* from Guyana, South America. – Life Sci. Contrib. Royal Ontario Mus., **73**: 1–15.

– (1971): The systematic status of the African molossid bats *Tadarida bemmelini* and *Tadarida cistura*. – Canadian J. Zool., **49**: 1347–1354.

– (1972): Systematic Status of the African Molossid Bats *Tadarida congica, T. niangarae* and *T. trevori*. – Life Sci. Contrib. Royal Ontario Mus., **85**: 1–32.

– (1974): Variation in the African Bat, *Tadarida, lobata*, with Notes on Habitat and Habits (Chiroptera: Molossidae). – Royal Ontario Mus. Life Sci. Occ. Pap., **24**: 1–8.

– (1981): Variation in the *tristis* group of the bentwinged bats of the genus *Miniopterus* (Chiroptera: Vespertilionidae). – Canadian J. Zool., **59**: 828–843.

– (1982): A new species of *Glauconycteris* from the east coast of Kenya (Chiroptera: Vespertilionidae). – Canadian J. Zool., **60**: 2521–2525.

– & D. A. SMITH (1973): A New Species of *Glauconycteris* (Vespertilionidae, Chiroptera). – Royal Ontario Mus. Life Sci. Occ. Pap., **22**: 1–9.

PHILLIPS, C. J. (1968): Systematics of Megachiropteran Bats in the Solomon Islands. – Univ. Kans. Publ. Mus. nat. Hist., **16**: 777–787.

PINE, R. H. (1972): The Bats of the Genus *Carollia*. – Texas A. and M. Univ., Texas Agric. Exp. Sta. Tech Monogr., **8**: 1–125.

– & A. RUSCHI (1976): Concerning Certain Bats Described and Recorded from Espirito Santo, Brazil. – An. Inst. Biol. Univ. Nal. Auton Mexico, **47**, Ser. Zoologia (2): 183–196.

RABOR, D. S. (1952): Two New Mammals from Negros Island, Philippines. – Nat. Hist. Miscellanea, Chicago Acad. Sci., **96**: 1–7.

REVILLIOD, P. (1911): Über einige Säugetiere von Celebes. – Zool. Anz., **37**: 513–517.

ROBBINS, C. B. (1978): Taxonomic identification and history of *Scotophilus nigrita* (SCHREBER) (Chiroptera, Vespertilionidae). – J. Mammal., **59**: 212–213.

ROOKMAAKER, L. C. & W. BERGMANS (1981): Taxonomy and Geography of *Rousettus amplexicaudatus* (GEOFFROY 1810) with Comparative Notes on Sympatric Congeners (Mammalia, Megachiroptera). – Beaufortia, **31**: 1–29.

ROSEVEAR, D. R. (1965): The Bats of West Africa. – Trustees Brit. Mus., London.

ROUK, C. S. & D. S. CARTER (1972): A New Species of *Vampyrops* (Chiroptera: Phyllostomatidae) from South America. – Occ. Pap. Mus. Texas Tech. Univ., **1**: 1–7.

ROZENDAAL, F. G. (1984): Notes on Macroglossine Bats from Sulawesi and the Moluccas, Indonesia, with the Description of a New Species of *Syconycteris* MATSCHIE 1899 from Halmahera (Mammalia: Megachiroptera). – Zool. Mededelingen, **58**: 187–212.

RUSCHI, A. (1951): Morcegos do Estado do Espirito Santo. – Bol. Mus. Biol. Prof. Mello-Leitao, Santa Teresa, E. E. Santo, Brasil, Zool., **6**: 1–22.

SANBORN, C. C. (1931): Bats from Polynesia, Melanesia, and Malaysia. – Field Mus. nat. Hist., (Zool. Ser.) **28**: 7–29.

– (1937): American Bats of the Subfamily Emballonurinae. – Zool. Ser. Field Mus. nat. Hist., **20**: 321–354.

– (1939): Eight New Bats of the Genus *Rhinolophus*. – Zool. Ser. Field Mus. nat. Hist., **24**: 37–43.

– (1943): External Characters of the Bats of the Subfamily Glossophaginae. – Zool. Ser. Field Mus. nat. Hist., **24**: 271–277.

– (1949): Bats of the Genus *Micronycteris* and its Subgenera. – Fieldiana Zool., **31**: 215–233.

– (1952): The Status of "*Triaenops wheeleri*" OSGOOD. – Nat. Hist. Miscellanea, Chicago Acad. Sci., **97**: 1–3.

– (1955): Remarks on the Bats of the Genus *Vampyrops*. – Fieldiana, Zool., **37**: 403–413.

– & J. A. CRESPO (1957): El Murciélago Blanquizco (*Lasiurus cinereus*) y sus Subspecies. – Bol. Mus. Argentino Cienc. Nat. "Bernardino Rivadavia", **4**: 1–12.

SCHLITTER, D. A., S. L. WILLIAMS, & J. E. HILL (1983): Taxonomic Review of Temminck's Trident Bat, *Aselliscus tricuspidatus* (TEMMINCK 1834) (Mammalia: Hipposideridae). – Ann. Carnegie Mus., **52**: 337–358.

SINHA, Y. P. (1969): A new pipistrelle bat (Mammalia: Chiroptera: Vespertilionidae) from Burma. – Proc. zool. Soc., Calcutta, **22**: 83–86.

– (1973): Taxonomic studies on the Indian horse-shoe bats of the genus *Rhinolophus* LACÈPEDE. – Mammalia, **37**: 603–630.

– & S. CHAKRABORTY (1971): Taxonomic Status of the Vespertilionid Bat, *Nycticejus emarginatus* DOBSON. – Proc. zool. Soc. Calcutta, **24**: 53–59.

SMITH, J. D. (1972): Systematics of the Chiropteran Family Mormoopidae. – Univ. Kansas Mus. nat. Hist. Misc. Publ., **56**: 1–132.

– & C. S. HOOD (1983): A New Species of Tube-Nosed

Fruit Bat (*Nyctimene*) from the Bismarck Archipelago, Papua New Guinea. – Occ. Pap. Mus. Texas Techn. Univ., **81**: 1–14.

– & G. MADKOUR (1980): Penial morphology and the question of chiropteran phylogeny. – *In* D. E. WILSON & A. L. GARDNER (eds.). Proceedings, Fifth Intenational Bat Research Conference. – Texas Tech Press, Lubbock.

STARRETT, A. & L. DE LA TORRE (1964): Notes on a Collection of Bats from Central America, with the Third Record for *Cyttarops alecto* THOMAS. – Zoologica, N. Y. Zool. Soc., **49**: 55–63.

STRAHAN, R. (ed.) (1983): The Australian Museum Complete Book of Australian Mammals. – Angus and Robertson, Sydney.

SWANEPOEL, P. & H. H. GENOWAYS (1979): Morphometrics. – *In* R. J. BAKER, J. K. JONES, & D. C. CARTER (eds) Biology of Bats of the New World Family Phyllostomatidae, Part III. – Spec. Publ. Mus. Texas Tech Univ., **16**: 13–106.

TADDEI, V. A. (1979): Phyllostomidae (Chiroptera) do Norte-Occidental do Estado de Sao Paulo, III-Stenodermatinae. – Cienc. e Cultura, **31**: 900–914.

– (1983): Uma Nova Espécie de *Lonchophylla* do Brasil e Chave para Identificaçao das Especies do Género (Chiroptera, Phyllostomidae). – Cienc. e Cultura, **35**: 625–629.

TAMSITT, J. R. & D. NAGORSEN (1982): *Anoura cultrata*. – Mamm. Spp., **179**: 1–5.

TATE, G. H. H. (1941a): Results of the Archbold Expeditions. No. 36. Remarks on Some Old World Leaf-Nosed Bats. – Amer. Mus. Novitates, **1140**: 1–11.

– (1941b): Results of the Archbold Expeditions. No. 37. Notes on Oriental *Taphozous* and Allies. – Amer. Mus. Novitates, **1141**: 1–5.

– (1941c): Results of the Archbold Expeditions. No. 39. Review of *Myotis* of Eurasia. – Bull. Amer. Mus. nat. Hist., **78**: 537–565

– (1941d): Results of the Archbold Expeditions. No. 40. Notes on Vespertilionid Bats. – Bull. Amer. Mus. nat. Hist., **78**: (567–597).

– (1942): Results of the Archbold Expeditions. No. 47. Review of the Vespertilionine Bats, with Special Reference to Genera and Species of the Archbold Collections. – Bull. Amer. Mus. nat. Hist., **80**: 221–297.

– (1943): Results of the Archbold Expeditions. No. 49. Further Notes on the *Rhinolophus philippinensis* Group (Chiroptera). – Amer. Mus. Novitates, **1219**: 1–7.

– & R. ARCHBOLD (1939a): Results of the Archbold Expeditions. No. 23. A Revision of the Genus *Emballonura* (Chiroptera). – Amer. Mus. Novitates, **1035**: 1–14.

– & R. ARCHBOLD (1939b): Results of the Archbold Expeditions. No. 24. Oriental *Rhinolophus* with Special Reference to Material from the Archbold Collections. – Amer. Mus. Novitates, **1036**: 1–12.

TAYLOR, E. H. (1934): Philippine Land Mammals. – Monogr. Bur. Sci Philippine Islands, **30**: 1–548.

THOMAS, O. (1913): On some rare Amazonien Mammals from the Collection of the Para Museum. – Ann. Mag. nat. Hist., (8) **11**: 130–136.

– (1922): The Generic Classification of the *Taphozous* Group. – Ann. Mag. nat. Hist., (9) **9**: 266–267.

THONGLONGYA, K. (1973): First record of *Rhinolophus paradoxolophus* (BOURRET 1951) from Thailand, with the description of a new species of the *Rhinolophus philippinensis* group (Chiroptera, Rhinolophidae). – Mammalia, **37**: 587–597.

– & J. E. HILL (1974): A New Species of *Hipposideros* (Chiroptera) from Thailand. – Mammalia, **38**: 285–294.

TOPAL, G. (1970a): The First Record of *Ia io* Thomas, 1902 in Vietnam and India and Some Remarks on the Taxonomic Position of *Parascotomanes beaulieui* BOURRET 1942, *Ia longimana* Pen, 1962, and the Genus *Ia* THOMAS 1902 (Chiroptera: Vespertilionidae). – Opusc. Zool. Budapest, **10**: 341–347.

– (1970b): On the systematic status of *Pipistrellus annectans* DOBSON 1871 and *Myotis primula* THOMAS 1920 (Mammalia). – Ann. Hist. nat. Mus. nation. Hungarica, pars Zool, **62**: 373–379.

TROUGHTON, E. L. G. (1925): A Revision of the Genera *Taphozous* and *Saccolaimus* (Chiroptera) in Australia and New Guinea, including a New Species and a Note on Two Malayan Forms. – Rec. Austral. Mus., **14**: 313–340.

TUPINIER, Y. (1977): Description d'un chauve souris nouvelle. *Myotis nathalinae* nov. sp. (Chiroptera-Vespertilionidae). – Mammalia, **41**: 327–339.

VREE, F. DE (1972): Description of a new form of *Pipistrellus* from Ivory Coast. – Rev. Zool. Bot. afr., **85**: 412–416.

WALLIN, L. (1969): The Japanese Bat Fauna. – Zool. Bidrag Uppsala, **37**: 223–440.

WEBSTER, W. D. & C. O. HANDLEY (1986): Systematics of MILLER'S Long-Tongued Bat, *Glossophaga longirostris*, with Descriptions of Two New Subspecies. – Occ. Pap. Texas Tech Univ., **100**: 1–22.

– & J. K. JONES (1980): Taxonomic and Nomenclatorial Notes on Bats of the Genus *Glossophaga* in North America, with Description of a New Species. – Occ. Pap. Texas Tech Univ., **71**: 1–12.

WILLIAMS, D. F. (1978): Taxonomic and Karyologic Comments on Small Brown Bats, Genus *Eptesicus*, from South America. – Ann. Carnegie Mus., **47**: 361–383.

WILLIAMS, S. L. & H. H. GENOWAYS (1980): Results of the Alcoa Foundation-Suriname Expeditions. IV. A New Species of Bat of the Genus *Molossops* (Mammalia: Molossidae). – Ann. Carnegie Mus., **49**: 487–498.

YENBUTRA, S. & H. FELTEN (1983): A new species of the fruit-bat genus *Megaerops* from SE-Asia. – Senckenbergiana biol., **84**: 1–11.

YOSHIYUKI, M. (1970): A New species of Insectivorous Bat of the Genus *Murina* from Japan. – Bull. Nation. Sci. Mus. Tokyo, **13**: 195–198.

– (1971): A New Bat of the *Leuconoe* Group in the Genus *Myotis* from Honshu, Japan. – Bull. Nation. Sci. Mus. Tokyo, **14**: 305–310.

– (1979): A New Species of the Genus *Ptenochirus* (Chiroptera, Pteropodidae) from the Philippine Islands. – Bull. Nation. Sci. Mus. Tokyo, (Ser. A, Zool.), **5**: 75–81.

– (1983): A New Species of *Murina* from Japan (Chiroptera, Vespertilionidae). – Bull. Nation. Sci. Mus. Tokyo, (Ser. A, Zool.), **9**: 141–150.

– (1984): A New Species of *Myotis* (Chiroptera, Vespertilionidae) from Hokkaido, Japan. – Bull. Nation. Sci. Mus. Tokyo, (Ser. A, Zool.), **10**: 153–158.

ZIEGLER, A. C. (1982): The Australo-Papuan Genus *Syconycteris* (Chiroptera: Pteropodidae) With the Description of a New Papua New Guinea Species. – Occ. Pap. Bernice P. Bishop Mus., **25** (5): 1–22.

Systematic Index

compiled by Eric Brothers, Scientific Assistant, Department of Mammalogy,
American Museum of Natural History

Key To Index:
underline   where the entry appears as a catalog listing.
bold   where the entry appears in a figure (caption).
normal   where the entry appears anywhere else

abae, Hipposideros 65
  Rhinolophus 58
abbotti, Myotis 106
abdita, Thyroptera 96
abei, Myotis 108
abramus, Pipistrellus 112
abrasus, Molossops 137
Acerodon 27
  alorensis 27
  arquatus 27
  celebensis 27
  floresi 27
  gilvus 27
  humilis 27
  jubatus 27
  leucotis 27, **148**
  lucifer 27
  mackloti 27
  prajae 27
acetabulosus, Mormopterus 136
achates, Taphozous 42
achilles, Rhinolophus 57
acrodonta, Pteralopex 28
acrotis, Rhinolophus 54
acuminatus, Rhinolophus 55
adami, Rhinolophus 54
adamsi, Pipistrellus 112
adana, Nycteris 50
admiralitatum, Pteropus 22
adversus, Myotis 106
aegyptiaca, Tadarida 139
aegyptiacus, Rousettus 19
aegyptius, Pipistrellus 114
aelleni, Myotis 107
Aello 71
aello, Nyctimene 37, **155**
aenea, Murina 133
aequalis, Aethalops 35
  Rhinolophus 53
aequatorialis, Artibeus 91
aequatorianus, Molossops 138
aero, Pipistrellus 114
aerosa, Kerivoula 100
Aethalops 35
  aequalis 35
  alecto 35, **154**
  ocypete 35
aethiopica, Nycteris 50
aethiops, Rhinolophus 58
afer, Triaenops 68
affinis, Lavia 52
  Noctilio 70
  Pipistrellus 113
  Rhinolophus 54, **162**
  Saccolaimus 44
afra, Coleura 45, **158**
africana, Kerivoula 98
  Tadarida 139
africanus, Miniopterus 134
  Pipistrellus 112
agnella, Kerivoula 98
aladdin, Pipistrellus 111
alascensis, Myotis 108

alaschanicus, Pipistrellus 114
alba, Ectophylla 90, **171**
albaniensis, Hipposideros 61
albatus, Myopterus 138
albescens, Myotis 108
albigula, Scotoecus 128
albipinnis, Taphozous 42
albiventer, Nyctimene 36
  Thyroptera 96
albiventris, Noctilio 70
albofuscus, Scotoecus 127
alboguttatus, Chalinolobus 125
albus, Dicliurus 48, **160**
alcyone, Rhinolophus 55
aldabrensis, Pteropus 23
alecto, Aethalops 35, **154**
  Cyttarops 48, **160**
  Emballonura 44
  Pteropus 26
alethina, Rhinophylla 84
alienus, Histiotus 122
Alionycteris 35
  paucidentata 35, **154**
alleni, Rhinolophus 57
  Rhogeessa 127
alongensis, Hipposideros 66
alorensis, Acerodon 27
aloysiisabaudiae, Chaerephon 140
altarium, Myotis 104
alticolus, Rhinolophus 53
alticraniatus, Myotis 104
altidudinis, Cynopterus 33
alvenslebeni, Scotophilus 129
amazonicus, Eumops 143
amboiensis, Hipposideros 61
Ametrida 93
  centurio 93, **173**
Amorphochilus 96
  schnablii 96, **175**
amotus, Myotis 108
amplexicaudatus, Rousettus 19
amplus, Artibeus 90
anatolicus, Eptesicus 120
anceps, Pteralopex 28, **148**
anchietai, Pipistrellus 114
  Plerotes 30, **150**
ancilla, Myotis 101
  Promops 144
adamanensis, Rhinolophus 54
anderseni, Artibeus 92
  Dobsonia 29
  Rhinolophus 53
andersoni, Eptesicus 120
anderssoni, Vespertilio 121
andinus, Eptesicus 120
andreinii, Rhinolophus 57
anetianus, Pteropus 24
angolensis, Epomophorus 31
  Hipposideros 63
  Laephotis 121, **179**
  Nycteris 50
  Rhinolophus 55
  Rousettus 20

angulatus, Cynopterus 33
  Pipistrellus 112
angustifolius, Rhinolophus 60
anjouanensis, Myotis 102
annectans, Myotis 104
annectens, Rhinolophus 53
  Pteropus 22
Anoura 81
  caudifera 82
  cultrata 82
  geoffroyi 82, **169**
  lasiopyga 82
  latidens 82
  peruana 82
ansorgei, Chaerephon 140
anthonyi, Nyctalus 118
Anthops 67
  ornatus 67, **162**
Anthorhina 76
antillarum, Glossophaga 80
antillularum, Tadarida 139
antricola, Hipposideros 61
Antrozoini 130
Antrozous 130
  bunkeri 130
  koopmani 130
  minor 130
  pacificus 130
  packardi 130
  pallidus 130, **183**
anurus, Epomophorus 31
aorensis, Pteropus 25
apache, Myotis 102
aphylla, Phyllonycteris 79
Aproteles 28
  bulmerae 28, **149**
aquilus, Vampyrops 87
arabicus, Pipistrellus 112
  Rousettus 19
arabium, Rhinopoma 41
aratathomasi, Sturnira 85
araxenus, Myotis 101
archipelagus, Cynopterus 33
arcuatus, Rhinolophus 59
Ardops 92
  koopmani 92
  luciae 92
  montserratensis 92
  nichollsi 92, **172**
arenarius, Miniopterus 134
arge, Nycteris 49
argentata, Kerivoula 99
argentatus, Chalinolobus 125
  Pteropus 23
argentinus, Eptesicus 120
argynnis, Casinycteris 32, **151**
ariel, Pipistrellus 115
  Plecotus 110
  Pteropus 26
Arielulus 115
Ariteus
  flavescens 93, **172**
armiger, Hipposideros 65

arnhemensis, Nyctophilus 131
arquatus, Acerodon 27
Artibeus **2**, 7, 90
   aequatorialis 91
   amplus 90
   anderseni 92
   aztecus 91
   bogotensis 92
   cinereus 92
   concolor 91
   fallax 91
   fraterculus 91
   fuliginosus 91
   harti 90
   hercules 91
   hesperus 92
   hirsutus 91
   inopinatus 91
   intermedius 90
   jamaicensis 91, **172**
   lituratus 90
   major 92
   minor 92
   nanus 92
   palatinus 92
   palmarum 90
   parvipes 91
   paulus 91
   phaeotis 92
   planirostris 91
   pumilio 92
   quadrivittatus 92
   ravus 92
   richardsoni 91
   rosenbergi 92
   schwartzi 91
   toltecus 92
   trinitatis 92
   triomylus 91
   watsoni 92
   yucatanicus 91
aruensis, Hipposideros 61
   Pteropus 24
   Rhinolophus 59
Asellia 67
   diluta 68
   italosomalica 68
   murraiana 68
   patrizii 68
   tridens 67, **163**
Aselliscus 67
   koopmani 67
   novaeguineae 67
   novaehebridensis 67
   stoliczkanus 67
   tricuspidatus 67, **162**
asirensis, Rhinopoma 40
astrolabiensis, Mormopterus 136
atacamensis, Myotis 105
ater, Hipposideros 61
   Molossus 144
   Myotis 104
   Pipistrellus 117
aterrimus, Pteropus 26
atrata, Emballonura 44
   Pteralopex 28
atrox, Hipposideros 61
   Kerivoula 100
audax, Rhinolophus 55
augur, Rhinolophus 54
aurantius, Melonycteris 39
   Rhinonycteris 68, **163**
aurarius, Vampyrops 87
aurata, Murina 132
auratus, Pteropus 23
aurex, Tylonycteris 122

auriculus, Myotis 102
auripendulus, Eumops 143
aurispinosus, Nyctinomops 143
aurita, Lonchorhina 74, **165**
   Myzopoda 97, **175**
   Nycteris 50
auritus, Chrotopterus 77, **166**
   Plecotus 110
austenianus, Pipistrellus 114
australis, Cloeotis 68
   Miniopterus 134
   Myotis 104
   Plecotus 111
   Syconycteris 38, **156**
   Tadarida 139
   Taphozous 42, **159**
austriacus, Plecotus 110
austroriparius, Myotis 108
aviator, Nyctalus 118
azoreum, Nyctalus 118
azteca, Carollia 84
aztecus, Artibeus 91
   Molossus 145
   Myotis 101–102

babi, Cynopterus 33
babu, Pipistrellus 113
Baeodon 127
bahamensis, Eptesicus 120
   Tadarida 139, **186**
bakeri, Glossophaga 80
   Pteropus 25
Balantiopteryx 46
   infusca 46
   io 47
   pallida 47
   plicata 47, **159**
Balionycteris 34
   maculata 34, **153**
   seimundi 34
balstoni, Nycticeius 126
balutus, Pteropus 22
banksianus, Pteropus 25
barbarus, Rhinolophus 56
Barbastella 5, 109
   barbastellus 109
   darjelingensis 110
   leucomelas 109, **176**
barbastellus, Barbastella 109
barbensis, Hipposideros 66
barnesi, Molossus 145
bartelsi, Myotis 101
Barticonycteris 74
basiliscus, Pteropus 23
bastiani, Nycteris 49
batchianensis, Hipposideros 63
batjanus, Glischropus 119
Bauerus 130
   dubiaquercus 130, **182**
baverstocki, Pipistrellus 116
beatrix, Chalinolobus 124
beatus, Hipposideros 64
beaulieui, Ia 118
beccarii, Emballonura 44
   Mormopterus 136
   Rhinolophus 60
bechsteini, Myotis 100
beckeri, Eumops 143
beddomei, Rhinolophus 58
behni, Micronycteris 74
bellieri, Pipistrellus 114
bellissima, Kerivoula 98
bemmelini, Chaerephon 139
bennettii, Mimon 76
bensoni, Eptesicus 120

benuensis, Nycteris 50
bernardinus, Eptesicus 120
berneri, Mimetillus 123
bernsteini, Coelops 69
bhaktii, Tylonycteris 122
bicolor, Kerivoula 97
   Hipposideros 60
   Pipistrellus 114
   Taphozous 42
bidens, Boneia 20, **147**
   Sturnira 86
   Tonatia 75, **165**
   Vampyressa 89
bifax, Nyctophilus 131
bilabiatum, Pygoderma 93, **173**
bilineata, Saccopteryx 46, **158**
bilobatum, Uroderma 86, **170**
bivittata, Chaerephon 140
blainvillii, Mormoops 72
blanfordi, Hesperoptenus 123
   Sphaerias 35, **153**
blasii, Rhinolophus 56
blepotis, Miniopterus 134
blossevillii, Lasiurus 129
blythi, Rhinolophus 56
blythii, Myotis 101
bobrinskoi, Eptesicus 119
bocagei, Myotis 106
   Tadarida 139
bocharicus, Rhinolophus 54
bodenheimeri, Pipistrellus 115
bogotensis, Artibeus 92
   Sturnira 86
bokermanni, Lonchophylla 78
boliviensis, Phylloderma 77
bombifrons, Erophylla 79, **168**
bombinus, Myotis 101
bonariensis, Eumops 143
bondae, Molossus 144
Boneia 20
   bidens 20, **147**
boothi, Pteronotus 71
borbonicus, Scotophilus 128
borealis, Lasiurus 129, **182**
borneensis, Rhinolophus 53
borneoensis, Myotis 106
   Pipistrellus 114
botswanae, Laephotis 121
bottae, Eptesicus 120
bougainville, Nyctimene 36
brachycephala, Myonycteris 20
brachycephalus, Vampyrops 87
brachygnathus, Rhinolophus 54
brachymeles, Molossops 137
brachyotis, Cynopterus 33
   Hipposideros 62
   Lasiurus 129
   Micronycteris 73
   Rousettus 20
Brachyphylla 79
   cavernarum 79, **167**
   intermedia 79
   minor 79
   nana 79
   pumila 79
Brachyphyllinae 78
brachypterus, Mops 141
   Philetor 122, **179**
brachysoma, Cynopterus 33
brandti, Myotis 103
brasiliense, Tonatia 75
brasiliensis, Eptesicus 120, **178**
   Tadarida 10, 139
bregullae, Chaerephon 140
brevicauda, Carollia 84
breviceps, Hipposideros 63

brevirostris, Cormura 47, **159**
    Glossophaga 80
brocki, Vampyressa 88
brockmani, Nycteris 50
    Rhinolophus 57
brooksi, Dyacopterus 34
browni, Myotis 104
brunneus, Pipistrellus 117
    Pteropus 22
bucharensis, Myotis 103
buettikoferi, Epomops 31
bulleri, Macrotus 74
bulmerae, Aproteles 28, **149**
bunkeri, Antrozous 130
    Rhinolophus 59
bureschi, Myotis 108
burius, Rhinolophus 59

cabrerai, Noctilio 70
Cabreramops 138
cadornae, Pipistrellus 114
caffer, Hipposideros 63
cagayanus, Pteropus 22
calcaratus, Hipposideros 61
caldwelli, Rhinolophus 57
calidus, Rhinolophus 56
californicus, Eumops 144
    Macrotus 74
    Myotis 104
caliginosus, Myotis 104
calypso, Rhinolophus 55
camerunensis, Hipposideros 64
camortae, Pipistrellus 112
campestris, Glossophaga 81
canescens, Murina 132
    Saccopteryx 46
caniceps, Pteropus 23
canus, Pteropus 22
canuti, Rhinolophus 59
capaccinii, Myotis 107
capensis, Nycteris 50
    Pipistrellus 117
    Rhinolophus 54
capistratus, Pteropus 25
capito, Saccolaimus 44
caprenus, Nycticeius 126
caraccioli, Vampyrodes 88, **171**
Carioderma 5, 51
    cor 51, **161**
carimatae, Myotis 106
carissima, Myotis 108
carolinae, Syconycteris 39
Carollia 84
    azteca 84
    brevicauda 84
    castanea 84
    perspicillata 84, **170**
    subrufa 84
    tricolor 84
Carolliinae 83
carrikeri, Tonatia 75
carteri, Eptesicus 120
    Mormoops 72
    Myotis 105
cartilagonodus, Otopteropus 35, **154**
Casinycteris 32
    argynnis 32, **151**
castanea, Carollia 84
castaneus, Lasiurus 129
    Molossus 144
    Scotophilus 128
caucasicus, Pipistrellus 114
caucensis, Myotis 105

caudatus, Cheiromeles 138
    Lasiurus 130
caudifera, Anoura 82
caurinus, Pipistrellus 116
cavaticus, Taphozous 42
cavernarum, Brachyphylla 79, **167**
celaeno, Nyctimene 37
celebensis, Acerodon 27
    Harpyionycteris 30
    Hipposideros 63
    Megaderma 51
    Miniopterus 134
    Rhinolophus 53
    Rousettus 20, 147
    Scotophilus 128
centralis, Centronycteris 46
    Hipposideros 63
    Promops 144, **187**
    Tonatia 75
centrasiaticus, Eptesicus 119
Centronycteris 46
    centralis 46
    maximiliani 46, **158**
Centurio 4, 5, 93
    greenhalli 93
    senex 93, **174**
centurio, Ametrida 93, **173**
cephalotes, Nyctimene 36
ceramicus, Hipposideros 66
cerastes, Molossops 137
certans, Nyctimene 36
cervinus, Hipposideros 62, 63
ceylonense, Megaderma 51
ceylonensis, Cynopterus 33
ceylonicus, Pipistrellus 114
Chaerephon 139
    aloysiisabaudiae 140
    ansorgei 140
    bemmelini 139
    bivittata 140
    bregullae 140
    chapini 140
    cisturus 140
    colonicus 140
    elphicki 140
    faini 140
    frater 140
    gallagheri 141
    gambianus 140
    hindei 140
    jobensis 140, **186**
    johorensis 140
    lancasteri 140
    langi 140
    leucogaster 140
    limbata 140
    major 140
    nigeriae 140
    nigri 140
    plicata 140
    pumila 140
    pusillus 140
    russata 140
    shortridgei 140
    solomonis 140
    spillmani 140
    webtseri 140
Chalinolobus 123
    alboguttatus 125
    argentatus 125
    beatrix 124
    dwyeri 124
    egeria 125
    gleni 125
    gouldii 124, **180**
    humeralis 124

    kenyacola 125
    machadoi 125
    morio 124
    neocaledonicus 124
    nigrogriseus 124
    phalaena 125
    picatus 124
    poensis 124
    rogersi 124
    sheila 125
    superbus 125
    tuberculatus 124
    variegatus 125
    venatoris 124
chapini, Chaerephon 140
chapmani, Dobsonia 29
    Eptesicus 120
chaseni, Rhinolophus 53
Cheiromeles 1, 138
    caudatus 138
    jacobsoni 138
    parvidens 138
    torquatus 138, **185**
chiloensis, Myotis 107
Chilonatalus 95
Chilonycteris 71
chinensis, Miniopterus 134
    Myotis 101
    Rhinolophus 58
chiralensis, Eptesicus 120
Chiroderma 2, 89
    doriae 89
    improvisum 89
    jesupi 89
    salvini 89
    scopaeum 89
    trinitatum 89
    villosum 89, **171**
Chironax 34
    melanocephalus 34, **153**
Chiroptera 1, 12, 14
Choeroniscus 82
    godmani 83
    inca 83
    intermedius 83, **169**
    minor 83
    persiosus 83
    ponsi 83
Choeronycteris 4, 83
    harrisoni 83, **170**
    mexicana 83
christiei, Plecotus 110
Chrotopterus 1, 77
    auritus 77, **166**
chrysauchen, Pteropus 26
chrysoproctus, Pteropus 24
ciliolabrum, Myotis 104
cineraceus, Hipposideros 61
cinerea, Otonycteris 129
cinereus, Artibeus 92
    Lasiurus 10, 129
circe, Rhinolophus 55
circumdatus, Pipistrellus 115
cirrhosus, Trachops 77, **166**
Cistugo 109
cisturus, Chaerephon 140
clarus, Pipistrellus 112
clavium, Emballonura 45
clinedaphus, Monophyllus 81
clivosus, Rhinolophus 54
Cloeotis 68
    australis 68
    percivali 68, **163**
cobanensis, Myotis 107
coburgiana, Mormopterus 136
coelophyllus, Rhinolophus 59

Coelops 69
   bernsteini 69
   formosanus 69
   frithi 69, **164**
   hirsuta 69
   inflatus 69
   robinsoni 69
   sinicus 69
Coelopsini 68
coffini, Trachops 77
cognatus, Pteropus 24
   Rhinolophus 56
coibensis, Molossus 145
Coleura 45
   afra 45, **158**
   gallarum 45
   kummeri 45
   seychellensis 45
   silhouettae 45
colias, Scotophilus 128
collinus, Pipistrellus 112
colombiae, Histiotus 122
colonicus, Chaerephon 140
colonus, Pteropus 22
commersoni, Hipposideros 67
commissarisi, Glossophaga 80
comorensis, Pteropus 23
compressus, Macrotus 74
concava, Lonchophylla 78
concolor, Artibeus 91
   Cynopterus 33
condylurus, Mops 141, **186**
congicus, Mops 142
consobrinus, Scotophilus 128
conspicillatus, Pteropus 26
continentis, Myotis 106
   Natalus 95
convexum, Uroderma 86
cor, Cardioderma 51, **161**
   Emballonura 45
cordofanicum, Rhinopoma 40
Cormura 47
   brevirostris 47, **159**
cornutus, Rhinolophus 56
coromandra, Pipistrellus 112
coronatus, Hipposideros 61
Corvira 86
corynophyllus, Hipposideros 64
coxi, Hipposideros 64
   Pipistrellus 116
cozumelae, Mimon 76
Craseonycteridae 2, 3, **10**, 41
Craseonycteris 41
   thonglongyai 41, **157**
crassa, Syconycteris 39
crassicaudatus, Molossus 145
crassulus, Pipistrellus 116
crassus, Saccolaimus 44
creaghi, Rhinolophus 59
crenulata, Dobsonia 29
crenulatum, Mimon 76, **165**
creticus, Rhinolophus 55
crumeniferus, Hipposideros 63
crypta, Kerivoula 99
crypticola, Rousettus 20
crypturus, Epomophorus 31
cubanus, Nycriceius 126
culex, Pipistrellus 113
cultrata, Anoura 82
cuneatus, Rhinolophus 55
cupidus, Hipposideros 61
cupreolus, Myotis 106
cuprosa, Kerivoula 98
cuprosus, Pipistrellus 115
curasoae, Leptonycteris 81
curtatus, Pipistrellus 115

curtus, Hipposideros 63
custos, Hipposideros 66
cuvieri, Mormoops 72
cyclops, Hipposideros 64, **162**
   Peropteryx 48
cyclotis, Murina 132
   Nyctimene 36
cynocephala, Tadarida 139
Cynomops 137
Cynopterina 33
Cynopterini 32
Cynopterus 33
   altitudinis 33
   angulatus 33
   archipelagus 33
   babi 33
   brachyotis 33
   brachysoma 33
   ceylonensis 33
   concolor 33
   harpax 33
   horsfieldi 33
   insularum 33
   javanicus 33
   lyoni 33
   major 33
   minor 33
   minutus 33
   pagensis 33
   persimilis 33
   princeps 33
   scherzeri 33
   serasani 33
   sphinx 33
   terminus 33
   titthaecheilus 33, **152**
cystops, Rhinopoma 41
Cyttarops 48
   alecto 48, **160**

dabbenei, Eumops 143
daedalus, Nyctophilus 131
daitoensis, Pteropus 23
damarensis, Nycteris 50
   Pipistrellus 117
   Rhinolophus 55
   Scotophilus 128
darioi, Stenoderma 93
darjelingensis, Barbastella 110
darlingi, Rhinolophus 55
darlingtoni, Pipistrellus 116
dasycneme, Myotis 109
dasymallus, Pteropus 23
Dasypterus 130
dasythrix, Miniopterus 134
daubentoni, Myotis 107
daubentonii, Myopterus 138, **185**
davidi, Myotis 103
daviesi, Micronycteris 74
davisi, Uroderma 86
davisoni, Promops 144
davyi, Pteronotus 71
debilis, Molossus 145
deckeni, Rhinolophus 55
degelidus, Lasiurus 129
deignani, Myotis 106
dekeyseri, Lonchophylla 78
delticus, Eumops 143
demissus, Eptesicus 119
   Hipposideros 66
demonstrator, Mops 141
denti, Rhinolophus 53
denticulus, Eudiscopus 111, **177**
Depanycteris 48
depressa, Kerivoula 99
Dermanura 91

desertorum, Myotis 102
Desmodontinae 6, 8, 9, 94
Desmodus 6, 94
   rotundus 94, **174**
diadema, Hipposideros 66
Diaemus 94
   youngi 94, **174**
dianae, Emballonura 45
Diclidurus 48
   albus 48, **160**
   ingens 48
   isabella 48
   scutatus 48
   virgo 48
diluta, Asellia 68
diminutus, Eptesicus 121
dinelli, Myotis 107
dingani, Scotophilus 128
dinops, Hipposideros 66
Diphylla 94
   ecaudata 94, **174**
Dirias 70
discifera, Thyroptera 96
discolor, Phyllostomus 76
diversus, Rhinolophus 58
dobsoni, Epomops 30, **150**
   Pteropus 23
Dobsonia 28
   anderseni 29
   chapmani 29
   crenulata 29
   emersa 29
   exoleta 29
   grandis 29
   inermis 30
   magna 29
   minimus 30
   minor 28
   moluccensis 29
   nesea 30
   pannietensis 29, **149**
   peroni 29
   praedatrix 29
   remota 29
   sumbana 29
   viridis 29
Dobsoniina 28
dogalensis, Myotis 106
dohrni, Rhinolophus 57
domincensis, Myotis 105
doriae, Chiroderma 89
   Hesperoptenus 123
   Hipposideros 62
   Mormopterus 136
dormeri, Pipistrellus 116
dorsalis, Vampyrops 87
douglasorum, Pipistrellus 116
downsi, Promops 144
draconilla, Nyctimene 36
drungicus, Pipistrellus 115
dryas, Myotis 106
dubiaquercus, Bauerus 130, **182**
dupreanum, Eidolon 19
durgadasi, Hipposideros 61
dutertreus, Eptesicus 120
dwyeri, Chalinolobus 124
Dyacopterus 34
   brooksi 34
   spadiceus 34, **152**
dyacorum, Hipposideros 62

ecaudata, Diphylla 94, **174**
ecaudatus, Megaerops 34, **152**
Ectophylla 90
   alba 90, **171**
edax, Rhinolophus 58

edulis, Pteropus 26
ega, Lasiurus 130
　Scleronycteris 82, **169**
egeria, Chalinolobus 125
egregius, Lasiurus 129
ehrhardti, Trachops 77
Eidolon 18
　helvum 19, **147**
　dupreanum 19
　sabaeum 19
eileenae, Murina 132
eisentrauti, Pipistrellus 114
elegans, Myotis 105
elongata, Glossophaga 81
elongatus, Phyllostomus 76
eloquens, Rhinolophus 57
elphicki, Chaerephon 140
emarginatus, Myotis 102
　Scotomanes 128
Emballonura 5, 44
　alecto 44
　atrata 44
　beccarii 44
　clavium 45
　cor 45
　dianae 45
　furax 45, **157**
　meeki 45
　monticola 44
　nigrescens 44
　palauensis 45
　palawanensis 44
　papuana 44
　raffrayana 45
　rivalis 44
　rotensis 45
　semicaudata 45
　solomonis 44
　stresemanni 45
　sulcata 45
Emballonuridae 3, 5, 6, 9, **11**, 41
Emballonurinae 2
Emballonuroidea 5, 40
emersa, Dobsonia 29
empusa, Rhinolophus 57
Enchisthenes 90
endoi, Pipistrellus 112
engana, Kerivoula 99
enganus, Hipposideros 66
　Pteropus 22
Eonycteris 7, 37
　major 38
　robusta 38
　rosenbergi 38
　spelaea 38, **155**
eotinus, Pteropus 25
Epomophorini 30
Epomophorus 31
　angolensis 31
　anurus 31
　crypturus 31
　gambianus 31, **150**
　haldemani 31
　labiatus 31
　minor 31
　parvus 31
　pousaguesi 31
　reii 31
　wahlbergi 31
Epomops 30
　buettikoferi 31
　dobsoni 30, **150**
　franqueti 31
　strepitans 31
Eptesicus 119
　anatolicus 120

　andersoni 120
　andinus 120
　argentinus 120
　bahamensis 120
　bensoni 120
　bernardinus 120
　bobrinskoi 119
　bottae 120
　brasiliensis 120, **178**
　carteri 120
　centrasiaticus 119
　chapmani 120
　chiralensis 120
　demissus 119
　diminutus 121
　dutertreus 120
　fidelis 121
　findleyi 121
　floweri 121
　furinalis 120
　fuscus 120
　gaumeri 120
　gobiensis 119
　guadeloupensis 120
　hingstoni 120
　hispaniolae 120
　horikawae 120
　hottentotus 120
　innesi 120
　innoxius 121
　isabellinus 120
　japonensis 119
　kashgaricus 119
　kobayashii 120
　matschiei 119
　melanopterus 120
　miradorensis 120
　montosus 120
　nasutus 119
　nilssoni 119
　ognevi 120
　omanensis 120
　osceola 120
　pachyomus 120
　pachyotis 119
　pallens 120
　pallidus 120
　parvus 119
　pashtonus 120
　pellucens 119
　peninsulae 120
　petersoni 120
　platyops 120
　portavernus 120
　serotinus 120
　shiraziensis 120
　tatei 119
　thomasi 120
　turcomanus 120
　wetmorei 120
epularius, Pteropus 26
erigens, Hipposideros 61
eriophora, Kerivoula 98
erongensis, Mormopterus 137
Erophylla 79
　bombifrons 79, **168**
　mariguanensis 79
　planifrons 79
　santacristobalensis 79
　sezekorni 79
　syops 79
erroris, Hipposideros 61
erythromos, Sturnira 85
escalerae, Rhinolophus 57
eschscholtzii, Miniopterus 134
esperitosantensis, Natalus 95

　Nyctinomops 142
Euderma 110
　maculatum 110, **177**
Eudiscopus 111
　denticulus 111, **177**
Eumops 143
　amazonicus 143
　auripendulus 143
　beckeri 143
　bonariensis 143
　californicus 144
　dabbenei 143
　delticus 143
　floridanus 143
　geijskesi 143
　gigas 144
　glaucinus 143
　hansae 143
　major 143
　maurus 143
　mederai 143
　nanus 143
　perotis 144
　sonoriensis 143
　trumbulli 144
　underwoodi 143, **187**
euotis, Hipposideros 66
europs, Nyctinomops 143
euryale, Rhinolophus 56
euryotis, Rhinolophus 58
evotis, Myotis 102
　Tonatia 75
exiguus, Rhinolophus 60
exoleta, Dobsonia 29
exsul, Rhinolophus 58
extremus, Myotis 105

faini, Chaerephon 140
falabae, Scotoecus 128
falcatus, Phyllops 92, **172**
fallax, Artibeus 91
　Rhinolophus 52
Falsistrellus 115
famulus, Rhinolophus 56
faunulus, Pteropus 22
feae, Rhinolophus 55
federatus, Myotis 106
femorosaccus, Nyctinomops 142
fernandezi, Lonchorhina 74
ferruginea, Nyctinomops 143
ferrumequinum, Rhinolophus 54
fidelis, Eptesicus 121
fimbriatus, Myotis 108
findleyi, Eptesicus 121
　Myotis 105
finlaysoni, Pipistrellus 116
finschi, Syconycteris 39
fischerae, Rhinophylla 84, **170**
fischeri, Haplonycteris 36, **155**
fitzsimonsi, Mormopterus 137
flavescens, Artieus 93, **172**
　Mesophylla 90
　Pipistrellus 117
flaviventris, Saccolaimus 43, **160**
flavomaculatus, Saccolaimus 44
flora, Kerivoula 99
floresi, Acerodon 27
floridanus, Eumops 143
　Lasiurus 130
　Pipistrellus 112
florium, Murina 132
floweri, Eptesicus 121
foetidus, Rhinolophus 58
formosae, Rhinolophus 58
formosanus, Coelops 69
formosus, Myotis 101

Otomops 142
Pteropus 23
fortidens, Myotis 106
fortis, Molossus 145
fosteri, Promops 144
fouriei, Pipistrellus 113
foxi, Rhinolophus 58
franqueti, Epomops 31
frantzii, Lasiurus 129
frater, Chaerephon 140
    Myotis 103
fraterculus, Artibeus 91
    Miniopterus 134
fraternus, Macroglossus 38
fretensis, Taphozous 42
frithi, Coelops 69, **164**
frons, Lavia 52, **161**
frosti, Neopteryx 28, **149**
fructivorus, Macroglossus 38
fujiensis, Myotis 103
fulginosus, Artibeus 91
    Hipposideros 63
    Miniopterus 134
    Pteronotus 71
fulminans, Tadarida 139
fulvida, Tylonycteris 122
fulvus, Hipposideros 61
    Pteronotus 71
fumigatus, Rhinolophus 57
fundatus, Pteropus 24
furax, Emballonura 45, **157**
furculus, Triaenops 68
furinalus, Eptesicus 120
Furipteridae 2, **14**, 96
Furipterus 96
    horrens 96, **175**
furvus, Nyctalus 118
fuscatus, Lasiurus 130
    Pipistrellus 114
fuscipes, Pipistrellus 116
fuscus, Eptesicus 120
    Miniopterus 134
    Murina 132
    Pteronotus 71

gailliardi, Pteropus 26
gairdneri, Scotophilus 128
galeritus, Hipposideros 62, 63
gallagheri, Chaerephon 141
gallarum, Coleura 45
gambianus, Chaerephon 140
    Epomophorus 31, **150**
gambiensis, Nycteris 50
garambae, Pipistrellus 117
gardneri, Myotis 107
gaskelli, Hesperoptenus 123, **180**
gaumeri, Eptesicus 120
geddiei, Pteropus 23
geijskesi, Eumops 143
geminus, Nyctimene 37
genovensium, Platalina 78, **167**
genowaysi, Rhogeesa 127
gentilis, Hipposideros 61
geoffroyi, Anoura 82, **169**
    Nyctophilus 131
georgianus, Taphozous 43
giganteus, Pteropus 26
gigas, Eumops 144
    Hipposideros 67
    Macroderma 51, **161**
gilberti, Hipposideros 61
gilvus, Acerodon 27
glaucinus, Eumops 143
Glauconycteris 124
gleni, Chalinolobus 125
Glischropus 118

batjanus 119
javanus 119
tylopus 119, **178**
Glossophaga 80
    antillarum 80
    bakeri 80
    brevirostris 80
    campestris 81
    commissarisi 80
    elongata 81
    handleyi 80
    hespera 80
    leachii 80
    longirostris 80
    major 81
    mexicana 80
    morenoi 80
    mutica 80
    reclusa 81
    rostrata 81
    soricina 80, **168**
    valens 80
Glossophaginae 6, 80
Glyphonycteris 73
gobiensis, Eptesicus 119
godmani, Choeroniscus 83
gonavensis, Pteronotus 71
goudoti, Myotis 102
gouldi, Nyctophilus 131, **183**
    Pteropus 26
gouldii, Chalinolobus 124, **180**
goweri, Pteropus 22
gracilior, Pipistrellus 117
gracilis, Myotis 103
    Rhinolophus 56
    Rhogeesa 127
grandidiere, Pipistrellus 117
grandis, Dobsonia 29
    Hipposideros 66
    Micropteropus 32
    Miniopterus 134
    Nycteris 49
    Pteropus 24
greenhalli, Centurio 93
    Molossops 137, **185**
greyii, Nycticeius 126
grisea, Murina 133
griseiventer, Molossops 138
grisescens, Myotis 10, **107**
griseus, Hipposideros 66
    Pteronotus 71
    Pteropus 22
grivaudi, Miniopterus 133
guadeloupensis, Eptesicus 120
guineensis, Hipposideros 63
    Pipistrellus 117
    Rhinolophus 55
gymnonotus, Pteronotus 71
gymnura, Saccopteryx 46

haagneri, Mormopterus 137
haedinus, Taphozous 42
hainanus, Rhinolophus 54
haitensis, Phyllops 92
haldemani, Epomophorus 31
halophyllus, Hipposideros 61
hamiltoni, Taphozous 43
handleyi, Glossophaga 80
    Lonchophylla 78, **167**
hansae, Eumops 143
Haplonycteris 35
    fischeri 36, **155**
harardai, Miniopterus 134
hardwickei, Kerivoula 99
    Rhinopoma 40, **157**
harpax, Cynopterus 33

harpia, Harpiocephalus 133, **184**
Harpiocephalus 6, 133
    harpia 133, **184**
    lasyurus 133
    madrassius 133
    mordax 133
    rufulus 133
Harpiola 133
Harpyionycterini 30
Harpyionycteris 30
    celebensis 30
    negrosensis 30
    whiteheadi 30, **149**
harrisoni, Choeronycteris 83, **170**
    Kerivoula 98
    Rhinopoma 40
harti, Artibeus 90
hasseltii, Myotis 106
hastatus, Phyllostomus 76, **165**
heathi, Scotophilus 128
hedigeri, Rousettus 20
heffernani, Pteropus 23
helios, Pipistrellus 113
helleri, Vampyrops 88, **170**
helvum, Eidolon 19, **147**
hemprichi, Otonycteris 129, **182**
hercules, Artibeus 91
herero, Scotophilus 128
hermani, Myotis 101
herrei, Myotis 104
hespera, Glossophaga 80
hesperia, Lonchophylla 78
Hesperoptenus 123
    blanfordi 123
    doriae 123
    gaskelli 123, **180**
    tickelli 123
    tomesi 123
hesperus, Artibeus 92
    Pipistrellus 113
hildebrandti, Rhinolophus 57
hildegardeae, Taphozous 42
hilgendorfi, Murina 132
hilli, Pteropus 26
    Rhinolophus 57
    Taphozous 42
himalayanus, Rhinolophus 54
hindei, Chaerephon 140
    Scotoecus 128
hingstoni, Eptesicus 120
Hipposiderina 60
Hipposiderinae 60
Hipposiderini 60
Hipposideros 1, 2, **4**, 60
    abae 65
    albaniensis 61
    alongensis 66
    amboiensis 61
    angolensis 63
    antricola 61
    armiger 65
    aruensis 61
    ater 61
    atrox 61
    barbensis 66
    batchianensis 63
    beatus 64
    bicolor 60
    brachyotis 62
    breviceps 63
    caffer 63
    calcaratus 61
    camerunensis 64
    celebensis 63
    centralis 63
    ceramicus 66

cervinus 62, 63
cineraceus 61
commersoni 67
coronatus 61
corynophyllus 64
coxi 64
crumeniferus 63
cupidus 61
curtus 63
custos 66
cyclops 64, **162**
demissus 66
diadema 66
dinops 66
doriae 62
durgadasi 61
dyacorum 62
enganus 66
erigens 61
erroris 61
euotis 66
fuliginosus 63
fulvus 61
galeritus 62, 63
gentilis 61
gigas 67
gilberti 61
grandis 66
griseus 66
guineensis 63
halophyllus 61
indus 66
inexpextatus 67
inornatus 66
insolens 62
jonesi 62
labuanensis 63
lamottei 63
lankadiva 66
larvatus 66
leptophyllus 66
lekaguli 66
longicaudus 62
lylei 65
macrobullatus 61
maggietaylorae 61
major 61
malaitensis 66
marisae 62
marungensis 67
masoni 66
maximus 64
megalotis 60
micropus 61
mirandus 66
misorensis 63
mixtus 66
muscinus 64
nanus 63
natunensis 66
neglectus 66
nequam 61
niangarae 67
niapu 63
nicobarensis 66
nicobarulae 61
obscurus 62
oceanitis 66
pallidus 61
papua 64
pelingensis 67
pendleburyi 65
pomona 61
poutensis 66
pratti 65
pullatus 66

pygmaeus 62
reginae 66
ridleyi 62
ruber 63
sabanus 62
saevus 61
schistaceus 66
schneideri 63
semoni 64
sinensis 61
speculator 66
speoris 66
stenotis 65
sumbae 66
tephrus 63
terasensis 65
thomensis 67
tranninhensis 65
trobrius 66
turpis 65
unitus 66
wollastoni 64
wrighti 61
hipposideros, *Rhinolophus* 57
hirsuta, *Coelops* 69
   Micronycteris 73
hirsutus, *Artibeus* 91
   Rhinolophus 57
hirundo, *Scotoecus* 127, **181**
hispaniolae, *Eptesicus* 120
hispida, *Nycteris* 50
Histiotus 122
   alienus 122
   colombiae 122
   inambarus 122
   laephotis 122
   macrotus 122
   magellanicus 122
   montanus 122, **179**
   velatus 122
hobbit, *Syconycteris* 39
homezi, *Micronycteris* 72
homochrous, *Plecotus* 110
hondurensis, *Sturnira* 86
horikawae, *Eptesicus* 120
horrens, *Furipterus* 96, **175**
horsfieldi, *Cynopterus* 33
   Megaderma 51
horsfieldii, *Myotis* 106
hosonoi, *Myotis* 103
hottentotus, *Eptesicus* 120
howensis, *Pteropus* 22
humbloti, *Pipistrellus* 117
humeralis, *Chalinolobus* 124
   Nycticeius 126, **181**
humilis, *Acerodon* 27
huttoni, *Murina* 132
Hylonycteris 82
   minor 82
   underwoodi 82, **169**
hypomelanus, *Pteropus* 22
Hypsignathus 3, 30
   monstrosus 30, **150**
Hypsugo 111

Ia 118
   beaulieui 118
   io 118, **178**
icarus, *Otomops* 142
ikonnikovi, *Myotis* 103
imaizumii, *Rhinolophus* 56
imbrensis, *Scotomanes* 128
imbricatus, *Pipistrellus* 113
imogene, *Pharotis* 131, **183**
importunus, *Rhinolophus* 53
improvisum, *Chiroderma* 89

inambarus, *Histiotus* 122
inca, *Choeroniscus* 83
incae, *Pteronotus* 71
incarum, *Vampyrops* 88
incautus, *Myotis* 107
indicus, *Pipistrellus* 114
indus, *Hipposideros* 66
inermis, *Donbsonia* 30
inexpextatus, *Hipposideros* 67
   Pipistrellus 114
inflata, *Pteronotus* 71
inflatus, *Coelops* 69
   Miniopterus 134
influatus, *Nycticeius* 126
infumatus, *Rousettus* 20
infusca, *Balantiopteryx* 46
infuscus, *Vampyrops* 87
ingens, *Diclidurus* 48
   Plecotus 111
innesi, *Eptesicus* 120
innoxius, *Eptesicus* 120
inopinatus, *Artibeus* 91
inops, *Rhinolophus* 59
inornatus, *Hipposideros* 66
insignis, *Tadarida* 139
insolens, *Hipposideros* 62
insularis, *Lasiurus* 130
   Miniopterus 134
   Pteropus 25
   Scotophilus 128
insularum, *Cynopterus* 33
   Myotis 103
interior, *Myotis* 108
intermedia, *Brachyphylla* 79
   Kerivoula 99
   Mormoops 72
   Nycteris 49
   Peropteryx 47
   Tadarida 139
intermedius, *Artibeus* 90
   Choeroniscus 83, **169**
   Lasiurus 130
   Micropteropus 32
   Pteropus 26
io, *Balantiopteryx* 47
   Ia 118, **178**
irani, *Rhinolophus* 55
isabella, *Diclidurus* 48
isabellinus, *Eptesicus* 120
italosomalica, *Asellia* 68

jacobsoni, *Cheiromeles* 138
jagorii, *Kerivoula* 100
   Ptenochirus 33, **152**
jamaicensis, *Artibeus* 91, **172**
   Macrotus 74
   Natalus 95
japonensis, *Eptesicus* 119
japoniae, *Miniopterus* 134
javanica, *Nycteris* 49
javanicus, *Cynopterus* 33
   Pipistrellus 112
   Rhinolophus 53
javanus, *Glischropus* 119
   Kerivoula 100
jeannei, *Myotis* 106
jesupi, *Chiroderma* 89
jin, *Otonycteris* 129
jobensis, *Chaerephon* 140, **186**
joffreyi, *Nyctalus* 118
johorensis, *Chaerephon* 140
jonesi, *Hipposideros* 62
jubatus, *Acerodon* 27
judaicus, *Rhinolophus* 56
jugularis, *Mormopterus* 136
juquiaensis, *Thyroptera* 96

kachensis, Taphozous 43
kaguyae, Myotis 103
kalinowskii, Mormopterus 136, **185**
kampenii, Taphozous 42
kapalgensis, Taphozous 42
kappleri, Peropteryx 47
kashgaricus, Eptesicus 119
keaysi, Myotis 105
keenani, Mimon 76
keenii, Myotis 102
kempi, Rousettus 20
keniensis, Rhinolophus 54
kenyacola, Chalinolobus 125
Kerivoula 1, 97
  aerosa, 100
  africana 98
  agnella 98
  argentata 99
  atrox 100
  bellissima 98
  bicolor 97
  crypta 99
  cuprosa 98
  depressa 99
  engana 99
  eriophora 98
  flora 99
  hardwickei 99
  harrisoni 98
  intermedia 99
  jagorii 100
  javanus 100
  lanosa 97
  lenis 99
  malayana 99
  malpasi 99
  minuta 99
  muscilla 98
  muscina 98
  myrella 99
  nidicola 99
  papillosa 99, **176**
  papuensis 100
  pellucida 99
  phalaena 98
  picta 98
  pusilla 97
  rapax 100
  smithi 98
  whiteheadi 97
  zuluensis 99
Kerivoulinae 97
keyensis, Pteropus 24
  Rhinolophus 52
  Syconycteris 39
kinabalu, Megaderma 51
kinneari, Rhinopoma 40
kitchneri, Pipistrellus 113
klossi, Rhinolophus 54
knorri, Rhinolophus 54
kobayashii, Eptesicus 120
koepckeae, Mimon 76
kolombatovici, Plecotus 110
koopmani, Antrozous 130
  Ardops 92
  Aselliscus 67
kuboriensis, Tadarida 139
kuhlii, Pipistrellus 114
  Scotophilus 128, **182**
kummeri, Coleura 45
kusnotoi, Megaerops 34

labiata, Nycteris 50
labiatus, Epomophorus 31
  Nyctalus 118
labuanensis, Hipposideros 63

Laephotis 121
  angolensis 121, **179**
  botswanae 121
  namibensis 121
  wintoni 121
laephotis, Histiotus 122
  Tonatia 75
lagochilus, Macroglossus 38
lambi, Molossus 145
  Myotis 108
Lamingtona 130
lamottei, Hipposideros 63
Lampronycteris 73
lancasteri, Chaerephon 140
landeri, Rhinolophus 55
lanei, Mops 142
lanensis, Pteropus 26
langi, Chaerephon 140
laniger, Myotis 107
lankadiva, Hipposideros 66
lanosa, Kerivoula 97
lanosus, Rhinolophus 58
  Rousettus 20
larensis, Myotis 105
larvatus, Hipposideros 66
Lasionycteris 109
  noctivagans 109, **176**
lasiopterus, Nyctalus 118
lasiopyga, Anoura 82
Lasiurini 129
Lasiurus 1, 3, 9, 129
  blossevillii 129
  borealis 129, **182**
  brachyotis 129
  castaneus 129
  caudatus 130
  cinereus 10, 129
  degelidus 129
  ega 130
  egregius 129
  floridanus 130
  frantzii 129
  fuscatus 130
  insularis 130
  intermedius 130
  minor 129
  panamensis 130
  pfeifferi 129
  seminolus 129
  semotus 129
  teliotis 129
  varius 129
  villosissimus 129
  xanthinus 130
lasyurus, Harpiocephalus 133
laticaudatus, Nyctinomops 142, **187**
Latidens 35
  salimalii 35, **154**
latidens, Anoura 82
latifolius, Phyllostomus 76
  Rhinolophus 54
latirostris, Myotis 104
lavellanus, Pteropus 24
Lavia 5, 51
  affinis 52
  frons 52, **161**
  rex 52
leachi, Rousettus 19
leachii, Glossophaga 80
lebonoticus, Nyctalus 118
leibii, Myotis 104
leisleri, Nyctalus 118
lekaguli, Hipposideros 66
lenis, Kerivoula 99
leonis, Kerivoula 99
leonis, Mops 141

lepidus, Natalus 96
  Pipistrellus 114
  Rhinolophus 55
leporinus, Noctilio 70, **164**
leptodon, Myonycteris 21
Leptonycteris 81
  curasoae 81
  nivalis 81, **169**
  sanborni 81
  tarlosti 81
  yerbabuenae 81
leptophyllus, Hipposideros 66
leptura, Saccopteryx 46
leschenaulti, Rousettus 19
lesueuri, Myotis 109
lesviacus, Myotis 101
leucocephalus, Pteropus 26
leucogaster, Chaerephon 140
  Murina 132, **183**
  Scotophilus 128
leucomelas, Barbastella 109, **176**
  Pipistrellus 116
Leuconoe 105
leucophaeus, Otonycteris 129
leucopleura, Taphozous 42
leucopterus, Peropteryx 48
  Pteropus 25
leucostigma, Mops 141
leucotis, Acerodon 27, **148**
levis, Myotis 107
Lichonycteris 81
  obscura 81, **169**
lilium, Sturnira 85
limbata, Chaerephon 140
lineatus, Vampyrops 87
Lionycteris 77
  spurrelli 77, **167**
liops, Pteropus 25
Liponycteris 43
Lissonycteris 20
lituratus, Artibeus 90
livingstonei, Pteropus 24
lobata, Tadarida 139
lobatus, Rhinolophus 55
lombocensis, Pteropus 24
Lonchophylla 77
  bokermanni 78
  concava 78
  dekeyseri 78
  handleyi 78, **167**
  hesperia 78
  mordax 78
  robusta 78
  thomasi 78
Lonchophyllinae 77
Lonchorhina 74
  aurita 74, **165**
  fernandezi 74
  marinkellei 75
  occidentalis 74
  orinocensis 74
longicaudatus, Myotis 103
longicaudus, Hipposideros 62
longicrus, Myotis 108
longifolium, Mimon 76
longimanus, Taphozous 42
longipes, Myotis 108
longirostris, Glossophaga 80
loochoensis, Pteropus 23
lophorhina, Nyctophilus 131
lophurus, Pipistrellus 113
loriae, Mormopterus 136
lucasi, Penthetor 35, **154**
luciae, Ardops 92
  Monophyllus 81
lucifer, Acerodon 27

lucifugus, Myotis 10, 108
luctus, Rhinolophus 58
ludovici, Sturnira 86
luisi, Sturnira 85
lullulae, Nyctimene 37
luteola, Nycteris 50
luteus, Pteropus 22
lutosus, Myotis 108
lylei, Hipposideros 65
　Pteropus 25
lyoni, Cynopterus 33
lyra, Megaderma 51, **161**
Lyroderma 51

macarenensis, Nyctinomops 143
macassaricus, Pteropus 22
macconnelli, Mesophylla 90, **171**
macdonaldi, Notopteris 39, **157**
macdougalli, Molossus 144
macellus, Myotis 106
macer, Natalus 95
machadoi, Chalinolobus 125
macinnesi, Rhinopoma 41
mackenziei, Pipistrellus 115
mackloti, Acerodon 27
maclaudi, Rhinolophus 57
macleayii, Pteronotus 71
macmillani, Mormopterus 137
　Pteropus 25
macrobullaris, Plecotus 110
macrobullatus, Hipposideros 61
macrocephalicus, Myotis 101
macrodactylus, Myotis 108
macrodens, Miniopterus 134
Macroderma 51
　gigas 51, **161**
Macroglossinae 37
Macroglossini 37
Macroglossus 38
　fraternus 38
　fructivorus 38
　lagochilus 38
　microtus 38
　minimus 38
　nanus 38
　pygmeus 38
　sobrinus 38, **156**
Macrophyllum 75
　macrophyllum 75, **165**
macrophyllum, Macrophyllum 75, **165**
macropus, Myotis 106
macrotarsus, Myotis 108
macrotis, Nycteris 50
　Nyctinomops 143
　Peropteryx 47, **159**
　Pipistrellus 115
　Plecotus 111
　Pteropus 26
　Rhinolophus 57
Macrotus 74
　bulleri 74
　californicus 74
　compressus 74
　jamaicensis 74
　mexicanus 74
　minor 74
　waterhousii 74, **165**
macrotus, Histiotus 122
macrurus, Rhinolophus 54
maculata, Balionycteris 34, **153**
maculatum, Euderma 110, **177**
madagascariensis, Nycteris 50
　Otomops 142
　Rousettus 20
maderensis, Pipistrellus 114
madrassius, Harpiocephalus 133

madurensis, Rhinolophus 53
magellanicus, Histiotus 122
maggietaylorae, Hipposideros 61
magna, Dobsonia 29
　Pygoderma 44
　Sturnira 86
magnamolaris, Myotis 107
magnater, Miniopterus 134, **184**
magnirostrum, Uroderma 86
magnus, Taphozous 43
mahaganus, Pteropus 27
major, Artibeus 92
　Chaerephon 140
　Cynopterus 33
　Eonycteris 38
　Eumops 143
　Glossophaga 81
　Hipposideros 61
　Natalus 95
　Nycteris 49
　Nyctimene 37
　Nyctophilus 131
　Syconycteris 39
　Vampyrodes 88
majori, Miniopterus 134
　Rhinolophus 57
majus, Megaderma 51
majusculus, Triaenops 68
malaccensis, Pteropus 26
malaitensis, Hipposideros 66
　Nyctimene 37
malayana, Kerivoula 99
　Tylonycteris 122
malayanus, Rhinolophus 53
maloneyi, Mimetillus 123, **180**
malpasi, Kerivoula 99
manavi, Miniopterus 133
marginatus, Pipistrellus 114
mariannus, Pteropus 23, **148**
marica, Nycteris 50
mariguanensis, Erophylla 79
marinkellei, Lonchorhina 75
maris, Pteropus 22
marisae, Hipposideros 62
maros, Rhinolophus 57
marrensis, Pipistrellus 114
marshalli, Rhinolophus 58
martiensseni, Otomops 142, **186**
martiniquensis, Myotis 105
marungensis, Hipposideros 67
masalai, Nyctimene 37
masoni, Hipposideros 66
mastersoni, Tadarida 139
mastivus, Molossops 137
　Noctilio 70
matroka, Pipistrellus 117
matschiei, Eptesicus 119
mattogrossensis, Molossops 138
mauritianus, Taphozous 43
maurus, Eumops 143
maximiliani, Centronycteris 46, **158**
maximus, Hipposideros 64
　Pipistrellus 113
mcintyrei, Rhinolophus 60
mearnsi, Pteropus 22
mecklenburzevi, Nyctalus 118
mederai, Eumops 143
media, Nycteris 50
medium, Megaderma 51
medius, Miniopterus 134
meeki, Emballonura 45
Megachiroptera 3, 6, 7, 8, 9, 17, 18
Megaderma 51
　celebensis 51
　ceylonense 51
　horsfieldi 51

　kinabalu 51
　lyra 51, **161**
　majus 51
　medium 51
　minus 51
　pangandarana 51
　sinensis 51
　spasma 51
　trifolium 51
Megadermatidae 2, 3, 5, 6, 7, 8, 9, **12**, 50
Megaerops 34
　ecaudatus 34, **152**
　kusnotoi 34
　niphanae 34
　wetmorei 34
Megaloglossus 38
　woermanni 38, **156**
megalophylla, Mormoops 71, **164**
megalotis, Hipposideros 60
　Micronycteris 72, **164**
　Paracoelops 69, **164**
megaphyllus, Rhinolophus 52
mehelyi, Rhinolophus 56
melanocephalus, Chironax 34, **153**
melanopogon, Pteropus 24
　Taphozous 42
melanops, Melonycteris 39, **156**
melanopterus, Eptesicus 120
melanorhinus, Myotis 104
melanotus, Pteropus 24
melckorum, Pipistrellus 117
melissa, Vampyressa 88
Melonycteris 39
　aurantius 39
　melanops 39, **156**
　woodfordi 39
meridionalis, Rhinolophus 56
mesoamericanus, Pteronotus 71
Mesophylla 89
　flavescens 90
　macconnelli 90, **171**
mexicana, Choeronycteris 83
　Glossophaga 80
　Micronycteris 72
　Tadarida 10, 139
mexicanus, Macrotus 74
　Molossops 137
　Natalus 95
　Nycticeius 126
　Plecotus 110
　Pteronotus 71
meyeni, Pipistrellus 112
meyeri, Tylonycteris 122
meyeroehmi, Rhinolophus 57
miarensis, Mops 142
Microchiroptera 3, 5, 6, 7, 8, 9, 17, **40**
microdon, Nyctophilus 131
Micronycteris 72
　behni 74
　brachyotis 73
　daviesi 74
　hirsuta 73
　homezi 72
　megalotis 72, **164**
　mexicana 72
　microtis 72
　minuta 73
　nicefori 73
　pusilla 73
　schmidtorum 73
　sylvestris 74
microphyllum, Rhinopoma 40
Micropterus 32
　grandis 32
　intermedius 32
　pusillus 32, **151**

micropus, Hipposideros 61
   Natalus 95
microtis, Micronycteris 72
   Nyctophilus 131
microtus, Macroglossus 38
midas, Mops 142
   Rhinolophus 57
Milithronycteris 123
milleri, Molossops 137
   Molossus 145
   Myotis 102
Mimetillus 122
   berneri 123
   maloneyi 123, **180**
   thomsi 123
Mimon 75, 76
   bennettii 76
   cozumelae 76
   crenulatum 76, **165**
   keenani 76
   koepckeae 76
   longifolium 76
   picatum 76
mimus, Pipistrellus 113
   Pteropus 22
minahassae, Pipistrellus 115
minimus, Dobsonia 30
   Macroglossus 38
Miniopterinae 133
Miniopterus 133
   africanus 134
   arenarius 134
   australis 134
   blepotis 134
   celebensis 134
   chinensis 134
   dasythrix 134
   eschscholtzii 134
   fraterculus 134
   fuliginosus 134
   fuscus 134
   grandis 134
   grivaudi 133
   harardai 134
   inflatus 134
   insularis 134
   japoniae 134
   macrodens 134
   magnater 134, **184**
   majori 134
   manavi 133
   medius 134
   minor 133
   natalensis 134
   newtoni 133
   oceanensis 134
   orianae 134
   pallidus 134
   parvipes 134
   paululus 133
   propritristis 134
   pusillus 134
   robustior 134
   rufus 145
   schreibersi 134
   shortridgei 133
   smitianus 134
   solomonensis 134
   tibialis 134
   tristis 134
   vicinior 134
   villiersi 134
   witkampi 134
minor, Antrozous 130
   Artibeus 92
   Brachyphylla 79

Choeroniscus 83
Cynopterus 33
Dobsonia 28
Epomophorus 31
Hylonycteris 82
Lasiurus 129
Macrotus 74
Miniopterus 133
Noctilio 70
Ptenochirus 33
Rousettus 20
minus, Megaderma 51
minuta, Kerivoula 99
   Micronycteris 73
minutilla, Rhogeessa 127
minutillus, Rhinolophus 56
minutus, Cynopterus 33
   Mormopterus 136
   Nyctimene 36
   Rhinolophus 57
mira, Rhogeessa 127
miradorensis, Eptesicus 120
mirandus, Hipposideros 66
misorensis, Hipposideros 63
mitratus, Rhinolophus 58
mixtus, Hipposideros 66
   Saccolaimus 43
modigliani, Pteropus 24
molaris, Uroderma 86
Molossidae 1, 2, 3, 5, 6, 8, 9, **16**, 135
molossinus, Pteropus 24
Molossops 137
   abrasus 137
   aequatorianus 138
   brachymeles 137
   cerastes 137
   greenhalli 137, **185**
   griseiventer 138
   mastivus 137
   mattogrossensis 138
   mexicanus 137
   milleri 137
   neglectus 137
   paranus 137
   planirostris 137
   sylvia 138
   temminckii 138
Molossus 144
   ater 144
   aztecus 145
   barnesi 145
   bondae 144
   castaneus 144
   coibensis 145
   crassicaudatus 145
   debilis 145
   fortis 145
   lambi 145
   macdougalli 144
   milleri 145
   molossus 144, **187**
   nigricans 144
   pretiosus 144
   pygmaeus 145
   sinaloae 144
   trinitatis 144
   tropidorhynchus
   verrillii 145
molossus, Molossus 144, **187**
moluccarum, Myotis 106
moluccensis, Dobsonia 29
monachus, Rhinolophus 52
monoceros, Rhinolophus 56
monoensis, Pteropus 24
Monophyllus 81
   clinedaphus 81

   luciae 81
   plethodon 81
   portoricensis 81
   redmani 81, **168**
monstrosus, Hypsignathus 30, **150**
montanus, Histiotus 122, **179**
   Nyctalus 118
   Rhinolophus 57
monticola, Emballonura 44
   Rhinolophus 55
montivagus, Myotis 106
montosus, Eptesicus 120
montserratensis, Ardops 92
Mops 141
   brachypterus 141
   condylurus 141, **186**
   congicus 142
   demonstrator 141
   lanei 142
   leonis 141
   leucostigma 141
   miarensis 142
   midas 142
   mops 142
   nanulus 141
   niangarae 141
   niveiventer 141
   orientis 141
   osborni 141
   petersoni 141
   sarasinorum 142
   spurrelli 141
   thersites 141
   trevori 141
   wonderi 141
mops, Mops 142
mordax, Harpiocephalus 133
   Lonchophylla 78
   Pipistrellus 114
   Sturnira 86
morenoi, Glossophaga 80
morio, Chaliolobus 124
   Pteropus 26
   Rhinolophus 58
Mormopterus 135, 136
   acetabulosus 136
   astrolabiensis 136
   beccarii 136
   coburgiana 136
   doriae 136
   erongensis 137
   fitzsimonsi 137
   haagneri 137
   jugularis 136
   kalinowskii 136, **185**
   loriae 136
   macmillani 137
   minutus 136
   norfolkensis 136
   petrophilus 137
   planiceps 136
   ridei 136
   setiger 137
   umbratus 137
Mormoopidae 2, 6, 8, **13**, 70
Mormoops 3, 5, 71
   blainvillii 72
   carteri 72
   cuvieri 72
   intermedia 72
   megalophylla 71, **164**
   tumidiceps 72
Mormopterus 5
morrisi, Myotis 102
Mosia 44
motalavae, Pteropus 25

moupinensis, Myotis 104
muricola, Myotis 103
Murina 131
  aenea 133
  aurata 132
  canescens 132
  cyclotis 132
  eileenae 132
  florium 132
  fuscus 132
  grisea 133
  hilgendorfi 132
  huttoni 132
  leucogaster 132, **183**
  peninsularis 132
  puta 132
  rozendaali 132
  rubella 132
  rubex 132
  silvatica 132
  suilla 132
  tenebrosa 132
  tubinaris 132
  ussuriensis 132
murina, Tadarida 139
Murininae 1, 131
murinus, Vespertilio 121, **179**
murraiana, Asellia 68
murrayi, Pipistrellus 112
muscatellum, Rhinopoma 41
musciculus, Pipistrellus 113
muscilla, Kerivoula 98
muscina, Kerivoula 98
muscinus, Hipposideros 64
muscula, Tadarida 139
Musonycteris 4, 83
mutica, Glossophaga 80
Myonycteris 20
  brachycephala 21
  leptodon 21
  relicta 21
  torquata 21, **147**
  wroughtoni 21
Myopterus 138
  albatus 138
  daubentonii 138, **185**
  whitleyi 138
Myotini 100
Myotis 1, **5**, 100
  abbotti 106
  abei 108
  adversus 106
  aelleni 107
  alascensis 108
  albescens 108
  altarium 104
  alticraniatus 104
  amotus 108
  ancilla 101
  anjouanensis 102
  annectans 104
  apache 102
  araxenus 101
  atacamensis 105
  ater 104
  auriculus 102
  australis 104
  austroriparius 108
  aztecus 101–102
  bartelsi 101
  bechsteini 100
  blythii 101
  bocagei 106
  bombinus 101
  borneoensis 106
  brandti 103

  browni 104
  bucharensis 103
  bureschi 108
  californicus 104
  caliginosus 104
  capaccinii 107
  carimatae 106
  carissima 108
  carteri 105
  caucensis 105
  chiloensis 107
  chinensis 101
  ciliolabrum 104
  cobanensis 107
  continentis 106
  cupreolus 106
  dasycneme 109
  daubentoni 107
  davidi 103
  deignani 106
  desertorum 102
  dinelli 107
  dogalensis 106
  dominicensis 105
  dryas 106
  elegans 105
  emarginatus 102
  evotis 102
  extremus 105
  federatus 106
  fimbriatus 108
  findleyi 105
  formosus 101
  fortidens 106
  frater 103
  fujiensis 103
  gardneri 107
  goudoti 102
  gracilis 103
  grisescens 10, 107
  hasseltii 106
  hermani 101
  herrei 104
  horsfieldii 106
  hosonoi 103
  ikonnikovi 103
  incautus 107
  insularum 103
  interior 108
  jeannei 106
  kaguyae 103
  keaysi 105
  keenii 102
  lambi 108
  laniger 107
  larensis 105
  latirostris 104
  leibii 104
  lesueuri 109
  lesviacus 101
  levis 107
  longicaudatus 103
  longicrus 108
  longipes 108
  lucifugus 10, 108
  lutosus 108
  macellus 106
  macrocephalicus 101
  macrodactylus 108
  macropus 106
  macrotarsus 108
  magnamolaris 107
  martiniquensis 105
  melanorhinus 104
  milleri 102
  moluccarum 106

  montivagus 106
  morrisi 102
  moupinensis 104
  muricola 103
  myotis 100
  mystacinus 103
  nathalinae 107
  nattereri 101
  nesopolus 105
  niasensis 104
  nigricans 105, **176**
  nipalensis 103
  nugax 104
  nyctor 105
  occultus 108
  omari 101
  oreias 104
  orientis 106
  oxalis 108
  oxygnathus 101
  oxyotus 107
  ozensis 103
  pacificus 102
  pahasapensis 101
  peninsularis 107
  pequinius 101
  pernox 108
  peshwa 106
  petax 107
  peytoni 106
  pilosotibialis 105
  pilosus 109
  planiceps 104
  primula 104
  pruinosus 108
  przewalskii 103
  punensis 105
  punicus 101
  relictus 108
  ricketti 109
  ridleyi 104
  riparius 106
  rosseti 104
  ruber 106
  rufoniger 101
  rufopictus 101
  saba 109
  saturatus 102
  schaubi 101
  scotti 104
  seabrai 109
  septentrionalis 102
  sicarius 101
  siligorensis 104
  simus 106
  sociabilis 108
  sodalis 105
  sogdianus 103
  sonoriensis 106
  sowerbyi 104
  stalkeri 109
  taiwanensis 106
  thaianus 104
  thysanodes 101
  transcaspicus 103
  tricolor 102
  tschuliensis 101
  tsuensis 101
  turcomanicus 102
  ussuriensis 107
  velifer 107
  vivesi 8, 109
  volans 108
  volgensis 107
  watasei 101
  weberi 101

welwitschii 101
yesoensis 103
yumanensis 108
myotis, Myotis 100
myrella, Kerivoula 99
Myropteryx 47
Mystacina 1, 6, 135
　　robusta 135, **184**
　　tuberculata 135
Mystacinidae **16**, 135
mystacinus, Myotis 103
Myzopoda 97
　　aurita 97, **175**
Myzopodidae 3, **14**, 96

naias, Syconycteris 39
najdiya, Nycteris 50
namibensis, Laephotis 121
nana, Brachyphylla 79
　　Nycteris 49
　　Sturnira 86
Nanonycteris 32
　　veldkampi 32, **151**
nanulus, Mops 141
　　Pipistrellus 116
nanus, Artibeus 92
　　Eumops 143
　　Hipposideros 63
　　Macroglossus 38
　　Pipistrellus 113
　　Rhinolophus 53
naso, Rhynchonycteris 46, **158**
nasutus, Eptesicus 119
　　Promops 144
natalensis, Miniopterus 134
　　Natalus 95
Natalidae 8, **14**, 95
natalis, Pteropus 24
Natalus 95
　　continentis 95
　　esperitosantensis 95
　　jamaicensis 95
　　lepidus 96
　　macer 95
　　major 95
　　mexicanus 95
　　micropus 95
　　natalensis 95
　　stramineus 95
　　tronchonii 95
　　tumidifrons 95
　　tumidorostris 95, **175**
nathalinae, Myotis 102
nathusii, Pipistrellus 112
nattereri, Myotis 101
natunae, Pteropus 26
natunensis, Hipposideros 66
nawaiensis, Pteropus 24
neglectus, Hipposideros 66
　　Molossops 137
negrosensis, Harpyionycteris 30
neocaledonica, Notopteris 39
neocaledonicus, Chalinolobus 124
neohibernicus, Pteropus 26
Neonycteris 73
Neoplatymops 138
Neopteryx 28
　　frosti 28, **149**
Neoromicia 117
nequam, Hipposideros 61
nereis, Rhinolophus 53
nesea, Dobsonia 30
nesites, Rhinolophus 54
Nesonycteris 39
nesopolus, Myotis 105
newtoni, Miniopterus 133

niadicus, Pteropus 24
niangarae, Hipposideros 67
　　Mops 141
niapu, Hipposideros 63
niasensis, Myotis 104
　　Rhinolophus 58
nicefori, Micronycteris 73
nichollsi, Ardops 92, **172**
nicobarensis, Hipposideros 66
nicobarulae, Hipposideros 61
nidicola, Kerivoula 99
nigellus, Vampyrops 87
niger, Pteropus 24
nigeriae, Chaerephon 140
nigrescens, Emballonura 44
　　Thoopterus 35, **153**
nigri, Chaerephon 140
nigricans, Molossus 144
　　Myotis 105, **176**
nigrita, Scotophilus 128
nigritellus, Scotophilus 128
nigrogriseus, Chalinolobus 124
nilssoni, Eptesicus 119
nipalensis, Myotis 103
niphanae, Megaerops 34
nippon, Rhinolophus 55
nitendiensis, Pteropus 25
nitidus, Pipistrellus 112
nivalis, Leptonycteris 81, **169**
niveiventer, Mops 141
Noctilio 2, 7, 70
　　affinis 70
　　albiventris 70
　　cabrerai 70
　　leporinus 70, **164**
　　mastivus 70
　　minor 70
　　rufescens 70
Noctilionidae 3, **13**, 69
Noctilionoidea 69
noctivagans, Lasionycteris 109, **176**
noctula, Nyctalus 9, 118, **178**
nordmanni, Rhinolophus 56
norfolkensis, Mormopterus 136
notius, Pipistrellus 117
Notopterini 39
Notopteris 18, 39
　　macdonaldi 39, **157**
　　neocaledonica 39
novaeguineae, Aselliscus 67
novaehebridensis, Aselliscus 67
nucella, Scotophilus 128
nudaster, Taphozous 43
nudicluniatus, Saccolaimus 44
nudiventris, Taphozous 43
nugax, Myotis 104
nux, Scotophilus 128
Nyctalus **5**, 117
　　anthonyi 118
　　aviator 118
　　azoreum 118
　　furvus 118
　　joffreyi 118
　　labiatus 118
　　lasiopterus 118
　　lebonoticus 118
　　leisleri 118
　　mecklenburzevi 118
　　montanus 118
　　noctula 9, 118, **178**
　　plancei 118
　　stenopterus 118
　　velutinus 118
　　verrucosus 118
Nycteridae 3, 5, 9, **12**, 49
Nycteris 49

adana 50
aethiopica 50
angolensis 50
arge 49
aurita 50
bastiani 49
benuensis 50
brockmani 50
capensis 50
damarensis 50
gambiensis 50
grandis 49
hispida 50
intermedia 49
javanica 49
labiata 50
luteola 50
macrotis 50
madagascariensis 50
major 49
marica 50
media 50
najdiya 50
nana 49
oriana 50
parisii 50
proxima 50
sabiensis 50
thebaica 50
tragata 49, **160**
villosa 50
vinsoni 50
woodi 50
Nycticeini 125
Nycticeinops 126
Nycticeius 125
　　balstoni 126
　　caprenus 126
　　cubanus 126
　　greyii 126
　　humeralis 126, **181**
　　influatus 126
　　mexicanus 126
　　rueppellii 126
　　sanborni 126
　　schlieffeni 126
　　subtropicalis 126
Nyctiellus 95
Nyctimene **2**, **3**, 36
　　aello 37, **155**
　　albiventer 36
　　bougainville 36
　　celaeno 37
　　cephalotes 36
　　certans 36
　　cyclotis 36
　　draconilla 36
　　geminus 37
　　lullulae 37
　　major 37
　　malaitensis 37
　　masalai 37
　　minutus 36
　　papuanus 36
　　rabori 37
　　robinsoni 37
　　sanctacrucis 37
　　scitulus 37
　　varius 36
　　vizcaccia 36
Nyctimenina 6, 36
Nyctinomops 142
　　aurispinosus 143
　　esperitosantensis 142
　　europs 143
　　femorosaccus 142

ferruginea 143
laticaudatus 142, **187**
macarenensis 143
macrotis 143
similis 143
yucatanica 143
Nyctophilini 3, 130
Nyctophilus 130
  arnhemensis 131
  bifax 131
  daedalus 131
  geoffroyi 131
  gouldi 131, **183**
  lophorhina 131
  major 131
  microdon 131
  microtis 131
  sherrini 131
  timoriensis 131
  walkeri 130
nyctor, Myotis 105
nymphea, Vampyressa 88

obliviosus, Rousettus 20
obscura, Lichonycteris 81, **169**
obscurus, Hipposideros 62
obtusa, Phyllonycteris 79
occidentalis, Lonchorhina 74
  Rousettus 19
  Sturnira 86
  Tonatia 75
occultus, Myotis 108
  Promops 144
oceanensis, Miniopterus 134
oceanitis, Hipposideros 66
ocularis, Pteropus 26
ocypete, Aethalops 35
ognevi, Eptesicus 120
omanensis, Eptesicus 120
omari, Myotis 101
ophiodon, Scotonycteris 32
oporophilum, Sturnira 86
oratus, Vampyrops 87
oreias, Myotis 104
oriana, Nycteris 50
orianae, Miniopterus 134
orientalis, Vespertilio 121
orientis, Myotis 106
orientis, Mops 141
orinocensis, Lonchorhina 74
ornatus, Anthops 67, **162**
  Pteropus 23
  Scotomanes 128, **181**
  Vampyrodes 88
osborni, Mops 141
osceli, Eptesicus 120
osgoodi, Rhinolophus 55
Otomops 142
  formosus 142
  icarus 142
  madagascariensis 142
  martiensseni 142, **186**
  papuensis 142
  secundus 142
  wroughtoni 142
Otonycteris 129
  cinerea 129
  hemprichi 129, **182**
  jin 129
  leucophaeus 129
  petersi 129
Otopteropus 35
  cartilagonodus 35, **154**
oxalis, Myotis 108
oxygnathus, Myotis 101

oxyotus, Myotis 107
ozensis, Myotis 103

pachyomus, Eptesicus 120
pachyotis, Eptesicus 119
pachypus, Tylonycteris 122
pacificus, Antrozous 130
  Myotis 102
packardi, Antrozous 130
paganensis, Pteropus 23
pagensis, Cynopterus 33
pagi, Rhinolophus 56
pahasapensis, Myotis 101
palatinus, Artibeus 92
palauensis, Emballonura 45
palawanensis, Emballonura 44
pallens, Eptesicus 120
pallescens, Plecotus 111
pallida, Balantiopteryx 47
pallidus, Antrozous 130, **183**
  Eptesicus 120
  Hipposideros 61
  Miniopterus 134
  Pteropus 22
  Scotoecus 127
palmarum, Artibeus 90
pamana, Promops 144
panamensis, Lasiurus 130
  Phyllostomus 76
panayensis, Scotophilus 128
pangandarana, Megaderma 51
pannietensis, Dobsonia 29, **149**
papillosa, Kerivoula 99, **176**
papua, Hipposideros 64
papuana, Emballonura 44
  Syconycteris 39
papuanus, Nyctimene 36
  Pipistrellus 112
  Pteropus 26
papuensis, Kerivoula 100
  Otomops 142
Paracoelops 69
  megalotis 69, **164**
paradoxolophus, Rhinolophus 58
paraguensis, Pteronotus 71
paranus, Molossops 137
Paranyctimene 36
  raptor 36, **155**
parcus, Rhinolophus 56
parisii, Nycteris 50
parnelli, Pteronotus 70, **164**
parvidens, Cheiromeles 138
  Sturnira 85
parvipes, Artibeus 91
  Miniopterus 134
parvula, Rhogeesa 127
parvus, Epomophorus 31
  Eptesicus 119
  Rhinolophus 53
pashtonus, Eptesicus 120
pastoris, Pteropus 25
paterculus, Pipistrellus 113
patrizii, Asellia 68
paucidentata, Alionycteris 35, **154**
paululus, Miniopterus 133
paulus, Artibeus 91
pearsoni, Rhinolophus 58
peguensis, Pipistrellus 112
pelewensis, Pteropus 23
peli, Saccolaimus 44
pelingensis, Hipposideros 67
pellucens, Eptesicus 119
pellucida, Kerivoula 99
pendleburyi, Hipposideros 65
peninsulae, Eptesicus 120
peninsularis, Myotis 107

  Murina 132
Penthetor 35
  lucasi 35, **154**
pequinius, Myotis 101
perauritus, Rhinolophus 57
percivali, Cloeotis 68, **163**
perforatus, Taphozous 42
periosus, Choeroniscus 83
permixtus, Pipistrellus 112
perniger, Rhinolophus 58
pernox, Myotis 108
peroni, Dobsonia 29
Peronymus 47
Peropteryx 47
  cyclops 48
  intermedia 47
  kappleri 47
  leucopterus 48
  macrotis 47, **159**
  phaea 47
  trinitatis 47
perotis, Eumops 144
persicus, Triaenops 68, **163**
persimilis, Cynopterus 33
personatus, Pteronotus 71
  Pteropus 25
perspicillata, Carollia 84, **170**
perspicillifer, Saccopteryx 46
peruana, Anoura 82
peshwa, Myotis 106
petax, Myotis 107
petersi, Otonycteris 129
  Pipistrellus 113
  Rhinolophus 54
petersoni, Eptesicus 120
  Mops 141
petrophilus, Mormopterus 137
peytoni, Myotis 106
pfeifferi, Lasiurus 129
phaea, Peropteryx 47
phaeocephalus, Pteropus 25
phaeotis, Artibeus 92
phalaena, Chalinolobus 125
  Kerivoula 98
Pharotis 131
  imogene 131, **183**
phasma, Pipistrellus 117
Philetor 122
  brachypterus 122, **179**
  rohui 122
  veraecundus 122
philippinensis, Rhinolophus 57
  Rousettus 20
  Taphozous 42
Phoniscus 99
Phygetis 21
Phylloderma 76
  boliviensis 77
  septentrionalis 77
  stenops 76, **166**
Phyllodia 70
Phyllonycterinae 79
Phyllonycteris 6, 79
  aphylla 79
  obtusa 79
  poeyi 79, **168**
Phyllops 92
  falcatus 92, **172**
  haitensis 92
Phyllostomidae 3, 4, 6, 7, 8, 9, **13**, 72
Phyllostominae 2, 9, 72
Phyllostomoidea 9
Phyllostomus 76
  discolor 76
  elongatus 76
  hastatus 76, **165**

211

latifolius 76
panamensis 76
verrucosus 76
phyllotis, Plecotus 110
picatum, Mimon 76
picatus, Chalinolobus 124
picta, Kerivoula 98
pilosotibialis, Myotis 105
pilosus, Myotis 109
  Pteropus 25
  Rhinolophus 59
Pipistrellus 2, 111
  abramus 112
  adamsi 112
  aegyptius 114
  aero 114
  affinis 113
  africanus 112
  aladdin 111
  alaschanicus 114
  anchietai 114
  angulatus 112
  arabicus 112
  ariel 115
  ater 117
  austenianus 114
  babu 113
  baverstocki 116
  bellieri 114
  bicolor 114
  bodenheimeri 115
  borneoensis 114
  brunneus 117
  cadornae 114
  camortae 112
  capensis 117
  caucasicus 114
  caurinus 116
  ceylonicus 114
  circumdatus 115
  clarus 112
  collinus 112
  coromandra 112
  coxi 116
  crassulus 116
  culex 113
  cuprosus 115
  curtatus 115
  damarensis 117
  darlingtoni 116
  dormeri 116
  douglasorum 116
  drungicus 115
  eisentrauti 114
  endoi 112
  finlaysoni 116
  flavescens 117
  floridanus 112
  fouriei 113
  fuscatus 114
  fuscipes 116
  garambae 117
  gracilior 117
  grandidieri 117
  guineesis 117
  helios 113
  hesperus 113
  humbloti 117
  imbricatus 113
  indicus 114
  inexpectatus 114
  javanicus 112
  kitchneri 113
  kuhlii 114
  lepidus 114
  leucomelas 116
  lophurus 113
  mackenziei 115
  macrotis 115
  maderensis 114
  marginatus 114
  marrensis 114
  matroka 117
  maximus 113
  melchorum 117
  meyeni 112
  mimus 113
  minahassae 115
  mordax 114
  murrayi 112
  musciculus 113
  nanulus 116
  nanus 113
  nathusii 112
  nitidus 112
  notius 117
  papuanus 112
  paterculus 113
  peguensis 112
  permixtus 112
  petersi 113
  phasma 117
  pipistrellus 111
  ponceleti 112
  portensis 112
  principulus 113
  pulcher 116
  pulveratus 113
  pumilus 116
  raptor 114
  rectitragus 117
  regulus 116
  rendalli 117
  rueppelli 116
  rusticus 114
  sagittula 116
  savii 114
  senegalensis 116
  sewelanus 112
  shanorum 115
  societatis 115
  somaliscus 117
  stampflii 113
  sturdeei 111
  subcanus 114
  subflavus 112, **177**
  subtilis 114
  subulidens 112
  tasmaniensis 115
  tenuipinnis 117
  tenuis 112
  tonfangensis 114
  tramatus 112
  ugandae 117
  veraecrucis 112
  vordermanni 115
  vulturnus 116
  wattsi 112
  westralis 112
  yunnanensis 113
  zuluensis 117
pipistrellus, Pipistrellus 111
plancei, Nyctalus 118
planiceps, Mormopterus 136
  Myotis 104
planifrons, Erophylla 79
planirostris, Artibeus 91
  Molossops 137
Platalina 78
  genovensium 78, **167**
Platymops 137
platyops, Eptesicus 120
Platyrrhinus 87
Plecotini 109
Plecotus **5**, 110
  ariel 110
  auritus 110
  australis 111
  austriacus 110
  christiei 110
  homochrous 110
  ingens 111
  kolombatovici 110
  macrobullaris 110
  macrotis 111
  mexicanus 110
  pallescens 111
  phyllotis 110
  rafinesquii 111
  sacrimontis 110
  teneriffae 110
  townsendii 110, **177**
  uenoi 110
  virginianus 111
  wardi 110
Plerotes 30
  anchietai 30, **150**
plethodon, Monophyllus 81
plicata, Chaerephon 140
  Balantiopteryx 47, **159**
pluto, Saccolaimus 44
pluton, Pteropus 26
poensis, Chalinolobus 124
poeyi, Phyllonycteris 79, **168**
pohlei, Pteropus 26
poliocephalus, Pteropus 26
pomona, Hipposideros 61
ponceleti, Pipistrellus 112
pondoensis, Scotophilus 128
ponsi, Choeroniscus 83
portavernus, Eptesicus 120
portensis, Pipistrellus 112
portoricensis, Monophyllus 81
  Pteronotus 71
pousarguesi, Epomophorus 31
poutensis, Hipposideros 66
praedatrix, Dobsonia 29
praestans, Rhinolophus 59
prajae, Acerodon 27
pratti, Hipposideros 65
pretiosus, Molossus 144
primula, Myotis 104
princeps, Cynopterus 33
  Rhinolophus 54
principulus, Pipistrellus 113
pronconsularis, Rhinolophus 60
Promops 144
  ancilla 144
  centralis 144, **187**
  davisoni 144
  downsi 144
  fosteri 144
  nasutus 144
  occultus 144
  pamana 144
propritristis, Miniopterus 134
proxima, Nycteris 50
proximus, Rhinolophus 55
pruinosus, Myotis 108
przewalskii, Myotis 103
pselaphon, Pteropus 25
psilotis, Pteronotus 71
Ptenochirus 33
  jagorii 33, **152**
  minor 33
Pteralopex 1, 27
  acrodonta 28
  anceps 28, **148**

atrata 28
Pterocyon 18
Pteronotus 70, 71
   boothi 71
   davyi 71
   fuliginosus 71
   fulvus 71
   fuscus 71
   gonavensis 71
   griseus 71
   gymnonotus 71
   incae 71
   inflata 71
   macleayii 71
   mesoamericanus 71
   mexicanus 71
   paraguensis 71
   parnelli 70, **164**
   personatus 71
   portoricensis 71
   psilotis 71
   pusillus 71
   quadridens 71
   rubiginosus 71
   suapurensis 71
   torrei 71
Pteropodidae 2, 3, 5, 6, 8, 9, **10**, 18
Pteropodina 21
Pteropodinae 18
Pteropodini 18
Pteropus **2**, **3**, 5, 21
   admiralitatum 22
   aldabrensis 23
   alecto 26
   anetianus 24
   annectens 22
   aorensis 25
   argentatus 23
   ariel 26
   aruensis 24
   aterrimus 26
   auratus 23
   bakeri 25
   balutus 22
   banksianus 25
   basiliscus 23
   brunneus 22
   cagayanus 22
   caniceps 23
   canus 22
   capistratus 25
   chrysauchen 26
   chrysoproctus 24
   cognatus 24
   colonus 22
   comorensis 23
   conspicillatus 26
   daitoensis 23
   dasymallus 23
   dobsoni 23
   edulis 26
   enganus 22
   eotinus 25
   epularius 26
   faunulus 22
   formosus 23
   fundatus 24
   gailliardi 26
   geddiei 23
   giganteus 26
   gouldi 26
   goweri 22
   grandis 24
   griseus 22
   heffernani 23
   hilli 26

   howensis 22
   hypomelanus 22
   insularis 25
   intermedius 26
   keyensis 24
   lanensis 26
   lavellanus 24
   leucocephalus 26
   leucopterus 25
   liops 25
   livingstonei 24
   lombocensis 24
   loochoensis 23
   luteus 22
   lylei 25
   macassaricus 22
   macmillani 25
   macrotis 26
   mahaganus 27
   malaccensis 26
   mariannus 23, **148**
   maris 22
   mearnsi 22
   melanopogon 24
   melanotus 24
   mimus 22
   modigliani 24
   molossinus 24
   monoensis 24
   morio 26
   motalavae 25
   natalis 24
   natunae 26
   nawaiensis 24
   neohibernicus 26
   niadicus 24
   niger 24
   nitendiensis 25
   ocularis 26
   ornatus 23
   paganensis 23
   pallidus 22
   papuanus 26
   pastoris 25
   pelewensis 23
   personatus 25
   phaeocephalus 25
   pilosus 25
   pluton 26
   pohlei 26
   poliocephalus 26
   pselaphon 25
   pumilus 22
   rayneri 24
   rennelli 24
   rodricensis 24
   rubianus 24
   rufus 23
   samoensis 24
   sanctacrucis 22
   satyrus 24
   scapulatus 27
   sepikensis 26
   seychellensis 23
   simalurus 22
   solitarius 24
   solomonis 22
   speciosus 22
   subniger 23
   tablasi 22
   temmincki 25
   tokudae 25
   tomesi 22
   tonganus 23
   tuberculatus 25
   tytleri 24

   ualanus 23
   ulthiensis 23
   vampyrus 26
   vetulus 25
   voeltzkowi 23
   woodfordi 27
   yapensis 23
pulcher, *Pipistrellus* 116
pullatus, *Hipposideros* 66
pulveratus, *Pipistrellus* 113
pumila, *Brachyphylla* 79
   Chaerephon 140
   Saccopteryx 46
pumilio, *Artibeus* 92
   Rhinophylla 84
pumilus, *Pipistrellus* 116
   Pteropus 22
punensis, *Myotis* 105
punicus, *Myotis* 101
punicus, *Myotis* 101
pusilla, *Kerivoula* 97
   Micronycteris 73
   Vampyressa 88, **171**
pusillum, *Rhinopoma* 41
pusillus, *Chaerephon* 140
   Micropteropus 32, **151**
   Miniopterus 134
   Pteronotus 71
   Rhinolophus 55
puta, *Murina* 132
pygmaeus, *Hipposideros* 62
   Macroglossus 38
   Molossus 145
Pygoderma 93
   bilabiatum 93, **173**
   magna 93

quadridens, *Pteronotus* 71
quadrivittatus, *Artibeus* 92

rabori, *Nyctimene* 37
raffrayana,
   Emballonura 45
rafinesquii, *Plecotus* 111
rapax, *Kerivoula* 100
raptor, *Paranyctimene* 36, **155**
   Pipistrellus 114
ravus, *Artibeus* 92
   Tomopeas 135, **184**
rayneri, *Pteropus* 24
recifinus, *Vampyrops* 87
reclusa, *Glossophaga* 81
rectitragus, *Pipistrellus* 117
redmani, *Monophyllus* 81, **168**
refulgens, *Rhinolophus* 55
reginae, *Hipposideros* 66
regulus, *Pipistrellus* 116
   Rhinolophus 55
reii, *Epomophorus* 31
Reithronycteris 79
relicta, *Myonycteris* 21
relictus, *Myotis* 108
remota, *Dobsonia* 29
rendalli, *Pipistrellus* 117
rennelli, *Pteropus* 24
rex, *Lavia* 52
   Rhinolophus 58
Rhinolophidae 2, 3, 5, 7, 8, **13**, 52
Rhinolophinae 52
Rhinolophoidea 48
Rhinolophus 1, 3, **4**, 52
   abae 58
   acrotis 54
   acuminatus 55
   adami 54
   aequalis 53

aethiops 58
affinis 54, **162**
alcyone 55
alleni 57
alticolus 53
andamanensis 54
anderseni 53
andreinii 57
angolensis 55
angustifolius 60
annectens 53
arcuatus 59
aruensis 59
audax 55
augur 54
barbarus 56
beccarii 60
beddomei 58
blasii 56
blythi 56
bocharicus 54
borneensis 53
brachygnathus 54
brockmani 57
bunkeri 59
burius 59
caldwelli 57
calidus 56
calypso 55
canuti 59
capensis 54
celebensis 53
chaseni 53
chinensis 58
circe 55
clivosus 54
coelophyllus 59
cognatus 56
cornutus 56
creaghi 59
creticus 55
cuneatus 55
damarensis 55
darlingi 55
deckeni 55
denti 53
diversus 58
dohrni 57
edax 58
eloquens 57
empusa 57
escalerae 57
euryale 56
euryotis 58
exiguus 60
exsul 58
fallax 52
famulus 56
feae 55
ferrumequinum 54
foetidus 58
formosae 58
foxi 58
fumigatus 57
gracilis 56
guineensis 55
hainanus 55
hildebrandti 57
hilli 57
himalayanus 54
hipposideros 57
hirsutus 57
imaizumii 56
importunus 56
inops 59
irani 55
javanicus 53
judaicus 56
keniensis 54
keyensis 52
klossi 54
knorri 54
landeri 55
lanosus 58
latifolius 54
lepidus 55
lobatus 55
luctus 58
maclaudi 57
macrotis 57
macrurus 54
madurensis 53
majori 57
malayanus 53
maros 57
marshalli 58
mcintyrei 60
megaphyllus 52
mehelyi 56
meridionalis 56
meyeroehmi 57
midas 57
minutillus 56
minutus 57
mitratus 58
monanchus 52
monoceros 56
montanus 57
monticola 55
morio 58
nanus 53
nereis 53
nesites 54
niasensis 58
nippon 55
nordmanni 56
osgoodi 55
pagi 56
paradoxolophus 58
parcus 56
parvus 53
pearsoni 58
perauritus 57
perniger 58
petersi 54
philippinensis 57
pilosus 59
praestans 59
princeps 54
proconsularis 60
proximus 55
pusillus 55
refulgens 55
regulus 55
rex 58
robertsi 57
robinsoni 54
rouxi 54
rubiginosus 54
rufus 59
ruwenzorii 57
sanborni 57
schwartzi 54
sedulus 58
septentrionalis 54
shameli 59
shortridgei 55
siamensis 57
silvestris 55
simplex 52
simulator 53
sinicus 54
sobrinus 58
solitarius 58
spadix 53
spurcus 58
stheno 53
subbadius 56
subrufus 59
sumatranus 55
superans 54
swinnyi 53
szechwanus 56
tatar 59
tener 54
thomasi 54
timidus 59
timoriensis 59
toxopeusi 60
tragatus 55
trifoliatus 58
truncatus 53
tunetae 56
vandeuseni 52
vespa 57
virgo 53
yunanensis 58
zuluensis 54
Rhinonycterina 68
Rhinonycteris 68
 aurantius 68, **168**
Rhinophylla 84
 alethina 84
 fischerae 84, **170**
 pumilio 84
Rhinopoma 1, 40
 arabium 41
 asirensis 40
 cordofanicum 40
 cystops 41
 hardwickei 40, **157**
 harrisoni 40
 kinneari 40
 macinnesi 41
 microphyllum 40
 muscatellum 41
 pusillum 41
 seianum 41
 senaariense 41
 sumatrae 40
 tropicalis 40
Rhinopomatidae 2, 3, 6, **10**, 40
Rhinopterus 121
rhodesiae, *Taphozous* 42
Rhogeessa 126
 alleni 127
 genowaysi 127
 gracilis 127
 minutilla 127
 mira 127
 parvula 127
 tumida 126, **181**
Rhynchonycteris 45
 naso 46, **158**
richardsoni, *Artibeus* 91
ricketti, *Myotis* 109
ridei, *Mormopterus* 136
ridleyi, *Hipposideros* 62
 Myotis 104
riparius, *Myotis* 106
rivalis, *Emballonura* 44
robertsi, *Rhinolophus* 57
robinsoni, *Coelops* 69
 Nyctimene 37
 Rhinolophus 54
robusta, *Eonycteris* 38
 Lonchophylla 78
 Mystacina 135, **184**

robustior, Miniopterus 134
robustula, Tylonycteris 122, **180**
robustus, Scotophilus 128
rodricensis, Pteropus 24
rogersi, Chalinolobus 124
rohui, Philetor 122
rosenbergi, Artibeus 92
    Eonycteris 38
rosseti, Myotis 104
rostrata, Glossophaga 81
rotensis, Emballonura 45
rotundus, Desmodus 94, **174**
Rousettina 18
Rousettus 1, 3, 5, 7, 18, 20, 19
    aegyptiacus 19
    amplexicaudatus 19
    angolensis 20
    arabicus 19
    brachyotis 20
    celebensis 20, **147**
    crypticola 20
    hedigeri 20
    infumatus 20
    kempi 20
    lanosus 20
    leachi 19
    leschenaulti 19
    madagascariensis 20
    minor 20
    obliviosus 20
    occidentalis 19
    philippinensis 20
    ruwenzorii 20
    seminudus 19
    shortridgei 19
    smithi 20
    spinalatus 20
    stresemanni 20
    unicolor 19
rouxi, Rhinolophus 54
rozendaali, Murina 132
rubella, Murina 132
ruber, Hipposideros 63
    Myotis 106
rubex, Murina 132
rubianus, Pteropus 24
rubiginosus, Pteronotus 71
    Rhinolophus 54
rueppelli, Pipistrellus 116
rueppellii, Nycticeius 126
rufescens, Noctilio 70
rufoniger, Myotis 101
rufopictus, Myotis 101
rufulus, Harpiocephalus 133
rufum, Stenoderma 93, **173**
rufus, Miniopterus 134
    Pteropus 23
    Rhinolophus 59
    Triaenops 68
russata, Chaerephon 140
rusticus, Pipistrellus 114
ruwenzorii, Rhinolophus 57
    Rousettus 20

saba, Myotis 109
sabaeum, Eidolon 19
sabanus, Hipposideros 62
sabiensis, Nycteris 50
Saccolaimus 2, 3, 43
    affinis 44
    capito 44
    crasus 44
    flaviventris 43, **160**
    flavomaculatus 44
    mixtus 43
    nudicluniatus 44

    peli 44
    pluto 44
    saccolaimus 43
saccolaimus, Saccolaimus 43
Saccopteryx 2, 46
    bilineata 46, **158**
    canescens 46
    gymnura 46
    leptura 46
    perspicillifer 46
    pumila 46
sacrimontis, Plecotus 110
saevus, Hipposideros 61
sagittula, Pipistrellus 116
salimalii, Latidens 35, **154**
salvini, Chiroderma 89
samoensis, Pteropus 24
sanborni, Leptonycteris 81
    Nycticeius 126
    Rhinolophus 57
sanctacrucis, Nyctimene 37
    Pteropus 22
santacristobalensis, Erophylla 19
sarasinorum, Mops 142
saturatus, Myotis 102, 108
satyrus, Pteropus 24
Sauromys 137
savii, Pipistrellus 114
scapulatus, Pteropus 27
schaubi, Myotis 101
scherzeri, Cynopterus 33
schistaceus, Hipposideros 66
schlieffeni, Nycticeius 126
schmidtorum, Micronycteris 73
schnablii, Amorphochilus 96, **175**
schneideri, Hipposideros 63
schreibersi, Miniopterus 134
schulzi, Tonatia 75
schwartzi, Artbeus 91
    Rhinolophus 54
scitulus, Nyctimene 37
Scleronycteris 82
    ega 82, **169**
scopaeum, Chiroderma 89
Scoteanax 126
Scoteinus 128
Scotoecus 127
    albigula 128
    albofuscus 127
    falabae 128
    hindei 128
    hirundo 127, **181**
    pallidus 127
    woodi 127
Scotomanes 128
    emarginatus 128
    imbrensis 128
    ornatus 128, **181**
    sinensis 128
Scotonycteris 32
    ophiodon 32
    zenkeri 32, **151**
Scotophilus 128
    alvenslebeni 129
    borbonicus 128
    castaneus 128
    celebensis 128
    colias 128
    consobrinus 128
    damarensis 128
    dingani 128
    gairdneri 128
    heathi 128
    herero 128
    insularis 128
    kuhlii 128, **182**

    leucogaster 128
    nigrita 128
    nigritellus 128
    nucella 128
    nux 128
    panayensis 128
    pondoensis 128
    robustus 128
    temmincki 128
    viridis 128
    watkinsi 128
Scotorepens 126
Scotozous 116
scotti, Myotis 104
scutatus, Diclidurus 48
seabrai, Myotis 109
secatus, Taphozous 42
secundus, Otomops 142
sedulus, Rhinolophus 58
seianum, Rhinopoma 41
seimundi, Balionycteris 34
Selysius 102
semicaudata, Emballonura 45
seminolus, Lasiurus 129
seminudus, Rousettus 19
semoni, Hipposideros 64
semotus, Lasiurus 129
senaariense, Rhinopoma 41
senegalensis, Pipistrellus 116
    Taphozous 42
senex, Centurio 93, **174**
sepikensis, Pteropus 26
septentrionalis, Myotis 102
    Phylloderma 77
    Rhinolophus 54
serasani, Cynopterus 33
serotinus, Eptesicus 120
setiger Mormopterus 137
sewelanus, Pipistrellus 112
seychellensis, Coleura 45
    Pteropus 23
sezekorni, Erophylla 79
shameli, Rhinolophus 59
shanorum, Pipistrellus 114
sheila, Chalinolobus 125
sherrini, Nyctophilus 131
shiraziensis, Eptesicus 120
shortridgei, Chaerephon 140
    Miniopterus 133
    Rhinolophus 55
    Rousettus 19
siamensis, Rhinolophus 57
sicarius, Myotis 101
silhouettae, Coleura 45
siligorensis, Myotis 104
silvatica, Murina 132
silvestris, Rhinolophus 55
silvicola, Tonatia 75
simalurus, Pteropus 22
similis, Nyctinomops 143
simplex, Rhinolophus 52
simulator, Rhinolophus 53
simus, Myotis 106
sinaloae, Molossus 144
sinensis, Hipposideros 61
    Megaderma 51
    Scotomanes 128
sinicus, Coelops 69
    Rhinolophus 54
smithi, Kerivoula 98
    Rousettus 20
smitianus, Miniopterus 134
sobrinus, Macroglossus 38, **156**
    Rhinolophus 58
sociabilis, Myotis 108
societatis, Pipistrellus 115

215

sodalis, Myotis 105
sogdianus, Myotis 103
solifer, Taphozous 42
solitarius, Pteropus 24
    Rhinolophus 58
solomonensis, Miniopterus 134
solomonis, Chaerephon 140
    Emballonura 44
    Pteropus 22
somalicus, Pipistrellus 117
sonoriensis, Eumops 143
    Myotis 106
soricina, Glossophaga 80, **168**
sowerbyi, Myotis 104
spadiceus, Dyacopterus 34, **152**
spadix, Rhinolophus 53
spasma, Megaderma 51
speciosus, Pteropus 22
spectrum, Vampyrum 77, **166**
speculator, Hipposideros 66
spelaea, Eonycteris 38, **155**
speoris, Hipposideros 66
Sphaerias 35
    blanfordi 35, **153**
Sphaeronycteris 93
    toxophyllum 93, **173**
sphinx, Cynopterus 33
spillmani, Chaerephon 140
spinalatus, Rousettus 20
spurcus, Rhinolophus 58
spurrelli, Lionycteris 77, **167**
    Mops 141
stalkeri, Myotis 109
stampflii, Pipistrellus 113
Stenoderma 93
    darioi 93
    rufum 93, **173**
Stenodermatinae 1, 2, 6, 9, 85
Stenodermatini 86
Stenonycteris 20
stenops, Phylloderma 76, **166**
stenopterus, Nyctalus 118
stenotis, Hipposideros 65
stheno, Rhinolophus 53
stoliczkanus, Aselliscus 67
stramineus, Natalus 95
strepitans, Epomops 31
stresemanni, Emballonura 45
    Rousettus 20
sturdeei, Pipistrellus 111
Sturnira 85
    aratathomasi 85
    bidens 86
    bogotensis 86
    erythromops 85
    hondurensis 86
    lilium 85
    ludovici 86
    luisi 85
    magna 86
    mordax 86
    nana 86
    occidentalis 86
    oporophilum 86
    parvidens 85
    thomasi 85
    tildae 85, **170**
Sturnirini 85
Styloctenium 28
    wallacei 28, **148**
suapurensis, Pteronotus 71
subbadius, Rhinolophus 56
subcanus, Pipistrellus 114
subflavus, Pipistrellus 112, **177**
subniger, Pteropus 23
subrufa, Carollia 84

subrufus, Rhinolophus 59
subtilis, Pipistrellus 114
subtropicalis, Nycticeius 126
subulidens, Pipistrellus 112
sudani, Taphozous 42
suilla, Murina 132
sulcata, Emballonura 45
sumatrae, Rhinopoma 40
sumatranus, Rhinolophus 55
sumbae, Hipposideros 66
sumbana, Dobsonia 29
superans, Rhinolophus 54
    Vespertilio 121
superbus, Chalinolobus 125
swinnyi, Rhinolophus 53
swirae, Taphozous 42
Syconycteris 38
    australis 38, **156**
    carolinae 39
    crassa 39
    finschi 39
    hobbit 39
    keyensis 39
    major 39
    naias 39
    papuana 39
sylvestris, Micronycteris 74
sylvia, Molossops 138
syops, Erophylla 79
szechwanus, Rhinolophus 56

tablasi, Pteropus 22
Tadarida 138
    aegyptiaca 139
    africana 139
    antillularum 139
    australis 139
    bahamensis 139
    bocagei 139
    brasiliensis 10, 139, **186**
    cynocephala 139
    fulminans 139
    insignis 139
    intermedia 139
    kuboriensis 139
    lobata 139
    mastersoni 139
    mexicana 10, 139
    murina 139
    muscula 139
    teniotis 139
    thomasi 139
    tragata 139
    ventralis 139
taiwanensis, Myotis 106
Taphozous 1, 2, 41, 42
    achates 42
    albipinnis 42
    australis 42, **159**
    bicolor 42
    cavaticus 42
    fretensis 42
    georgianus 43
    haedinus 42
    hamiltoni 43
    hildegardeae 42
    hilli 42
    kachensis 43
    kampenii 42
    kapalgensis 42
    leucopleura 42
    longimanus 42
    magnus 43
    mauritianus 43
    melanopogon 42
    nudaster 43

    nudiventris 43
    perforatus 42
    philippinensis 42
    rhodesiae 42
    secatus 42
    senegalensis 42
    solifer 42
    sudani 42
    swirae 42
    theobaldi 42
    troughtoni 43
    zayidi 43
tarlosti, Leptonycteris 81
tasmaniensis, Pipistrellus 115
tatar, Rhinolophus 59
tatei, Eptesicus 119
teliotis, Lasiurus 129
temmincki, Pteropus 25
    Scotophilus 128
temminckii, Molossops 138
tenebrosa, Murina 132
tener, Rhinolophus 54
teneriffae, Plecotus 110
teniotis, Tadarida 139
tenuipinnis, Pipistrellus 117
tenuis, Pipistrellus 112
tephrus, Hipposideros 63
terasensis, Hipposideros 65
terminus, Cynopterus 33
thaianus, Myotis 104
thebaica, Nycteris 50
theobaldi, Taphozous 42
thersites, Mops 141
thomasi, Eptesicus 120
    Lonchophylla 78
    Mimetillus 123
    Rhinolophus 54
    Sturnira 85
    Tadarida 139
    Uroderma 86
thomensis, Hipposideros 67
thonglongyai, Craseonycteris 41, **157**
Thoopterus 34
    nigrescens 35, **153**
thyone, Vampyressa 88
Thyroptera 9, 96
    abdita 96
    albiventer 96
    discifera 96
    juquiaensis 96
    tricolor 96, **175**
Thyropteriade 14, 96
thysanodes, Myotis 101
tibialis, Miniopterus 134
tickelli, Hesperoptenus 123
tildae, Sturnira 85, **170**
timidus, Rhinolophus 59
timoriensis, Nyctophilus 131
    Rhinolophus 59
titthaecheilus, Cynopterus 33, **152**
tokudae, Pteropus 25
toltecus, Artibeus 92
tomesi, Hesperoptenus 123
    Pteropus 22
Tomopeas 135
    ravus 135, **184**
Tomopeatinae 135
Tonatia 9, 75
    bidens 75, **165**
    brasiliense 75
    carrikeri 75
    centralis 75
    evotis 75
    laephotis 75
    occidentalis 75
    schulzi 75

silvicola 75
tonfangensis, Pipistrellus 114
tonganus, Pteropus 23
torquata, Myonycteris 21, **147**
torquatus, Cheiromeles 138, **185**
torrei, Pteronotus 71
townsendii, Plecotus 110, **177**
toxopeusi, Rhinolophus 60
toxophyllum, Sphaeronycteris 93, **173**
Trachops 3, 9, 77
  cirrhosus 77, **166**
  coffini 77
  ehrhardti 77
tragata, Nycteris 49, **160**
  Tadarida 139
tragatus, Rhinolophus 55
tramatus, Pipistrellus 112
tranninhensis, Hipposideros 65
transcaspicus, Myotis 103
trevori, Mops 141
Triaenops 68
  afer 68
  furculus 68
  majusculus 68
  persicus 68, **163**
  rufus 68
tricolor, Carollia 84
  Myotis 102
  Thyroptera 96, **175**
tricuspidatus, Aselliscus 67, **162**
tridens, Asellia 67, **163**
trifoliatus, Rhinolophus 58
trifolium, Megaderma 51
trinitatis, Artibeus 91
  Molossus 144
  Peropteryx 47
trinitatum, Chiroderma 89
  Uroderma 86
Trinycteris 73
triomylus, Artibeus 91
tristis, Miniopterus 134
trobrius, Hipposideros 66
tronchonii, Natalus 95
tropicalis, Rhinopoma 40
tropidorhynchus, Molossus 145
troughtoni, Taphozous 43
trumbulli, Eumops 144
truncatus, Rhinolophus 53
tschuliensis, Myotis 101
tsuensis, Myotis 101
tuberculata, Mystacina 135
tuberculatus, Chalinolobus 124
  Pteropus 25
tubinaris, Murina 132
tumida, Rhogeesa 126, **181**
tumidiceps, Mormoops 72
tumidifrons, Natalus 95
tumidorostris, Natalus 95, **174**
tunetae, Rhinolophus 56
turcomanicus, Myotis 102
turcomanus, Eptesicus 120
turpis, Hipposideros 65
Tylonycteris 5, 9, 122
  aurex 122
  bhaktii 122
  fulvida 122
  malayana 122
  meyeri 122
  pachypus 122
  robustula 122, **180**
tylopus, Glischropus 199, **178**
tytleri, Pteropus 24

ualanus, Pteropus 23
uenoi, Plecotus 110
ugandae, Pipistrellus 117

ulthiensis, Pteropus 23
umbratus, Mormopterus 137
  Vampyrops 87
underwoodi, Eumops 143, **187**
  Hylonycteris 82, **169**
unicolor, Rousettus 19
unitus, Hipposideros 66
Uroderma 86
  bilobatum 86, **170**
  convexum 86
  davisi 86
  magnirostrum 86
  molaris 86
  thomasi 86
  trinitatum 86
ussuriensis, Murina 132
  Myotis 107
  Vespertilio 121
valens, Glossophaga 80

Vampyressa 88
  bidens 89
  brocki 88
  melissa 88
  nymphea 88
  pusilla 88, **171**
  thyone 88
Vampyrodes 88
  caraccioli 88, **171**
  major 88
  ornatus 88
Vampyrops 87
  aquilus 87
  aurarius 87
  brachycephalus 87
  dorsalis 87
  heller 88, **170**
  incarum 88
  infuscus 87
  lineatus 87
  nigellus 87
  oratus 87
  recifinus 87
  umbratus 87
  vittatus 87
  zarhinus 88
Vampyrum 77
  spectrum 77, 166
vampyrus, Pteropus 26
vandeuseni, Rhinolophus 52
Vansonia 116
variegatus, Chalinolobus 125
varius, Lasiurus 129
  Nyctimene 36
velatus, Histiotus 122
veldkampi, Nanonycteris 32, **151**
velifer, Myotis 107
velutinus, Nyctalus 118
venatoris, Chalinolobus 125
ventralis, Tadarida 139
veraecrucis, Pipistrellus 112
veraecundus, Philetor 122
verrillii, Molossus 145
verrucosus, Nyctalus 118
  Phyllostomus 76
vespa, Rhinolophus 57
Vespadelus 116
Vespertilio 121
  anderssoni 121
  murinus 121, **179**
  orientalis 121
  superans 121
  ussuriensis 121
Vespertilionidae 3, 5, 8, 9, **15**, 97
Vespertilioninae 100
Vespertilionini 111

Vespertillionoidea 5, 94
vetulus, Pteropus 25
vicinior, Miniopterus 134
villiersi, Miniopterus 134
villosa, Nycteris 50
villosissimus, Lasiurus 129
villosum, Chiroderma 89, **171**
vinsoni, Nycteris 50
virginianus, Plecotus 111
virgo, Diclidurus 48
  Rhinolophus 53
viridis, Dobsonia 29
  Scotophilus 128
vittatus, Vampyrops 87
vivesi, Myotis 8, 109
vizcaccia, Nyctimene 36
voeltzkowi, Pteropus 23
volans, Myotis 108
volgensis, Myotis 107
vordermanni, Pipistrellus 115
vulturnus, Pipistrellus 116

wahlbergi, Epomophorus 31
walkeri, Nyctophilus 130
wallacei, Styloctenium 28, **148**
wardi, Plecotus 110
watasei, Myotis 101
waterhousii, Macrotus 74, **165**
watkinsi, Scotophilus 128
watsoni, Artibeus 92
wattsi, Pipistrellus 112
weberi, Myotis 101
websteri, Chaerephon 140
welwitschii, Myotis 101
westralis, Pipistrellus 112
wetmorei, Eptesicus 120
  Megaerops 34
whiteheadi, Harpyionycteris 30, **149**
  Kerivoula 97
whitleyi, Myopterus 138
wintoni, Laephotis 121
witkampi, Miniopterus 134
woermanni, Megaloglossus 38, **156**
wollastoni, Hipposideros 64
wonderi, Mops 141
woodfordi, Melonycteris 39
  Pteropus 27
woodi, Nycteris 50
  Scotoecus 127
wrighti, Hipposideros 61
wroughtoni, Myonycteris 21
  Otomops 142

xanthinus, Lasiurus 130
Xenoctenes 73

Yangochiroptera 3, 69
yapensis, Pteropus 23
yerbabuenae, Leptonycteris 81
yesoensis, Myotis 103
Yinochiroptera 3, 40
youngi, Diaemus 94, **174**
yucatanica, Nyctinomops 143
yucatanicus, Artibeus 91
yumanensis, Myotis 108
yunanensis, Rhinolophus 58
yunnanensis, Pipistrellus 113

zarhinus, Vampyrops 88
zayidi, Taphozous 43
zenkeri, Scotonycteris 32, **151**
zuluensis, Kerivoula 99
  Pipistrellus 117
  Rhinolophus 54

Informationen für Autoren

Handbuch der Zoologie, Band VIII Mammalia

Verlag: Walter de Gruyter, Berlin
Herausgeber: D. Starck, H. Schliemann, J. Niethammer
Schriftleiter: H. Wermuth

Das Handbuch der Zoologie, Band VIII Mammalia (Säugetiere) behandelt das gesamte Gebiet der Speziellen Zoologie der Säugetiere. Es wird der gegenwärtige Stand der Kenntnisse in kurzer Form dargestellt, wobei vor allem die wesentlichen Tatsachen Berücksichtigung finden und auf offene, noch zu klärende Fragen hingewiesen wird.

Der Band VIII erscheint in Teilbänden mit fortlaufender Numerierung entsprechend dem Eingang der Beitragsmanuskripte.

Der Umfang der Manuskripte und der Termin für deren Ablieferung wird gemeinsam mit Autoren, Herausgebern und Verlag vertraglich festgelegt.

Jeder Teilband umfaßt
   Inhaltsübersicht
   Text
   Abbildungen (fortlaufend numeriert)
   Abbildungslegenden
   Tabellen (fortlaufend numeriert)
   Verzeichnis der Gattungs- und Artnamen mit Autor und Jahreszahl
   Sachregister mit Seitenzahl
   Literaturverzeichnis

Text: Das Manuskript soll mit Schreibmaschine (2facher Zeilenabstand) auf DIN-A-4-Blättern geschrieben sein. Links soll ein 3 cm breiter Rand frei bleiben. Wissenschaftliche Namen der Gattungen, Arten und Unterarten sind im Manuskript gewellt zu unterstreichen und werden kursiv gesetzt. Vorschläge für Kleindruck werden durch einen senkrechten Strich am linken Rand mit dem Zusatz „Petit" gekennzeichnet. Im Text sollen Autorennamen der Taxa vermieden werden. Literaturhinweise erfolgen durch Nennung des Autors in eckigen Klammern evtl. mit Zusatz der Jahreszahl und bei Mehrfachzitierung verschiedener Arbeiten des gleichen Autors mit dem Zusatz a, b, c. Für die endgültige Vorbereitung des Manuskriptes für den Satz ist der Schriftleiter verantwortlich.

Information for Authors

Handbook of Zoology, Volume VIII Mammalia

Publishers: Walter de Gruyter, Berlin
Editors: D. Starck, H. Schliemann, J. Niethammer
Managing Editor: H. Wermuth

The Handbook of Zoology, Vol. VIII, Mammalia treats the whole field of the zoology of the mammals. It briefly outlines current knowledge, laying particular emphasis on the most essential facts. Reference is made to important questions that are still open.

Volume VIII will be published in parts, numbered consecutively according to the receipt of the manuscripts.

The size of the manuscripts as well as the data of submittance will be determined by the editors together with the contributors and the publisher.

Each part comprises
   Table of contents
   Text
   Illustrations (numbered consecutively)
   Legends to illustrations
   Tables (Numbered consecutively)
   Index of generic and species names with author and year
   Subject Index in alphabethical order
   References

Text: The manuscript should be typed double-spaced on DIN A4 paper (21 × 29,7 cm) with a left margin of 3 cm. Scientific names of genera and species should be marked with a wavy underline and will be set in italics. Suggestions for small type may be made by a vertical line in the left margin with the note "Petit". The names of authors of taxa should be avoided in the text. Literature citations should be made by giving the name of the author in square brackets, together with the year of publication, if desired. (In the case of citations for several works of the same author, they are to be distinguished by the addition of a, b, c, etc.) The Managing Editor is responsible for the final preparation of the manuscript for typesetting.

Das Literaturverzeichnis enthält alle Veröffentlichungen, die im Text zitiert werden, sie werden nach dem folgenden Muster in alphabetischer Reihenfolge aufgeführt.

Zeitschriftenbeiträge: Schildknecht, H., Maschwitz, V. & Winkler H. (1968): Zur Evolution der Carabiden-Wehrdrüsen-Sekrete. − Die Naturwissenschaften, Berlin, 3: 112−117.

Bücher oder andere selbständige Veröffentlichungen: Riedl, R. (1975): Die Ordnung des Lebendigen. Systembedingungen der Evolution. Paul Parey, Hamburg und Berlin.

Handbuchbeiträge: Stell, F.F. (1971): Mechanism of synaptic transmission. In: Neurosciences Research. (S. Ehrenpreis, ed.). Academic Press, New York, London. pp. 1−27

Unveröffentlichte Arbeiten sollen nur zitiert werden, wenn sie zur Veröffentlichung angenommen sind, und zwar unter Angabe der Zeitschrift, die die Arbeit angenommen hat:
Kuhn, H.-J. (1976): Antorbitaldrüse und Tränennasengang von Neotragus pygmaeus. − Z. Säugetierkunde (im Druck).

Periodica sollen im Literaturverzeichnis nach der „World list of scientific periodicals, published in the years 1900−1960" und folgende (Butterworths, London) zitiert werden.

Abbildungsvorlagen (Zeichnungen, Photos) sollen in reproduktionsfähigem Zustand geliefert werden. Die Abbildungen werden soweit wie möglich verkleinert, entweder auf die Breite einer Spalte (76 mm) oder auf die gesamte Satzbreite (157 mm). Die Höhe einer Spalte ist 238 mm. Für Abbildungen, die auf ¼ ihrer Größe (½ Breite oder Höhe) verkleinert werden sollen, werden Strichstärken von 0,5 bis 0,8 mm und eine Schriftgröße von 8 mm empfohlen. Es sollten Schriftschablonen benutzt werden (keine Schreibmaschinenschrift). Photos müssen kontrastreich sein, da im Druck etwas Kontrast verlorengeht.

Das Manuskript soll in drei gesonderten Teilen eingereicht werden:
 1. Text
 2. Abbildungen, Tabellen und Diagramme
 3. Abbildungslegenden

The reference section must contain an alphabetical list of all references mentioned in the text. Please note the following examples.

Articles in journals: Schildknecht, H., Maschewitz, V. & Winkler, H. (1968): Zur Evolution der Carabiden-Wehrdrüsen-Sekrete. − Die Naturwissenschaften, Berlin, 3: 112−117

Books or other publications: Riedl, R. (1975): Die Ordnung des Lebendigen. Systembedingungen der Evolution. Paul Parey, Hamburg und Berlin.

Articles in reference works: Stell, F.F. (1971): Mechanism of synaptic transmission. In: Neurosciences Research. (S. Ehrenpreis, ed.). Academic Press, New York, London. pp. 1−27.

Unpublished material should only be cited when it has been accepted for publication. The name of the journal where the paper is to appear must be given: Kuhn, H.-J. (1976): Antorbitaldrüse und Tränennasengang von Neutragus pygmaeus. − Z. Säugetierkunde (in press)

In the reference section scientific periodicals should be cited according to the "World list of Scientific Periodicals, Published in the Years 1900−1960" (Butterworths, London) and later listings.

Illustrations (drawings and photos) must be submitted in camera-ready form. The figures will be reduced as far as possible, either to the width of one column (76 mm) or two columns (157 mm). The length of a column is 238 mm. For a figure that is to be reduced to ¼ its size (½ length of side), lines of 0,5 to 0,8 mm and letters 8 mm high are recommended. A lettering device should be used (no typing). Photographs must be of good contrast as there is a loss of contrast in printing.

The manuscript should be submitted in 3 separate sections:
 1. complete text
 2. illustrations, tables and diagrams
 3. legends to the illustrations

Das Manuskript für das Sachregister, das alle sachdienlichen Hinweise auf den Text enthalten soll, ist nach Erhalt der umbrochenen Seiten zu erstellen.

Der Autor wird gebeten, eine Manuskriptkopie zurückzubehalten.

Als Sprache sind Deutsch und Englisch zugelassen.

Manuskripte, korrigierte Fahnen und Umbruch sind an den Herausgeber zu senden.

The manuscript of the subject index recording all pertinent statements made within the body of the text is to be prepared on receipt of the page proofs.

The author is requested to retain a copy of the manuscript.

The manuscripts are to be written in English or German.

Manuscripts, galley-proofs and page proofs should be sent to the editor.

Herausgeber/Editors

Professor
Dr. Jochen Niethammer
Zoologisches Institut der
Universität Bonn
Poppelsdorfer Schloß
D-53115 Bonn
Tel. (0228) 73 54 57

Professor
Dr. Harald Schliemann
Zoologisches Institut und
Zoologisches Museum
Martin-Luther-King-Platz 3
D-20146 Hamburg
Tel. (040) 41 23 39 17

Professor
Dr. med. Dr, phil. h. c.
Dietrich Starck
Balduinstraße 88
D-60599 Frankfurt

Schriftleiter/Managing Editor

Dr. Heinz Wermuth
Falkenweg 1
D-71691 Freiberg
Tel. (07141) 7 49 77

Verlag/Publisher

Walter de Gruyter & Co.
Genthiner Straße 13
D-10785 Berlin
Tel. (030) 26005-0

Walter de Gruyter, Inc.
Scientific Publishers
200 Saw Mill River Road
Hawthorne, N.Y. 10532
U.S.A.
Tel. (914) 747-0110

Handbuch der Zoologie

Band VIII Mammalia

Band VIII Mammalia wird von nun an in Teilbänden mit einer fortlaufenden Numerierung erscheinen, die an die bisherige Zählung der Lieferungen anschließt. Jeder Teilband wird von 55 an gebunden geliefert.

Handbook of Zoology

Volume VIII Mammalia

Volume VIII Mammalia will appear from now on in parts with consecutive numeration in connection with the previous numeration of the instalments. Each part will be delivered bound from part 55 on.

Bisher sind erschienen/Already published

Lieferung instalment	Autor, Titel / Author, Title	Erscheinungsjahr Publication date
1	K. Herter: Winterschlaf G. Lehmann: Das Gesetz der Stoffwechselreduktion.	1956
2	M. Meyer-Holzapfel: Das Spiel bei Säugetieren W. Fischel: Haushunde E. Mohr: Das Verhalten der Pinnipedier H. Pilters: Das Verhalten der Tylopoden.	1956
3	H. Mies: Physiologie des Herzens und des Kreislaufs H. v. Hayek: Die Lunge.	1956
4	W. Schoedel: Die Atmung.	1956
5	F. Tischendorf: Milz H.E. Voß: Der Einfluß endokriner Drüsen auf den Stoffwechsel der Säugetiere.	1956
6	C. Heidermanns: Physiologie der Exkretion E. Heinz & H. Netter: Wasserhaushalt.	1956
7	G.P. Baerends: Aufbau des tierischen Verhaltens P. Leyhausen: Das Verhalten der Katzen H. Frick: Morphologie des Herzens. Vergriffen.	1956
8	K. Lorenz: Methoden der Verhaltensforschung I. Eibl-Eibesfeldt: Ausdrucksformen der Säugetiere. M. Meyer-Holzapfel: Das Verhalten der Bären. Vergriffen.	1957
9	K. Herter: Das Verhalten der Insektivoren. G. Tembrock: Das Verhalten des Rotfuchses. Vergriffen.	1957
10	L. v. Bertalaffny: Wachstum.	1957
11	G. Siebert & K. Lang: Energiewechsel A. Kuritz: Das Autonome Nervensystem.	1958
12	I. Eibl-Eibesfeldt: Das Verhalten der Nagetiere.	1958
13	W. Krüger: Der Bewegungsapparat.	1958

Lieferung instalment	Autor, Titel / Author, Title	Erscheinungsjahr Publication date
14	W. Krüger: Der Bewegungsapparat. Abschluß von Lieferung 13.	1958
15	W. Krüger: Bewegungstypen E.J. Slijper: Das Verhalten der Wale (Cetacea).	1958
16	Th. Haltenorth: Klassifikation (Monotremata) Th. Haltenorth: Klassifikation (Marsupialia).	1958
17	P.O. Chatfield: Physiologie der peripheren Nerven H. Brune: Rohstoffe der Haussäugetiere.	1958
18	L.S. Crandall: Über das Verhalten des Schnabeltieres in der Gefangenschaft H. Hediger: Verhalten der Marsupialier C.R. Carpenter: Soziologie und Verhalten freilebender nichtmenschlicher Primaten.	1958
19	H. Bartels: Physiologie des Blutes E.H. Hess: Lernen und Engramm.	1959
20	J.H. Schuurmans Stekhoven: Biologie der Parasiten der Säugetiere P. Cohrs & H. Köhler: Tod und Todesursachen bei Säugetieren.	1959
21	Th.H. Schiebler: Morphologie der Nieren.	1959
22	D. Starck: Ontogenie und Entwicklungsphysiologie der Säugetiere.	1959
23	G. Birukow: Statischer Sinn M. Watzka: Superfecundatio, Superfedatio, multiple Ovulation, Zwillinge, Mehrlinge bei Säugetieren.	1959
24	B. Kummer: Biomechanik des Säugetierskeletts.	1959
25	H. Grau & Boessneck: Der Lymphapparat E.J. Slijper: Die Geburt der Säugetiere.	1960
26	R. Ortmann: Die Analregion der Säugetiere.	1960
27	H. Hediger & H. Kummer: Das Verhalten der Schnabeligel H. Krieg & U. Rahm: Das Verhalten der Xenarthren (Ameisenbären, Faultiere und Gürteltiere); Das Verhalten der Schuppentiere U. Rahm: Das Verhalten der Erdferkel O. Kalela: Wanderungen.	1961
28	H.W. Matthes: Verbreitung der Säugetiere in der Vorzeit.	1962
29	K. Neubert & E. Wüstenfeld: Morphologie des akustischen Organs.	1962
30	J. Aschoff: Spontane lokomotorische Aktivität.	1962
31	J. Eibl-Eibesfeldt: Technik der vergleichenden Verhaltensforschung.	1962
32	Th. Haltenorth: Klassifikation: Artiodactyla.	1963
33	H. Elias: Leber- und Gallenwege.	1963
34	H. Hofer & J. Tigges Makromorphologie des Zentralnervensystems 1. Teil.	1964
35	R. Schneider: Larynx der Säugetiere.	1965
36	F. Strauss: Weibliche Geschlechtsorgane, 1. Teil.	1965
37	F. Goethe: Das Verhalten der Musteliden U. Rahm: Das Verhalten der Klippschliefer.	1965

Lieferung instalment	Autor, Titel Author, Title	Erscheinungsjahr Publication date
38	G. Dücker: Das Verhalten der Viverriden.	1965
39	J.F. Eisenberg: The Social Organization of Mammals.	1966
40	F. Strauss: Weibliche Geschlechtsorgane, 2. Teil.	1966
41	H. Schriever: Physiologie des akustischen Organs.	1967
42	H. Frädrich: Das Verhalten der Schweine (Suidae, Tayassuidae) und Flußpferde (Hippotamidae).	1967
43	D. Müller-Using & R. Schloeth: Das Verhalten der Hirsche (Cervidae).	1967
44	M. Montjé: Physiologie des Auges.	1968
45	L. Róka: Intermediärer Stoffwechsel.	1968
46	R. Schenkel & E.M. Lang: Das Verhalten der Nashörner.	1969
47	E. Thenius: Stammesgeschichte der Säugetiere (einschließlich der Hominiden), 1. Teil.	1969
48	E. Thenius: Stammesgeschichte der Säugetiere (einschließlich der Hominiden), 2. Teil.	1969
49	H. Klingel: Das Verhalten der Pferde (Equidae).	1972
50	W. Platzer: Morphologie der Kreislauforgane.	1973
51	M.R.N. Prasad: Männliche Geschlechtsorgane.	1975
52	W. Schober & K. Brauer: Makromorphologie des Zentralnervensystems, 2. Teil.	1975
53	W. Schultz: Der Magen-Darm-Kanal der Monotremen und Marsupialier.	1976
54	F.R. Walther: Das Verhalten der Hornträger (Bovidae).	1979
55	F. Strauss: Der weibliche Sexualzyklus	1986
56	E. Thenius: Zähne und Gebiß der Säugetiere	1989
57	Z. Halata: Die Sinnesorgane der Haut und der Tiefensensibilität	1993
58	M.S. Fischer: Hyracoidea	1992
60	K.F. Koopman: Chiroptera (Systematics)	1994